AF842282

Decolonizing Western-Indigenous Dialogues

Decolonizing Western-Indigenous Dialogues

Interwoven Epistemologies for Multiple Modernities

Edited by
Gregory Cajete, Kefilwe Johanna Ditlhake,
Maria do Carmo dos Santos Gonçalves,
Karsten Kiewitt, Ronald Lutz

BLOOMSBURY ACADEMIC
LONDON • NEW YORK • OXFORD • NEW DELHI • SYDNEY

BLOOMSBURY ACADEMIC
Bloomsbury Publishing Plc
50 Bedford Square, London, WC1B 3DP, UK
1359 Broadway, New York, NY 10018, USA
29 Earlsfort Terrace, Dublin 2, Ireland

BLOOMSBURY, BLOOMSBURY ACADEMIC and the Diana logo are trademarks of Bloomsbury Publishing Plc

First published in Great Britain 2025

Copyright © Gregory Cajete, Kefilwe Johanna Ditlhake, Maria do Carmo dos Santos Gonçalves, Karsten Kiewitt, Ronald Lutz, 2025

Gregory Cajete, Kefilwe Johanna Ditlhake, Maria do Carmo dos Santos Gonçalves, Karsten Kiewitt, Ronald Lutz have asserted their right under the Copyright, Designs and Patents Act, 1988, to be identified as Editors of this work.

Series design by Grace Ridge
Cover images © Dmitrijs Novikovs and Joanna McCarthy via Getty Images

All rights reserved. No part of this publication may be reproduced or transmitted in any form or by any means, electronic or mechanical, including photocopying, recording, or any information storage or retrieval system, without prior permission in writing from the publishers.

Bloomsbury Publishing Plc does not have any control over, or responsibility for, any third-party websites referred to or in this book. All internet addresses given in this book were correct at the time of going to press. The author and publisher regret any inconvenience caused if addresses have changed or sites have ceased to exist, but can accept no responsibility for any such changes.

A catalogue record for this book is available from the British Library.

Library of Congress Cataloging-in-Publication Data available

ISBN: HB: 978-1-3504-2520-0
ePDF: 978-1-3504-2522-4
eBook: 978-1-3504-2521-7

Typeset by RefineCatch Limited, Bungay, Suffolk

To find out more about our authors and books visit www.bloomsbury.com and sign up for our newsletters.

Contents

List of Illustrations vii
Notes on Contributors viii
Preface xv

Introductions

1 On the Decolonization of Dialogues *Karsten Kiewitt* 3
2 What Is Native Science? *Gregory Cajete* 9

Part One Dialogues

3 To Know from Within *Benno Glauser* 17
4 To Be Open for Constructive Strangeness: Experiences with Intercultural Knowledge Exchange in Development Cooperation and Science *Peter Strack* 25
5 Embodied Knowledge: Decolonizing Philosophy by Turning to the Nature that We Are *Barbara Schellhammer and Stan Wilson* 39
6 Perhaps We Didn't Need a Bridge: In Dialogue with Indigenous Mathematics *Edward Doolittle and Amber Hughes* 51
7 Western and Indigenous Knowledge Systems: Attainment of the Sustainable Development Goals and the Umunthu Developmental Approach *Chance Chagunda* 69
8 Kitheka ki matu: A Forest Has Ears *Mark Lawrence* 81

Part Two Challenges

9 Conceded Land Rights, Indigenous Territoriality, and Cooperation: Contradictions and Challenges *Volker von Bremen* 103
10 Spirited Five Fingers Model: A Way of Reconnecting to the Spirit of Reciprocity in Social and Community Work *Ksenija Napan* 117
11 Mobilities in the Global South and New Circuits of Affects *Maria do Carmo dos Santos Gonçalves* 129
12 Indigenous Social Security Systems for Endogenous Social Development in Southern Africa *Ndangwa Noyoo* 141

13 Indigenous in Europe: The Challenge in Bringing Back Livonian *Valts Ernštreits and Gunta Kļava* 153
14 The Hijab and Faith: Fashion and Modesty *Samira Adel Osman* 169

Part Three Perspectives

15 Thirty Years of Democracy in South Africa: Reflecting on Social Work and Climate Change Challenges through a Ubuntu Lense *Kefilwe Johanna Ditlhake* 183
16 Decoloniality as Total Disentanglement with Modern Capitalist and Neo-liberal Complex: Social Work Perspective *Mbazima Simeon Mathebane* 195
17 Indigenous Youth and the Recomposition of Community Life: Some Experiences in Oaxaca, Mexico *Eduardo Bautista Martínez and María Leticia Briseño Maas* 203
18 "Not about us, but with us!" Perspectives of Insurgent Research with Children of the Global South *Manfred Liebel* 215
19 From Alternative Epistemologies to Alternative Aesthetics: A Plea for a Decolonial Environmental Aesthetics *Werther Gonzales León* 231
20 Buen Vivir and Sustainability: Relation between Indigenous Knowledge and Western Concepts *Ronald Lutz* 243

Conclusion

21 Conclusion: Decolonizing Knowledge *Ronald Lutz* 255

Index 263

Illustrations

1 Königsberg Bridges map view 52
2 Euler's map of Königsberg 54
3 Pregel River drainage basin 55
4 Concept map of our conversations, with clustering and mapping 62
5 A holistic medicine wheel 65
6 The modified medicine wheel in the Ontario Mathematics Curriculum, with spirituality pushed to the personal domain and out of the public domain 65
7 Old and new railways with respect to indigenous socioecologies of Kibwezi Forest and Maimu Pond 87
8 Contested indigenous spaces within the sisal plantation of the Dwa Estate 89
9 Time-space in Ayoreo cosmovision 107

Contributors

Volker von Bremen has a PhD in anthropology (Berlin, Germany); fieldwork with indigenous people in Paraguay (Ayoreo); since 1979, he has accompanied indigenous peoples in the Gran Chaco (Argentina, Bolivia, Paraguay) and other regions of South America-Lowlands. He is a consultant of local and international organizations cooperating with indigenous communities, and networks with indigenous organizations and those working with indigenous peoples in topics of conflict and development; lectures, workshops, and seminars in different universities, academies, and congresses related to indigenous peoples (in Latin America and Europe). His focal points of research are indigenous land rights, "Gestión Territorial Indígena" ("indigenous territorial land-shaping"), indigenous cosmovisions and spirituality, protection of nature and development, and monitoring in transcultural conflicts.

Gregory Cajete (PhD) is Professor for Native American Studies and Language, Literacy and Sociocultural Studies in the Department of Native American Studies at the University of New Mexico, Albuquerque, USA. He is a Native American scholar and educator (Tewa from Santa Clara Pueblo, New Mexico), an artist in ceramic, pastel, and metal, and a scholar of herbalism and holistic health. He served as Dean of the Center for Research and Cultural Exchange, Chair of Native American Studies and Professor of Ethno-Science and he is the former Director of Native American Studies (eighteen years) at the University of Mexico. Gregory Cajete designs culturally responsive curricula geared to the special needs and learning styles of Native American students. He has pioneered reconciling Indigenous perspectives in sciences with a Western academic setting.

Chance Chagunda (Dr.) is a senior associate researcher of the Southern African Policy and Development Nexus (SAPDN), a research-based and policy-driven consultancy and think tank based in Cape Town, South Africa. Among other things, SAPDN aims to foster the growth of Africa for Africans and is grounded in Indigenous African Knowledge and seeks to arrive at appropriate solutions to African challenges. Dr. Chagunda is also the South African country manager for the Whitaker Peace and Development Initiative (WPDI), which works with youth to develop their skills as leaders, peace builders, and community builders, including various indigenous groups across the globe. Dr. Chagunda is from the Ngoni tribe from Matiasi Village in the Traditional Authority of Ndawambi in Mchinji District, Malawi. He is an activist for indigenous knowledge systems. His research interests include public policy, social development, and Indigenous Knowledge Systems.

Kefilwe Johanna Ditlhake is a social worker and associate professor at the Department of Social Work and Community Development, University of Johannesburg, South

Africa. She holds bachelor's and master's degrees in social work from the University of Witwatersrand and RAU University, respectively, and a PhD from Wits University. Her research interest is in ubuntu, indigenization, and decolonization of social work, community development education, and practice are fundamental for epistemic diversity, legitimacy, and recognition of indigenous epistemologies. The dominant worldviews, knowledge production, and claim for universality constrained a deeper examination of the African worldviews, philosophies, or epistemologies. An agenda for transformation and dialogue could offer viable perspectives to develop knowledge applicable to contextual and unique experiences.

Edward Doolittle is Kanyen'kehake (Mohawk) from Six Nations in southern Ontario. Edward has a PhD in mathematics (partial differential equations) from the University of Toronto. He is currently Associate Professor of Mathematics in the Department of Indigenous Knowledge and Science at First Nations University of Canada in Regina, Saskatchewan, Canada, where he is also Associate Dean of Research and Graduate Programs. First Nations University occupies the atim kâ-mihkosit (Red Dog) Urban Reserve, which is situated on Treaty 4 Territory in Canada. In his effort to serve his community and Indigenous people in general, Edward conducts research not only in Indigenous mathematics (which includes mathematics of practical interest to Indigenous people) but also in Indigenous mathematics education.

Valts Ernštreits (Dr. phil.) is a Livonian-born linguist and language activist who received his doctorate at the University of Tartu with a dissertation on the Livonian standard language. He specializes in Livonian lexicography, grammar, development of the language, and electronic resources. His research and community work have been widely acknowledged in Latvia and abroad, and he has received many recognitions. V. Ernštreits is also a member of the Global Task Force for the UN International Decade of Indigenous Languages (2022–32), a member and representative of the Livonian community for the Board of the Historical Lands and Cultural Spaces of Latvia, and a corresponding member of the Latvian Academy of Science.

Benno Glauser is a Swiss-born philosopher, writer, and activist who lives in Paraguay since the 1970s, having committed his work to the politically persecuted and to repressed and invisibilized social groups—street children, campesinos, and indigenous peoples—seeking to contribute to their cultural vitality, autonomy, and self-determination. Since the 1990s, conscious of the unique value of indigenous cosmovisions, he has been exploring the situation and life culture of indigenous peoples without any contact with our modern civilization. Since 2002, has also been active in carrying out long-term, institutional fieldwork in defense of their integrity and continuity. He shares his insights and learnings through teaching and writing, as an incentive for a necessary questioning and rethinking of the life culture of modern societies.

Werther Gonzales León is currently a PhD candidate in philosophy at the Friedrich Schiller University Jena, Germany. He holds an MA in futures studies from the Free University of Berlin and a BA in philosophy from the National University of San Marcos,

Lima. During his doctoral studies, he carried out research stays at the Ibero-American Institute (IAI) and the Institute for Latin American Studies (LAI), both located in Berlin. A native of Peru, since his childhood he has been in contact with people from different origins, among which the Andean stands out. In order to make visible the forms of life of his birthplace, he develops in his doctoral research an environmental aesthetics with a transcultural and decolonial orientation. This implies questioning the dominant European patterns and demanding other ways of being in the world.

Maria do Carmo dos Santos Gonçalves is a coordinator of social projects at Associação Educadora São Carlos, a civil society organization working in the areas of health, education, and social work, based in Porto Alegre, Brazil. She holds a PhD in social sciences from the Pontifical Catholic University of Rio Grande do Sul (PUCRS). Her research interests include biographical and ethnographic narrative research, studies on contemporary diasporas, migration and refuge, the Middle East, and Islam. She is an Afro-Brazilian woman and a member of the Congregation of the Scalabrinian Sisters, who works directly with migrants and refugees in 26 countries. Starting from a decolonial perspective on migration in the Global South, she considers the limits of migratory categories, highlighting the way migrants narrate their mobilities and the political affects they mobilize throughout their migratory process.

Amber Hughes is an Australian, non-Indigenous woman who works at the University of Newcastle. Her family is of predominantly Anglo-Celtic descent and have resided in Awabakal Country (Newcastle, NSW) for generations. Amber would like to pay her respects to the traditional owners of the lands she calls home, the Awabakal and Worimi peoples. Amber has completed her doctoral degree on Indigenous Mathematics Education in Australia. Her research focus has been on deconstructing prominent mathematics education policy discourses in Australia and ways of framing equity and social justice for Indigenous learners therein. Amber's current research interests are in further understanding the role of settler-colonizer educators/researchers in the advancement of social justice and equity for and *with* Indigenous learners in mathematics, and the theorizing of social justice to reflect Indigenous ontologies within education broadly.

Karsten Kiewitt was born and raised in Germany, with ancestral roots to the Baltic Prūsai people. He is a music therapist and social worker and worked as a professor in the field of social inclusion at the Clara Hoffbauer University of Applied Sciences in Potsdam, Germany. Karsten is currently working as a freelance scientist, author, and guest lecturer at the Humanistic University of Applied Sciences in Berlin, Germany. His research interests include social inclusion, decolonization of social work, and Indigenous knowledge in the context of culturally responsive social work and ethics. Outside of his academic work, he supports the survival of the traditions and language of his Baltic ancestors.

Gunta Kļava (Dr. philol.) is a sociolinguist with extensive experience in researching language situations and language learning. She obtained her PhD in General Linguistics from the University of Latvia in 2019 with a dissertation devoted to the sociolinguistic aspects of migration processes. Her research interests include language policy, endangered languages, language and identity, and language acquisition. For more than fifteen years, she has been researching the language situation in Latvia that serves as the foundation for language policy in the country and has been the editor-in-chief and author of several reviews of the situation of the Latvian language.

Mark Lawrence is Professor Emeritus of Geography at the Center for Sustainability Studies, Bemidji State University, and a frequent lecturer for graduate social work courses at the Fachhochschule Erfurt in Germany. With a teaching emphasis on regional planning and political ecology, his interests have included development-induced displacement and resettlement, urban biodiversity, emergent urbanism in slums, participatory GIS, and new spatial semiotics. For twenty years, he worked with villagers in Kenya's Kibwezi Division mapping sacred landscapes for purposes of wood and water conservation and in resistance to a foreign-owned plantation's seizure of local people's lands.

Manfred Liebel is Professor Emeritus of Sociology at the Technical University of Berlin and honorary professor of intercultural studies on childhood and children's rights at the University of Applied Sciences Potsdam, Germany; born in Germany as the child of a baker and working as a baker before his studies in sociology and public law; due to self-experienced discrimination as a working-class child, early engagement with bourgeois class arrogance and social injustice; later engagement in the popular education (*educación popular*) and as action researcher in Latin America, primarily with working and indigenous children and youth; collaborator and consultant of the movements of working children and adolescents in Latin America, Africa and India; convinced that all educational practice and research must be based on solidarity and contribute to the liberation and emancipation of disadvantaged and oppressed young people.

Ronald Lutz is a non-Indigenous professor emeritus and doctor in sociology and cultural anthropology at the University of Applied Sciences, Germany. He was born and grew up in Germany. His research interests include the sociology of modernity, international relations, globalization, postcolonial theories, indigenization, and epistemologies of the South.

María Leticia Briseño Maas (Dra.) is a researcher attached to the Universidad Autónoma "Benito Juárez" of Oaxaca. With twenty-three years of academic experience, she has consolidated the following lines of research: education, youth, and gender. She is the founder and leader of the CA Education and Knowledge Construction, and manages university policies for gender mainstreaming and eradication of violence against women. Throughout her career, she has maintained a deep social commitment and responsibility in Oaxaca, an entity characterized by high rates of poverty, violence, and educational backwardness.

Eduardo Bautista Martínez has a doctorate in social sciences from the Universidad Autónoma Metropolitana (UAM) and is a professor-researcher at the Instituto de

Investigaciones Sociológicas of the Universidad Autónoma "Benito Juárez" de Oaxaca (UABJO). Born in the community of Villa de Mitla, Bautista Martínez has Zapotec and Afro-descendant roots, which have nurtured his research approach from a perspective deeply connected to the socio-cultural realities in Oaxaca. During the twenty-seven years of his academic career, his work has focused on issues related to education, interculturality, identities, and social actors, as well as processes of struggles and resistance to power.

Mbazima Simeon Mathebane is a professor and is Head of Graduate Studies or Higher Degrees at the College of Human Sciences, University of South Africa (UNISA). He also supervises postgraduate students. He previously chaired the Departmental Scientific Research Committee. His research interests include African indigenous knowledge systems, Afrocentrism, and decoloniality. He obtained his BA and MA in social work from the University of the Witwatersrand, Johannesburg, and a PhD from UNISA. He is a recipient of the Association of South African Social Work Education Institutions (ASASWEI)'s 2017 Social Work Up-and-Coming Educator of the Year Award. He previously served on the Executive Committee of ASASWEI as treasurer. He is a member of the editorial board of the Journal of Indigenous Social Development (JISD) as well as the Journal of Environment and Social Psychology. He serves on the Editorial Advisory Board of the Journal of Social Work/Maatskaplike Werk and is a reviewer for several local and international journals. Professor Mathebane lives in Pretoria and is a member of the Xitsonga-speaking Shangaan nation in South Africa. He was born and raised in a village known as Shawela, found in Bulobedu in the former homeland of Lebowa, now part of the province of Limpopo. His chapter is written from the vantage point of being an indigenous South African and his lived experiences of coloniality and apartheid in South Africa.

Ksenija Napan, RSW (registered social worker), MSocPsych (Master of Social Psychiatry), PhD (in social work education), Associate Professor at Massey University, Tāmaki Makaurau (Auckland), Aotearoa (New Zealand). Ksenija was born in Zagreb, Croatia, and Aotearoa is the country of her choice, where, inspired by Māori wisdom, she learned more about her own culture and her Indigenous background than she had during thirty-three years of living in Zagreb. Politically aware of injustices perpetrated against Indigenous people and the global majority around the world, she managed to stay a hopeless optimist, believing that there is still a possibility of co-creating a socially just and culturally and spiritually respectful sustainable world. She is passionate about bridging seemingly opposing polarities and hopes that within a context of reciprocity and mutual respect, Indigenous wisdom can work in partnership with the advances of modern technology. Ksenija is passionate about inventing novel teaching and learning methods that reach people personally, professionally, politically, and culturally. She is a social worker by profession, a teacher and activist by calling, and an endlessly curious lifelong learner committed to the peaceful resolution of wicked problems.

Ndangwa Noyoo is a professor and director of the Southern African Policy and Development Nexus (SAPDN), a research-based and policy-driven consultancy and think tank that is based in Cape Town, South Africa. He is also a Research Fellow in the

Department of Social Work at the University of the Free State, South Africa. Among other things, SAPDN aims to foster the development of Africa for Africans and is anchored in Indigenous African Knowledge and seeks to arrive at relevant solutions to African challenges. He is originally from Zambia, from the Lozi ethnic group. The Lozi and their pre-colonial kingdom of Bulozi or Barotseland and its indigenous governance systems still exist today, in present-day Zambia, albeit in diluted form. Ndangwa Noyoo's research interests are social policy, public policy, social development, comparative social welfare regimes typologies, Corporate Social Responsibility (CSR), and indigenous social work and social welfare and knowledge systems. He studied at the universities of Zambia, Cambridge, and the Witwatersrand. He was a postdoctoral fellow at the Fondation Maison des Sciences de l'Homme (FMSH) in Paris, France (2005–6).

Samira Adel Osman is Professor of Asian History at Federal University of São Paulo (UNIFESP), Brazil. Born in Brazil, in a Lebanese and Muslim family. She has been working on topics related to the Middle East, the Arab world and Islam, and their diasporas in Brazil and Latin America. The point of view from which she approached the research and her chapter is related to the perspective of authors such as Fatima Mernissi, Leila Ahmed, and Lila Abu-Lughod, as well as Edward Said and Ella Shohat. These authors present a counterpart to how the others of the Middle East have been viewed by the West. Although she is Muslim, she does not wear the hijab.

Barbara Schellhammer is a professor at the Munich School of Philosophy; she holds the research chair for Intercultural Social Transformation and is the director of the Center for Social and Development Studies. She lives in Tirol, Austria, and used to live in Canada, where she conducted research in the Arctic (Northwest Territories). Some of her main research interests are intercultural philosophy (with a focus on Indigenous thought), philosophical anthropology, and phenomenology of the alien. She writes from the perspective of a German philosopher; however, due to her background in social work and family therapy and her international experiences, she comes from "in-between" places: between different cultures and different disciplines, also incorporating artistic and bodily experiences in her way of doing philosophy.

Peter Strack was born in 1958 in Dillenburg, Germany. A former health worker, he came first in 1978 to Bolivia for a social service in SOS Children's Village, San José de Chiquitos, and obtained a diploma and doctorate in social sciences at the University of Bielefeld. He worked from 1991 to 2015 at the press department and as a coordinator for the Andean Regional Office of child rights NGO Terre des Hommes Germany until retirement, as well in Cochabamba as the Bolivia country program manager of Suisse NGO INTERTEAM in the organization of knowledge transfer and capacity building in the area of prevention of violence and therapy. Actually assessing grassroots initiatives in Bolivia and publishing in German ila-journal of Center for Information on Latin América and at "Latinorama" Blog of German newspaper taz. Intercultural dialogue was a daily reality not only in office work and contact with partner projects, but also in family relations being partnered with a Bolivian Quechua musician and songwriter, media, and recently also rural worker with a traditional miner's background.

Stan Wilson (born 1938) has a PhD from the University of California (1989) and is a member of the Opaskwayak Cree Nation, where he spent his early years speaking and living his family's language, culture, and traditions of Ainisininewak. Stan works from within an Indigenous paradigm using what he refers to as *Indigegogy*, promoting and encouraging Indigenous people to honor and utilize their own unique knowledge. After teaching for many years and working as an educational administrator, Stan returned to university to complete his PhD. He has taught at the University of Alaska in Fairbanks, California State University, Sacramento, and finally at the University of Alberta. Stan and his wife, Peggy, are now retired from the University of Alberta, Edmonton, where they co-founded the Ph.D. Specialization in First Nations Education and established the Masters' Degree in Land Based Education which is now at the University of Saskatchewan. He is often asked to speak as a guest on university courses to discuss issues in Indigenous education. He serves as an Elder for the Opaskwayak Cree Nation Council.

Preface

The texts in this book point to a demand for recognition of narratives about the world based on Indigenous, decolonized knowledge and worldviews. Perhaps more than reparation for a distant past, there is this desire to provide new ways of, as Paulo Freire pointed out, "saying the world, saying the future."

In this sense, it might be important to say that many of the global challenges and problems are connected to a "monologue" that doesn't recognize this plurality of worldviews. We must acknowledge.

The current and terrible wars are a monologue, all of them. Environmental exploitation and degradation are a monologue. The condemnation of entire populations to oblivion is a type of monologue.

We hope the book's contribution is to encourage off-center dialogues. To promote listening to everything that lives in this world, to give visibility to the beauty of the encounter.

We decided to choose an image for the cover of the book that shows the sea. Many creation stories around the world say that all life comes from the sea. The sea and water are often attributed to the power to purify the mind. In this sense, the sea can be seen as a symbol of decolonization if decolonization is understood as purifying the mind from colonial logic.

We are very thankful to all the authors; we appreciate their efforts to participate in this project and to develop the ideas of dialogues as a contradiction to monologues. We hope it will continue.

We are also very grateful to Kyra Emilia Lutz for her great editorial support, which helped us a lot in preparing the manuscript.

Gregory Cajete, Kefilwe Johanna Ditlhake, Maria do Carmo dos Santos Gonçalves, Karsten Kiewitt, Ronald Lutz

Introductions

1

On the Decolonization of Dialogues

Karsten Kiewitt

A personal story

Some years ago, at the suggestion of a friend who is a member of the Piikani First Nation in Canada, I read a book by an Indigenous author for the first time. This book, *Blackfoot Ways of Knowing. The Worldview of the Siksikaitsitapi*, by the social worker and professor Betty Bastien (2004), highlights the importance of the memory of traditional ways for future survival, cultural integrity, and renewal. I was surprised to find that this book had an afterword by a German psychologist, Jürgen W. Kremer, which begins with the question: "Why should a social worker, educator, psychologist, or health professional from a different cultural background or tradition be immersed in *Siksikaitsitapi* ways of knowing?" (2004: 184).

This question can be extended: why should Western academics and researchers be concerned with Indigenous knowledge systems? "At the heart of any answer to these questions lies the challenge to establish egalitarian and respectful exchanges among cultures" (Bastien 2004: 184). That is what I am looking for. But how can this be done? For Jürgen W. Kremer, creating a relationship to one's own cultural roots provides a central foundation for the dialogue with Indigenous people. He recommended I should read the book *Ethnoautobiography. Stories and Practices for Unlearning Whiteness, Decolonization and Uncovering Ethnicities* (2014), which he wrote with Robert Jackson-Patton.

At the beginning of the book is a key question: "Who am I?" (Kremer and Jackson-Patton, 2014: 15). This question moved me to look more closely into my own cultural background. I was helped in this by detailed information about my family that my grandfather had gathered together, and today I know where my cultural roots lie. My ancestors are from the *Prusai* people, an ancient Baltic ethnic group that was colonized in the thirteenth century by the Knights of the Teutonic Order. Descendants are now working to revive the religion, traditions, and language. The *Prusai* belonged to the same cultural and linguistic group as their neighbors, the Lithuanians and the Latvians. This helps me to find out more about my cultural roots and even makes it possible to learn the reconstructed language.

Exchanges with Indigenous people have shown me the importance of ties to ancestors, to locations, to traditions, and to nature. The interest in my own cultural background seems to me to be crucial for the process of my self-decolonization. Having said this, I would now like to present the following thoughts.

First thoughts

The world is facing enormous challenges in the shape of growing global inequality, threats of climate change, and violent conflicts and wars. It is becoming increasingly clear that the roots of these challenges lie primarily in the economic models and the imperial lifestyles that have developed in the societies of the Global North. The separation of nature and culture (Descola, 2021) has not only resulted in the destruction of the ecosystems of our planet but has also led to a colonization of the world by the Global North. Along with this, non-Western worldviews, traditions, and epistemologies have been devalued and marginalized.

With the colonial, imperial, and hegemonial structures that the Global North established globally, a Eurocentric science was established which views itself as universal and objective. Only knowledge generated in accordance with the criteria of this modern science is held to be legitimate and acceptable. In this context, non-Western epistemologies and worldviews have largely been rejected as unscientific knowledge.

However, alongside Western science, there are also various "Epistemologies of the South" (De Sousa Santos, 2016). Indigenous knowledge systems are based on experiences and insights gained by Indigenous peoples over thousands of years in direct relationship with the nature around them (Cajete, 2000; see also *Native Science* by Gregory Cajete in this book). Indigenous knowledge systems and Indigenous or Native sciences are unique and legitimate.

In Europe and in the Eurocentric societies, there has been hardly any discussion about cooperation with Indigenous peoples and how a respectful exchange about their knowledge and their sciences can be made easier. One of the reasons for this is that the science of the Global North still always draws on colonial and imperial practices, which hinder a dialogue with Indigenous thinkers and scholars. There are now a large number of publications by Indigenous authors on this topic, but little attention has been paid to them by the majority of Western scientists.

It seems that there have been very few attempts to understand Indigenous epistemologies and to enter into a dialogue with the preservers of this knowledge. This gap must be closed. Intensification of the dialogue between Indigenous and non-Indigenous thinkers seems to be urgently necessary in order to promote a change of perspectives and show ways to overcome the global problems of the Anthropocene. Such a dialogue can be a driving force for the urgently necessary decolonization of hegemonic positions of the North.

But what is decolonization, and where does it begin? The discourse about decolonization is currently growing in intensity, as shown by increasing numbers of publications. In particular, people who have been impacted by colonial practices and their consequences are quite rightly formulating demands for decolonization. Specific sectors are often named that need to be decolonized, e.g., the Western paradigm of research (Tuhiwai Smith, 2012), sociology (Meghji, 2020), social work (Gray et al., 2016), or the (Western) understanding of solidarity (Land, 2015).

In the continuation of this discourse, a book is presented here that aims to contribute to the decolonization of the dialogue between non-Indigenous and Indigenous people. According to Paulo Freire (1972), a dialogue cannot occur between anyone denying others the right to speak and those whose right to speak is being denied. For Freire, dialogue is "encounter between men, mediated by the world, in order to name the world" (Freire, 1972: 61). Such a dialogue is only possible if people listen to each other, learn from each other, and are willing to act together. When people with a colonial heritage, whether the colonized or the colonizers, enter into a dialogic exchange, this requires a profound acceptance of decolonization. This applies in particular to people with a Western socialization, who have long been acculturated to the hegemony of the West over the Rest (cf. Hall, 1996).

The colonization of the mind

The so-called European Enlightenment began a process of indoctrination that rapidly spread all over the world. "Mental slavery, or the colonization of the mind, is the process of subjugating the mind of people so they willingly accept the political narratives of the colonial powers and behave accordingly. It is the process of manipulating the minds of people" (Hira, 2023: 83). This process has two central functions: it indoctrinates people with a Western socialization with an idea of superiority and those impacted by colonization with a feeling of inferiority (Hira, 2023: 83).

A central colonial tenet of the Enlightenment is its universalism, and Western science still claims today to be the only correct way to understand, explain, and administer the world. Such a claim already shows that this science is indeed prejudiced. Its reductionism and its methods of decontextualizing lead to a distorted view of nature and society. They play a central role in the suppression and discrediting of Indigenous knowledge systems and traditions as mere superstition and presenting these as inferior and uncivilized (Deloria, 2002).

An important element in the colonial implementation of Western science is language. Colonialized peoples all over the world have been forced to speak the languages of the colonial powers. In order to reach an international readership, Indigenous and non-Indigenous authors, thinkers, and scholars have to publish in English. The Kenyan writer Ngũgĩ wa Thiong´o pointed out the dual character of language as a means of communication and as a carrier of culture (2008). The rejection of speaking and writing in one's own language is a colonial strategy. While it enables communication, it suppresses culture. The result is that the communication is fragmentary.

The outcome of this indoctrination is the loss of memories by both the colonizers and the colonized. This "organization of amnesia" (Hira, 2023: 107) involves false information, trivialization, and silence regarding colonial crimes. The colonial amnesia contributes to barbaric acts of the colonial powers being excluded from the consciousness of the colonizers so that a public acknowledgment of colonial crimes is prevented (Yellow Bird, 2016). Withholding knowledge about how Indigenous populations were shaped by the colonial powers is a mechanism that can lead to colonization amnesia (Porter, 2005). A colonialized mind is a fundamental block for dialogue. The central task is therefore to decolonize the mind.

Decolonizing the mind

Who am I? Answering this question is fundamental for decolonization. "Decolonization is a process that begins with the understanding that one is colonized (at whatever level that may be). It is creating and consciously using various strategies to liberate oneself from or adapt to or survive in oppressive conditions. It is the restoration of cultural practices, thinking, beliefs, and values that were taken away or abandoned but are still relevant or necessary for survival and well-being" (Yellow Bird, 2016: pos. 6448). In order to decolonize the mind, the colonial and the colonization amnesia must be overcome. For people with a Western socialization, this means acquiring knowledge about the colonial practices. This includes becoming aware of the processes and mechanisms of colonization and how they influence people. But it is also necessary to understand how the colonization impacted their ancestors.

Many people in Europe have ancestors who experienced colonization. European and Eurocentric people also come from societies that had connections with nature and location. In the process of decolonization, an awareness must be developed that this knowledge is also present within them, even though it has been suppressed within the framework of Western socialization. The Indigenous musician, poet, and political activist John Trudell said in 1998: "We are all descendants of tribes [. . .] We all carry the collective experience of 'before' when we understood that life was about spirit—the 'being part', who we are" (as quoted by Seguine, 1998). He explained that in the course of the civilization process, people became more and more distanced from what they are. Along with the "mining of minerals," there has also been a "mining of the spirit," with destructive consequences (Seguine, 1998). The modern, colonialized Western mind denies the sources of the equilibrium in us and the importance of our background for the future. "The viewpoint of western modernity [. . .] is the distancing inspection of people and nature [. . .] such a worldview has its origin in a dissociative and objectifying understanding of self and reality" (Kremer and Jackson-Paton, 2014: 22). Thus, the decolonization of the mind is the precondition and the foundation for the dialogue with people in different cultural contexts.

In my opinion, the recollection and revival of the experience of the "before" is fundamental. Learning about the life, the culture, and the religion of the ancestors and understanding the importance of this learning are the first steps toward self-decolonization. Re-establishing and revitalizing the relationship to the ancestors, their community, ceremonies, traditional stories, locations, and nature is a key step in order to expand the mind and the feeling for oneself. It is the way to the decolonization of the Western mind and, in the end, enables a real decolonized dialogue. This "ethnoautobiography" (Kremer and Jackson-Patton, 2014) helps to change the Eurocentric modes of thought that get in the way of the dialogue. "Looking at Indigenous cultures can also break open the modern individualistic senses of a well-boundaried and defended self that sees itself in opposition to and separate from community and nature" (Kremer and Jackson-Patton, 2014: 24). This is not only about finding traces of the personal cultural past in us but rather about taking down the hegemonic and imperial habitus toward the world. People will still engage in colonial deeds, however subtle these may be, until they understand their own cultural roots and relationships. A real dialogue only becomes possible when the dialogue partners know who they are as cultural persons, even if the knowledge is only fragmentary. The decolonization of the mind is a prerequisite for people from different cultural contexts to meet in an "ethical space of engagement" (Ermine, 2007) and create a dialogue.

This book sees itself as such a space, in which the authors can share their knowledge, their thoughts and ideas in dialogue, and develop visions for shaping a just and varied world. Indigenous and non-Indigenous authors, thinkers, and scholars—people with varied and individual cultural backgrounds around the world express themselves here. The authors present various perspectives about the creation and shaping of dialogues and about challenges and opportunities of dialogic exchanges. They develop perspectives and utopias about how a decolonized dialogue will be possible.

At the same time, this volume's contributions express a contradictory complexity in a diverse and differentiated way. It arises from the different perspectives and approaches of the authors and illustrates the diversity of ways of thinking. This complexity cannot be resolved and must be accepted as it is. It offers the opportunity and the challenge of dialogue. Each contribution is unique and legitimate. The goal of decolonization is combined with the endeavor to listen and learn from each other.

References

Bastien, B. (2004), *Blackfoot Ways of Knowing. The Worldview of the Siksikaitsitapi*, Calgary: University of Calgary Press.

Cajete, G. (2000), *Native Science. Natural Laws of Interdependence*, Santa Fe: Clear Light Publishers.

Deloria, V., Jr. (2002), *Evolution, Creationism, and Other Modern Myths. A Critical Inquiry*, Wheat Ridge, Colorado: Fulcrum Publishing.

De Sousa Santos, B. (2016), *Epistemologies of the South. Justice against Epistemicide*, Abingdon: Routledge.

Ermine, W. (2007), "The ethical space of engagement," *Indigenous Law Journal*, 10(6): 193–207. https://jps.library.utoronto.ca/index.php/ilj/article/view/27669/20400 (accessed August 13, 2020).

Freire, P. (1972), *Pedagogy of the Oppressed*, London and New York: Penguin Books.

Gray, M., J. Coates, M., Yellow Bird, and T. Hetherington (eds.) (2016), *Decolonizing Social Work*, Abingdon: Routledge.

Hall, S. (1996), "The West and the rest. Discourse and power," in S. Hall, D. Held, D. Hubert, and K. Thompson, *Modernity: An Introduction to Modern Societies*, Oxford: Blackwell Publishers.

Hira, S. (2023), *Decolonizing the Mind. A Guide to Decolonial Theory and Practice*, The Hague: Amrit Publisher.
Kremer, J. W. (2004), "Afterword: remembering ancestral conversations," in B. Bastien, *Blackfoot Ways of Knowing. The Worldview of the Siksikaitsitapi*, Calgary: University of Calgary Press.
Kremer, J. W. and R. Jackson-Patton (2014), *Ethnoautobiography. Stories and Practices for Unlearning Whiteness, Decolonization, Uncovering Ethnicities*, Sebastopol: ReVision.
Land, C. (2015), *Decolonizing Solidarity. Dilemmas and Directions for Supporters of Indigenous Struggles*, London: Zed Books.
Meghji, A. (2020), *Decolonizing Sociology. An Introduction*, Cambridge: Polity Press.
Ngũgĩ wa Thiong´o (2008), *Decolonizing the Mind. The Politics of Language in African Literature*, Martlesham: Boydell and Brewer.
Porter, R. (2005), "The decolonization of Indigenous governance," in W. A. Wilson and M. Yellow Bird (eds.), *For Indigenous Eyes Only: A Decolonization Handbook*, Santa Fe: School of American Research Press.
Seguine, J. (1998), "Progress has produced 'mining of the spirit' that we must work individually, collectively to reclaim, Trudell says," https://ratical.org/many_worlds/JohnTrudell/MiningSpirit.html (accessed October 9, 2023).
Tuhiwai Smith, L. (2012), *Decolonizing Methodologies. Research and Indigenous Peoples* (2nd. ed.), London: Zed Books.
Yellow Bird, M. (2016), "Postscript terms of endearment: a brief dictionary for decolonizing social work with Indigenous peoples," in M. Gray, J. Coates, and M. Yellow Bird (eds.), *Indigenous Social Work around the World. Towards Culturally Relevant Education and Practice*, Abingdon: Routledge (Kindle edition).

2

What Is Native Science?

Gregory Cajete

Introduction

The word "science" is derived from the Greek word for "knowledge." In this perspective, *science* is used in terms of the most inclusive of its meanings, that is, as a story of the world and a practiced way of living it. *Native science* is used as a metaphor for *Native knowledge* and its creative participation with the natural world in both theory and practice. This essay celebrates and acknowledges the Native contribution to an evolving and inclusive world philosophy of science as well as ecological awareness. This is not an essay about Native religion; in fact, religion will be discussed only as a reflection of environmental philosophy. The essence of Native spirituality is not religion in the Western sense of the word, but rather a set of core beliefs in the sanctity of personal and community relationships to the natural world, which are creatively acted upon and expressed at both the personal and communal levels.

Indeed, Native science is a product of a different creative journey and a different history than that of Western science. Native science is not quantum physics or environmental science, but it has come to similar understandings about the workings of the natural laws through experience and participation with the natural world. In researching Native science, the groundwork for a fruitful dialogue and exchange of knowledge has the potential for being created. But it must be a dialogue in which Native cultures have the opportunity to gain as much as they share about their understanding of natural laws.

Native science reflects the unfolding story of a creative universe in which human beings are active and creative participants. When viewed from this perspective, science is evolutionary—its expression unfolds through the general scheme of the creative process of first insight, immersion, creation, and reflection. Native science is a reflection of the metaphoric mind and is embedded in creative participation with nature. It reflects both the intellectual and the sensual capacities of humans. It is tied to spirit and is both ecological and integrative.

Note: the terms *Native* and *Indigenous* are used interchangeably in this essay to refer to tribal and cultural groups who continue to live close to the environment on ancestral aboriginal homelands.

What is Native science?

What is Native science? In order to address this question, it is first important to define the boundaries of Native science. In reality, "Native science" is a broad and inclusive term that can include categories such as metaphysics and philosophy; art and architecture; practical technologies and agriculture; and ritual and ceremony practiced by indigenous peoples both past and present. More specifically, Native science encompasses such areas as astronomy, farming, plant domestication, plant medicine, animal husbandry, hunting, fishing, metallurgy, geology, and an array of other studies related to plants, animals, and natural phenomena. Yet Native science extends beyond even these areas to include spirituality, community, creativity, and appropriate technologies that sustain environments and support essential aspects of human life. In addition, Native science may also include exploration of basic questions such as the nature of language, thought, and perception; the movement of time-space; the nature of human knowing and feeling; the nature of proper human relationship to the cosmos; and other such questions related to natural reality. Native science is the collective heritage of human experience with the natural world and, in its most essential form, a map of natural reality drawn from the experiences of thousands of human generations that have given rise to the diversity of human technologies and even to the advent of modern mechanistic science. In very profound ways, Native science can be said to be "inclusive" of modern science, although properly conditioned Western scientists would go to great lengths to deny such inclusivity.

Why is Native science important?

In Native languages, there is no word for science, as there is no word for art or for philosophy, psychology, or any of the other linguistic labels for foundational ways of *coming to know* and understand the nature of life and our relationships therein.

Not having, or more accurately *not needing*, a word for science or art or psychology did not diminish either its practice or importance in Native life. For Native people, *seeking life* was the encompassing task of their lives, and every member of the tribe, in their own time and through their own unique capacity, was a scientist, an artist, a storyteller, and a participant in the great web of life around them. While there were tribal "specialists" with particular knowledge, technologies, and ritual understanding for specific cultural practices and societal needs, the creative spirit and focus on essential life-sustaining relationships guided the processes of Native science.

Native science is a metaphor for a wide range of tribal processes of perceiving, thinking, acting, and "coming to know" that have evolved over millennia of human experience with the natural world. Native science is born of a lived and storied participation with a natural landscape and reality. To gain a sense for the essence of Native science, one must also participate with the natural world; to understand the foundations of Native science, one must become open to the roles of sensation, perception, imagination, emotion, symbols, and spirit, as well as concepts, logic, and rational empiricism. Much of the essence of Native science is beyond words and literal description. Indeed, misapplied words many times destroy the real and holistic experience of nature as a direct participatory act around which Native science has evolved. In the literal terms of biology, Native science may be seen as an exemplification of "biophilia," or the innate instinct that we and other living things have for affiliation with other life and with the animate world.

In anthropological terms, Native science may be viewed as "animism," or the worship of nature. In the conceptual framework of philosophy, Native science may be said to be based upon perceptual phenomenology. There can be numerous other definitions, but at its core, Native science is based on the natural perceptive wisdom and knowledge gained from using the whole body of our senses and on direct participation with the natural world.

In today's world, the appreciation and understanding of the nature of Native science is essential to the re-creation of a "participatory" science of life that is so desperately needed to balance the imbalance of science and technology and its continuing social and economic crisis consequences. For Native Peoples themselves, the revitalization of Native science is an essential component of cultural revitalization and preservation. It is the participation with the communal tribal landscape, which evolved from a direct relationship to "a place or places" in their history, that defines them as a people.

Today, modern societies are wholly estranged from the natural world. Yet, cut off from direct experience of and participation with the natural world, modern peoples collectively suffer a kind of sensory and emotional starvation, which at times reaches epidemic proportions and leads to the many and various forms of "acting out mind "at the individual and collective level.

According to ecological philosopher Henryk Skolimowski, the modern mind has lost its sensate bearings, its orientations to its roots and to the natural world of process of which it is a part. To address this perceptual blindness, a "new science" must be developed that includes at its very core the reintroduction of the *participatory mind*. It must once again become a mind that cyclically renews its orientation to the primal world of life and living processes. It must be a mind that recalls and re-establishes the basis of human awareness in the larger ecology of the world. Science as a whole must once again become the story of this awareness and relationship to the animate living world. And language must once again be understood as the body-based vehicle that refers to the animate landscape that first gave rise to its expression. The thought that guides our actions and comprises our contemplation must be understood as a part of the dynamic expression of the physical as well as metaphorical AIR that animates all nature. The intelligence that guides, focuses, and expresses our unique nature must be understood to be present in all things in nature. Therefore, a plant, an animal, a mountain, or a place may be metaphorically said to have "intelligence," its own mind and psyche that is unique to it and with which our human intelligence continually interacts (Skolimoski, 1994).

> The sense of being immersed in a sentient world is preserved in the oral stories and songs of indigenous peoples—in the belief that all sensible phenomena are all alive and aware, in the assumption that all things have the capacity of speech. Language for oral peoples, is not a human invention but a gift from the land itself.
>
> Abram 1996: 262–96

It is the "ecologically aware and participatory mind" that modern science must encompass, for it is that mind, that way of thinking, understanding, and participating with the natural world, that holds the best and most life-sustaining solutions to the current disconnection of science from the ground of its own being.

Sociobiologist E. O. Wilson advocates that everything that we consider human is reducible to our essential biological nature, which in turn is predicated upon the laws of physics and nature. Evolutionary philosopher Ken Wilber believes that there can be no return to pre-modern modes of life in the evolutionary flow of human consciousness, although prior forms of consciousness continue to exist in human societies today. Yet both agree that some kind of integration of science and spiritual practice must begin if human knowledge is to advance to its next level of development. The integration of the concentric rings of the "great nest of being," or matter, life, mind, soul, and spirit, and their associated realms of study, i.e., physics, biology, psychology, theology, and mysticism, must be unified as the field of human knowledge and meaning if we are to move beyond the current mono-dimensional view of mechanistic science. A new balance must be created for a new evolutionary phase to occur.

The unity of knowledge, or the integration of science and religion, has become one of the major theoretical debates in the philosophy of science and in theology. Essentially, science philosophers and theologians alike have come to realize that the integration of science and spirit, or the lack thereof, will determine the direction of contemporary society in the twenty-first century.

Native science in its diverse forms of historic as well as contemporary expression presents templates for the kind of reconstruction and evolution of modern science required to address the issues and needs of a twenty-first-century world in the throes of ecological crisis and for which it shares responsibility for creating. Ecologically responsible science must become the hallmark of twenty-first-century technology if we are to face the monumental environmental and social challenges of the next millennium. The new story of science must once again assist us in:

> living in storied relation and reciprocity with the myriad things, the myriad beings, that perpetually surround us. . ..Only if we can renew that reciprocity- grounding our newfound capacity for literate abstraction in those older, oral forms of experience- only then will the abstract intellect find its real value. It is surely not a matter of "going back," but rather of coming full circle, uniting our capacity for cool reason [science] with the more sensorial and mimetic ways of knowing, letting the vision of the common world root itself in our direct, participatory engagement with the local and the particular.
>
> Abram 1997: 270

Native science offers both challenges and opportunities for Western science since its' insights, forms of knowledge, and processes parallel the insights of evolving areas of Western science such as quantum mechanics, chaos theory, string theory, and others.

The development, simultaneous exploration, and comparison between Native science and Western science can provide the foundation for the flexibility and creative orientation to thinking and application in science that is essential to the future of human societies. Native scientists, though few in number, have the opportunity to become leading advocates for such rethinking and transformation of Western science.

How is Native science viewed?

There are those who would argue that there is no such thing as "Native science." That science is an invention of modern Western society and that so-called "indigenous" peoples have a body of cultural folklore, living practices, and thought that cannot be considered a rational and ordered system of theory and investigation comparable to anything found in Western science. Whether Native science exists in Western terms is largely an incestuous argument of semantics. Using Western orientations and definitions to measure the credence of non-Western peoples ways of knowing and being in the world has been historically applied to deny their lived reality. The fact remains that Indigenous people *are*; they exist and do not need an external authority to validate their long existence in the world. Attempts to define Native science, which is by its nature alive, dynamic, and ever evolving through generations, fall short since it is characteristically a high-context inclusive system of knowledge.

Native science may be viewed as little more than primitive animism, sentimentalism, and an object to be studied by science rather than as a "real" science by practitioners of Western science. Regardless of these biases, the perspective of Native science has great potential for developing insight and guidance for the creation of the kind of environmental ethics and deep understanding that must be gained in the critical times ahead. The serious study of environmental relationships to land, plants, animals, and natural resources provides much-needed models for understanding environmental sustainability as an emergent process.

Native science engenders in its very process and content the revitalization of our human "biophilic" sensibilities, our instinctual predisposition to seek and be nourished by natural relationships. It also provides a way to connect with the indigenous soul deep within each of us. As a system of thought and process of application, it can provide an expansive paradigm for applying scientific understanding. Indeed, the "perception" of science must expand to make it the whole and comprehensive form of human knowledge it needs to become.

For Indigenous people themselves, the revitalization of Indigenous knowledge through a truly self-determined education provides the most direct route for Native sovereignty. Nowhere is the path for this more direct than in the connection that Indigenous people feel for their homelands.

The research of Native science is only in its initial stages as an area of serious study. Nevertheless, preliminary exploration has already revealed tremendous data in such areas as plant medicines, archeo-astronomy, and philosophical areas related to quantum theory. This is only the beginning of a new and creative evolution of science. Yet, it must be a *dialogue* and not the replay of the past in which Native peoples have been exploited for their resources and ideas, receiving very little in return. The issues of intellectual and cultural property rights as they relate to plant medicines provide one case in point. There are many others. The ideas and processes of Native science are equally important as conceptual well-springs for helping to bring about the integration of science and spirit, which is essential to the marriage of "truth," the ideal goal of science, with "meaning," the ideal goal of spiritual practice. Native peoples must be given credit for their intellectual achievements and benefit from the tremendous research that has been accumulated with regard to Native science.

References

Abram, David (1996), *The Spell of the Sensuous: Perceptions and Language in a More-Than-Human World*, New York: Vintage Books.

Cajete, Gregory A. (2000), *Native Science: Natural Laws of Interdependence*, Santa Fe: Clear Light Publishers.

Skolimowski, Henryk (1994), *The Participatory Mind: A New Theory of Knowledge and the Universe*, New York: Arkana Books.

Wilber, Ken (1981), *Up From Eden: A Transpersonal View of Human Evolution*, Garden City: Anchor/Doubleday.

Wilson, E. O. (1984), *Biophilia*, Cambridge: Harvard University Press.

Excerpts from: Cajete, Gregory A. (2000). *Native Science: Natural Laws of Interdependence*, Santa Fe: Clear Light Publishers, pp. 2–5.

Part One

Dialogues

3

To Know from Within

Benno Glauser

Introduction

Some 600 or 800 kilometers north of where I write these lines, several small groups of indigenous people live in voluntary isolation in the still remaining forests of South America's Gran Chaco ecoregion. They belong to the Ayoreo peoples, and they share their very special condition—a life without contact with our modern world—with a good hundred other ethnic groups of the Amazon ecoregion. This chapter focuses on and explores the ways the isolated Ayoreo have with what in modern Western-type life culture that we call *knowledge*. They seem radically different from ours.

There are limits to the very possibility of describing features of another human culture, to any "cross-cultural" description, even when made with a "decolonial intention." It is not possible to describe the others and their realm or world without applying the ways of gaining knowledge, the concepts, and words of our own culture. Cultures are mutually incommensurable; we are all part of a culture—at least one—and unable to assume an outside, "neutral" standpoint from which to describe, let alone value or judge, another culture different to ours.[1]

Thus, the above-stated attempt—to speak about the isolated Ayoreo's ways of dealing with knowledge—is sure to fail, and, practicing an attitude of radical respect, I should stop writing right here. But I nevertheless continue, as I have experienced that exploring limits, even when they seem totally impassable, and going as close as possible, even if there doesn't seem to be any point in doing so, at the very least produces new, unexpected outlooks and insights; at times, there are even surprises and breakthroughs.

I am personally used to such experiences: from 2002 onward, my work, which I shared with my working team colleagues of many years, consisted in protecting the integrity of the lives of the mentioned isolated groups in their territories. "Integrity" meant they should be able to continue living their own lives in their own way, without the interference of an expanding, persistently colonial modernity.[2] In the early years,

[1] See Raimon Panikkar on intercultural dialogue.

[2] The institutional vessel founded for this work is the NGO Iniciativa Amotocodie (www.iniciativa-amotocodie.org).

we said we protected their territories; today we understand that in reality, it is their "life realm" or "world" we protected. They have a word for their world or realm: "eami." Eami includes everything known that exists, all beings in nature: the soil, the plants, the animals, humans, colors, weather conditions, human-made objects, etc.—anything material, immaterial, and spiritual, tangible or not. Our objective was to make sure nobody—ourselves included—would penetrate their realm and endanger the delicate balance of their existence through what we call "contact." Meaning disturbing them, addressing them, or prompting or forcing them to alter and abandon the way they live. "Ourselves included": we had to learn to perceive them in their incessant roaming migrations well enough to protect them but no more, making sure our "seeing" where they were—with the help of traces or trails we happened upon, or of left-behind objects, or testimonies by other people who had spotted them from afar, etc.—would never ever touch their realm. Over the many years and combining diverse approaches and ways of observation, we developed a complex "monitoring methodology" and brought it to a level of the highest possible subtlety and near-perfection. It was a little like balancing a very breakable, virtually invisible being, a fragile living sphere with a complete life inside—"holding" it somehow without ever damaging or pinching it. Even the smallest mistake would mean breaking the coherence of their vital isolation and making them irremediably—and literally—fall into our world.

As a part of our work of protection, we tried to communicate and describe their life—what we knew about it—to others, in particular to the surrounding scarce population in the Gran Chaco, and in general to our own expanding society menacing them, including the government, and even further beyond, when realizing that the delicate relationship with the groups without contact was not a local matter but one of a paradigmatic quality concerning the entire humanity.

Thus we mentioned, or "said," them—in our descriptions and reports—making sure our words would not disturb them; we took pictures and filmed their forest, and we regularly scrutinized the satellite views so we would in time detect new deforestations nearing their territories and migration routes—pictures in which they undoubtedly were somewhere. We did so hoping our appropriation through our looking at them would not interfere in any way with them, indeed would not even be noticed by them. We seem to have succeeded insofar as no voluntary or involuntary contact from either side ever ensued from our careful gazing around in their direction and in their vicinity. We seem to have succeeded in granting their lives some degree of protection. Still, often we asked ourselves whether thinking about them, or talking about them with others, or dreaming about them at night could disrupt their normal existence or might somehow be perceived by them.

Exploring the limits and pushing them

So far the frame, the limitations, and the conditions in which my scrutinizing of their ways of knowing may produce some result. I was tempted to stop my writing several times when facing once more the prison of our own conceptual framework. We only have our own words, and they fail to penetrate the alien realm and gain some degree of hold on it unless we admit and allow the words to deploy their colonial effect, but then they invade and transform. So, I attempt a description for which I have no words. Beyond that, my perception is charged by biases of many different kinds: I naturally tend to use, when describing their world, the logic inherent to the descriptions of ours.

I use two main sources of knowledge about the isolated Ayoreo. The first is the already mentioned many years of listening intently to the forests where they live somewhere. The second are life stories shared with us by Ayoreo men and women who spent years of their youth and sometimes decades of their adult lives in the forest before they were contacted[3] and forced to enter our world.

Of course, when telling us about their lives before contact, they use the language of *our* world. Yet, many features of *their* world shine through and can be perceived somehow and felt. A wealth of insights into a life related to "eami" and the territories emerges, a profoundly spiritual way of life based on a natural, permanent rootedness and connectedness with it all. It seems characteristic of the Ayoreo that this becomes manifest not so much by explicit mentions and deeds, but implicitly; a delicately balanced partnership of equals between the Ayoreo and nature and the world renews itself constantly through a daily collective life, lived in finely and invisibly articulated family groups.

My writing necessarily uses the words of our language and our type of discourse with its cold rationality, abstractions, generalizations, etc. However, I grant myself certain liberties: I use a form of "active imagination"; instead of holding back at the limit, I actively enter the alien realm with an open curiosity, aware of the limitations or even the impossibility of a clear perception. What do I feel, and what could such feelings point at? What could they mean? What does my intuition say if I take it dead seriously instead of discarding it? I also force myself to admit impreciseness consciously and try to formulate even though I don't fully understand what I perceive; I leave myself open to simply grasping and plain rendering without much classifying; I admit the unresolved and leave it in the bag of possible knowledge I carry along just the same.

Finally, I allow myself a not-going-any-further when the limits I encounter seem really overwhelming; I withdraw and let it stand. I wonder whether this receding and leaving what-is-there and merely witnessing is a source of knowing also, "seeing" that there is something although it remains beyond my and our reach.

Quite characteristically, it seems, the mentioned methodological "liberties" are not alien to stances and attitudes that can be found in the toolbox of cognitive devices of the Ayoreo themselves.

[3] Of some 5,000 Ayoreo roaming until the mid-twentieth century in their immense traditional territory, more than 95% were contacted from 1945 onward and prompted/forced to abandon their traditional life. The last such contacts took place in 1979, 1986, 1998, and 2004. Many accounts and life stories told by those contacted were listened to and recorded by researchers and NGOs over the last decades. Seven such—particularly rich—life stories range among the ones received by Bolivian anthropologist Rosa Maria Quiroga (see References).

Life in that other world

I consciously use the verb form *knowing* rather than the substantive *knowledge*; it seems to suit an active, nomadic life better, a life in perpetual movement and only carrying along the strictly essential: something one *does* rather than something one *has*, like an object to be held or to be stored away outside ourselves. This also matches another quality that is present: the constant renewal, a basic characteristic of nature itself. The Ayoreo have to be actively knowing all the time, as knowledge acquired in the past may already be outdated and should not be taken for granted.

The Ayoreo are—and have been since "immemorial times"—hunters and gatherers, and only seasonally also planters (planting the seeds while continuing the nomadic migration, returning only some time later to harvest). Always on the go, they roam the respective group territories and sometimes wider parts of the entire, immense Ayoreo territory incessantly. They may get up in the mornings and—following directions given by somebody in the group fulfilling the role of scout—walk toward the next campsite along one of the many migration routes; some routes are used more often than others. The daily moving from place to place may be a matter of a few hours only; as a consequence, there is time to sit, to be freely together, to talk, to think, to be silent, to sing, to observe, and to exchange. Times of concentrated, prolonged, and intensive activity like hunting are followed by long periods of recovery and rest. The accomplished daily distance can be between two and five kilometers (the thick, thorny Chaco bush and its forest have countless challenges in stock and oblige one to go slow). When imagining this wandering from one campsite to the next, it springs to mind that one can only *have* extremely limited material belongings: only the strictly essential, only what can be carried along. As very small children must be carried, a natural regulation of the frequency of births is necessary. The old and the sick must be able to walk; if not, they stay—or are left—behind; such circumstances determine and define for the Ayoreo what in modern life culture we call "health" and "being unwell," and what meaning and place dying occupies in life. The group on its wanderings may return to a given campsite only every three or four years; this implies that what they will encounter when moving is only partly known to them, the rest is unknown and cannot be anticipated: the places on the route can have changed, surprises can lie in waiting, menaces by wild animals or enemies, but also chances for good hunting, for gathering fruit or harvesting honey; heavy weather may create sudden obstacles and may even force them to change the planned route; a sudden flood can mean the loss of vital seeds for the crops in summer. Finally and centrally important, the availability of water resources on the way—lagoons, rivers, waterholes—remains a permanent difficulty and can affect and even determine the group's wanderings to a large extent, especially in the dry winter season.

A matter of delicate balance becomes visible, a daily feeling anew and balancing out what nature and the world need to receive or are asked to give; also, how much they will give and what consequences this implies for the life of an Ayoreo group living with eami. They have long learned the correct measures between taking, receiving, and giving or yielding. What we call proportion comes into it, along with the quality of sufficiency:[4] knowing one's place in the world, knowing that humans cannot ask for just anything, that there are natural limits for what can be asked or taken from nature, and knowing that eami is a living being that—although it is enormous—has finite dimensions.

In addition to that, there is another quality present in the way of being of the Ayoreo, as in that of many other indigenous peoples; it is decisive in the present context: the certainty that the world is good as it is, and also, that the humans—the Ayoreo—are good as they are. There is no need to improve anything—or to develop—there is no original sin or debt that has to be purged, there is no guilt. The world can be trusted. The only task there for humans is gauging the mentioned balance, assuring the continuity. This is beyond any evolutionary concept, and interestingly, it places cultures like the Ayoreo way of life on the same level as, for instance, our own modern life culture, which we consider highly developed and—normally—superior. There is a chance we may finally be able to look across and recognize a new kind of equality, which also opens new chances for learning. A learning we may dearly be in need of if we want to survive.

Knowing from within

What would *knowing* in the Ayoreo world be like?

What springs to mind immediately for us in modern life culture is the *absence of writing*. To appreciate more accurately what this means, I visualize how writing in our own lives can be seen as an act of *giving expression to* something we *know*—an act that resembles delimiting and cutting out a piece of reality and transposing it into a given language, as if we were pouring it into a pre-existing form; by this, we become able to *handle* it and *place* it somewhere outside of ourselves (a piece of paper, a digital code, a library, a poster, an envelope, etc.).

In contrast, the Ayoreo language *zamuco* is an oral language; there are no books, no libraries. Traditional Ayoreo life does not use anything comparable to writing except in artistic, creative expressions, like the ornaments used in the weavings (they carry clear meanings known to all). Apart from that, knowledge may materialize when two separate groups in the forest need to communicate their whereabouts and intentions by way of rudimentary installations left on the path, which may use little sticks, parts of branches, seeds, pieces of string, and similar.

Thus, in Ayoreo culture, *knowing*/what is known can be expressed in words and can be said, but it remains immaterial without becoming objectified in any way comparable to what is written. It is "held"—somehow "carried along"—*within*, by everybody, individually and collectively; it has to be noted that for the Ayoreo, the collective "we"—in the first place the large family group—is not an abstraction, but appears to be felt and lived with at least equal force and awareness as we do with the individual "I."

[4] See Leopold Kohr and Iván Illich on the concept of proportion.

This *knowing*, and the resulting knowledge, is open and freely accessible to all; there is specialized knowledge that will be used mainly by determined "knowers," according to explicit or spontaneous roles distributed in the group: people with knowledge about planting, hunters, women gatherers and weavers, elders, scouts with their knowledge of the group territory, etc.; there are only certain, very special domains that are exclusive, like the formulas used for healing or the knowledge used by shamans.

The *knowing* must be practiced anew constantly, the knowledge renewed: every day is a new day and a new beginning; the new content of knowing is always at least a little different from the previous one; the daily roaming confirms and also adjusts what is known, according to the constant movements and changes in the life of the world; there is an inner attitude in everyone of trusting expectation and readiness to adjust rapidly.

What the constantly renewed *knowing* produces is an incredibly vast and detailed body of knowledge. Its character remains *local* and *related to the given world*: it is remains centered in what needs to be known in order to live a good life with eami, the Ayoreo world. Spoken, sung, thought, and remembered language—names, words, stories, formulas, songs, clan denominations—refer to and make it possible to relate to everything that exists, to anything in the world.

Knowledge and (Ayoreo) territory are inseparable; knowing is only possible through the connectedness within the territory. Leaving the territory cuts the possibility of partaking of the shared knowing; the applicability of already gained knowledge fades. Connectedness (or rootedness) seems to manifest itself in dreams and seems to habilitate the possibility of taking part in a dream field.

Equally inseparable are knowledge and community (or social group); being together habilitates the vitally needed connectedness.

Thus, knowledge has a participatory quality: I partake of what is known—by way of staying in communication with my kin and also by dwelling in the territory in a connected mode that provides "messages" by inspiration (in feelings, proprioception, dreams, visions, songs, etc.).

Knowing and knowledge become themselves a part of the world (instead of being separated); its role and function in eami is making it possible not to control but to come by, to flourish and carry fruits but not at the cost of others and the other.

In our world

Let us imagine—trying to feel—what is known as something that is both inside ourselves individually and inside the social body we belong to (the closest permanent human surroundings).

"What is known" is the totality of knowing, irrational and sometimes rational, conscious and not conscious, about how to live, act, where to go, what to watch out for, what to seek, etc.

Let's imagine that we are connected permanently to our life realm; the moment we aren't, we lose the knowledge.

This connection is felt: it gives longing, warning, urge, bliss, belonging, alarm, sadness, need, and so on.

Even when we don't "think" of it—or we aren't aware—it is there, just the same, whether we sleep—and dream—or we are "awake."

We don't need to "have" it, "keep" it, or even "hoard" it; as long as we are connected to the world, we can safely leave it and later return to find it—and we will spontaneously return to its newest update.

What is known doesn't belong to anyone—it is the other way round: we belong to it, we are part of it.

We are known. We are known, therefore we are—I know, therefore I am.

Conclusion

If what we found in our exploration attracts us and creates a certain added awareness—although plagued with doubts and uncertainty, if it may prompt us to modify something of our concepts and doings with regard to *knowing* and *knowledge*, we would get a little closer to "them": to the remaining many outside our own cultural framework.

Yet to get even closer—really close—we would have to change our life culture—only quite radical changes would open us and give us access to the privilege of their ways of knowing.

References

Bremen, Volker von (1991), *Zwischen Anpassung und Aneignung: Zur Problematik von Wildbeuter- Gesellschaften im modernen Weltsystem am Beispiel der Ayoréode*, Munich: Anacon.

Bremen, Volker von (1987), *Moderne Jagd- und Sammelgründe - Entwicklungshilfeprojekte unter Indianern des Gran Chaco*. Frankfurt/M: epd-Entwicklungspolitik Materialien III/87.

Fischermann, Bernd (1986), *Zur Weltsicht der Ayoréode Ostboliviens.* Bonn, Rheinische Friedrich-Wilhelm-Universität. See also Spanish edition: Bernd Fischermann, "La cosmovisión de los ayoréode del Chaco Boreal", Biblioteca del Bicentenario, La Paz, Bolivia, 2022, ISBN: 978-9917-33-021-9 (foreword by Benno Glauser).

Glauser, Benno (2023), *Trails from a World Beyond - The Ayoreo and their quest to return.* ISBN 9 783757 543778 - epubli, Berlin (available on platforms).

Quiroga, Rosa María with Fischermann, Bernd: Hablamos porque nos escuchan. *Siete historias de vida ayoreo* ("We speak because you listen. Seven ayoreo life stories"). (2018) Asunción: Iniciativa Amotocodie, ISBN 978-99953-898-4-0.

4

To Be Open for Constructive Strangeness: Experiences with Intercultural Knowledge Exchange in Development Cooperation and Science

Peter Strack[1]

When I proposed Jürgen Habermas' theory of communicative action as a main paradigm for a study on functional changes of religious rites in the former Jesuit reductions of the Bolivian Chiquitos region (Strack, 1991), my doctoral supervisor was skeptical. Why use a theory of the differentiation of modern societies, developed in Germany, to analyze indigenous cultures in Eastern Bolivia? I replied that if a theory claims to be universally valid, it must also be applicable to the Chiquitanía. Furthermore, wasn't San José de Chiquitos, even if it seemed so marginalized at that time, already part of globalized economic and social relations? Didn't the indigenous Chiquitano move quite flexibly in non-traditional contexts such as schools and trade unions, as well as in traditional religious ritual processes strongly influenced by the European Baroque, whose outward manifestations resemble Alpine cultures?

Every culture is mixed, but power relations make a difference

"This is nothing new; it is quite obviously mestizaje" is what we were told when we (Strack and Kühne, 1989) presented the results of our research at a Latin Americanist congress in Germany. From the perspective of the Chiquitano, I perceived, it was an authentic indigenous culture of its own, and *mestizaje* (racial or cultural mixing) was a foreign attribution. When I once showed photos of a Holy Week procession in Wiedenbrück/Westfalia to Chiquitano people of San José, they were surprised that apparently there were also indigenous people in Germany (Strack, 1994).

The absorption and appropriation of originally foreign cultural elements are, after all, phenomena in all living cultures. Some cultures experience this less, as seen among

1 I thank Julia Gabriela Strack Diaz for comments, translations from Spanish to English, and correction of style and grammar of the English written parts of the text.

the Aymara, the largest ethnic group in the Bolivian Altiplano, but others experience it more, as I discovered in my later work in development cooperation, particularly among the Quechua, who underwent more migration processes. Given the diverse cultural roots, aren't central European cultures also *mestizo* cultures? This has been true not only since the introduction of Andean potatoes or Asian tofu into German cuisine but dates back to much earlier times.

Hugo Celso Felipe Mansilla, in his critical analysis of Andean Philosophy (2019:8), points out that many "imported concepts" from Western societies are widely accepted by the indigenous population in the South, especially within the younger generation. "Nearly all peasant and rural communities in the Andean region, long ago, were subjected to processes of acculturation, mestizaje, and modernization" (ibid. 11). This is no less valid in urban environments. He considers current debates in Andean philosophy, which emphasize the principles of "vivir bien" (living well) in harmony with nature, as a reaction to this profane reality.

At some point, in conversation with project partners who advocated for strengthening Andean agrarian culture and who, with well-founded reasons, rejected the destruction brought about by Western science and production methods, I, from a structuralist perspective, pointed out similarities between Andean agrarian cultures and small-scale farming cultures in Europe. I suspected that this was due to the agro-centered view on both sides of the ocean. Equally, a multilateral exchange of information on agrarian production methods could be another form of *mestizaje*, as occurred when Andean practices were introduced to Europe.

The response I received from Grimaldo Rengifo, a leader of the PRATEC Network for Andean technologies (see CEDI/PRATEC, 2007), was that he could not recognize similarities because it was seen as a destructive relationship to nature, treating nature only as an object, brought from the West to Peru. Some even defended the idea that both worlds are irreconcilable, and mutual understanding is not possible without giving up one's own culture. Or, they would state that one can communicate, but only within one cultural framework at a time. When a Pijao representative returned to his indigenous community after one of our partner meetings in Bogotá, Colombia, he would not enter his community without a prior purification ritual to protect the community from Western influences. The aforementioned German Latin American scholars similarly considered European and indigenous cultures as separate, with similarities explained through *mestizaje* as the adaptation of indigenous cultures to European ones, generally not the other way around.

This brings us back to the concerns of my doctoral supervisor and to Critical Theory, specifically to the reflection on the conditionality of one's views and approaches, which are shaped by one's social and cultural background. Instead of considering this within a positivistic tradition as a deficit or obstacle in the search for scientific truth, it can be viewed as a path to a better understanding of society or reality, seen as socially constructed or as "relational ontologies" (Escobar, 2018: 74), not far from the concept of relational truth in indigenous "truth speaking" described by Barbara Schellhammer (2019:66) in processes about cultural genocide in Canada.

All of this underscores the importance of considering diversities. To contribute to the nuance of the conversation, we should refrain from generalizing indigenous knowledge as a singular entity but rather acknowledge the significant diversity and richness inherent in these various ways of cultivating wisdom in the world.

If there can be no universally valid theory and science, there is at least the potential for productive strangeness. Exposing oneself to different points of view or cultures, as well as sharing one's own ways of thinking, questions, and comments. Even if they fall outside the norm and are strange, they can stimulate reflections or generate new ideas about what appeared to be self-evident. However, this curious openness to alternative paths to knowledge requires mutual respect and the dismantling of structural power imbalances to be productive rather than destructive. This principle should apply to development cooperation as well as the academic world, where one perspective often holds disproportionate resources and dominates both the rules of academic or development action and the discourse (see Reiter, 2018: 315). It requires striving for as much "communication free of domination" (Habermas' "herrschaftsfreie Kommunikation") as possible, minimizing the impact of power relations (and defining power) and economic imbalances. The following text provides examples of how "epistemologies for a New Modernity" could be interwoven. In other words, it describes some opportunities and constraints of Western-indigenous dialogues, based on my professional career and interviews conducted for this essay.

Dialogue in development cooperation: to use both hands (and feet)

Once I read in a report of the partner organization Asociación Bartolomé Aripaylla (ABA) about my former monitoring visit, "We never had a visitor who asked us so many questions." ABA is working in the Quispillaccta community of Ayacucho, Peru, on the recovery of Andean agricultural knowledge (see also CEDI/PRATEC, 2007). They achieve what is called impact in development cooperation, namely, converting desertified landscapes back into green and productive soils, making villages more hospitable, and simultaneously generating water for the town of Ayacucho (Huamanga).

They formulate and explain their strategy using the concept of the soul of plants or stones, with the Putaqa plant nurturing water, or the explanation that concrete and water don't get along. They also explain it through storytelling. All this is done while staying away from logical frameworks or modern impact chains, which have been established as methodologies in development cooperation. ABA agronomists had left their rural community to study at universities, but they returned to their community to relearn from their fathers and mothers and to guide their actions in a Quechua communitarian way. This includes making offerings to Mother Earth, asking the stars in the sky, frogs, or plants when, where, and what to sow, and even convincing ants not to consume all the fruit of an acre. They do this with a ritual that recognizes the importance of nature and the need for all living creatures to eat, including ants.

Once, when I had to ask ABA staff to monitor and report their project activities in a new format provided by the donor organization, they thought it wouldn't necessarily help improve their work or make it more efficient (which was the argument for introducing the format). They replied that they could do all this in their free time at night, much like people who play chess for leisure. However, they asked me, as the representative of the donor, to accept that they would not apply it in their daily project work. They would continue with their own methodology to achieve the desired impacts. They used the Western methodology of planning and monitoring primarily to secure the funding, which, due to power dynamics in development cooperation, they wouldn't have received with their culturally based methods.

So, the hundreds of questions from this stranger were posed with the intention of understanding a bit more of the agricultural knowledge accumulated over hundreds of years in the Andean landscape. Some explanations showed me a path to interpreting their explanations within a Western logic, such as the indicator plants that can provide better and cheaper information about specific soil conditions for seeds than laboratory analysis. Sometimes I received complimentary information from Western scientists who confirmed the Andean knowledge, such as the fact that with certain soil, water does indeed disappear if you construct channels with concrete. However, many aspects remained, from a Western scientific perspective, somewhat of a cultural mystery.

But if a culture lacks a Western academic explanation for a certain phenomenon, it does not necessarily mean that there is no explanation, especially when another culture offers one. In dialogue with project partners like the Machaca Sisters of ABA, I also learned that some aspects of our German or NGO institutional culture may appear strange and mysterious to our partners. For instance, planning logic intended to improve efficiency often resulted in more bureaucracy and less impact.

Although Terre des Hommes Germany tried to mitigate such impositions and power imbalances with partner organizations through systematic dialogue and co-decision-making processes, questioning our standards was the exception rather than the rule. So, the reality of international interconnectivity led partners like ABA to think that in order to strengthen Andean culture, one must first weaken the imposition of strange Western thinking and return confidently to one's roots. They also argue that in the long run, it is necessary to manage both knowledge systems. They call it Iskay Yachay (double knowledge). The argument is that it is better to work with both hands, not just the left or right one. Thus, rural school curricula development promoted by ABA and other members of the PRATEC network aims to integrate Andean knowledge with an equal level of recognition into theoretical and practical classes. The participation of indigenous knowledge-bearers in regular classes is one essential element. Another element is the participation of students and regular teachers in community and agricultural activities outside the school building. To promote more professionals with this methodology, the PRATEC network implemented postgraduate courses with various universities in Peru and Bolivia. Although they use a fairly traditional scholastic format rather than dialogue to achieve what they call a "de-professionalizing" attitude, the content mainly presents and contrasts Western and Andean concepts on different issues.

Making information relevant for a changing reality

"It is necessary for the university to once again come into contact with the reality of rural communities," says the sociologist Fernando Antezana, who is both a professor at the Universidad San Simón of Cochabamba and director of the NGO Pusisuyu, which works on rural development in the mountain region in the North of Potosí. Just like ABA, Pusisuyu also focuses on the recuperation of Andean knowledge and the strengthening of Andean agriculture. "Indigenous cosmovision gives hope to achieve a new societal contract with nature in order to preserve the natural bases of production. Andean cultures have a different view of nature, which, from their practice, challenges the common development ideology," says Antezana.

But how does he integrate this into his own academic work? Indigenous contributions appear to be useful not only as a source of agricultural knowledge, where Andean rural workers are experts, but also to validate his academic theory in the field and gather information about a social reality that has changed dramatically since the beginning of the new century. "There is high rural-urban mobility. People have two or three homes. I can take this into account in my teaching," says Antezana. He is aware that indigenous people change their relation to nature as they have more contact with urban society, but this, too, is a topic to be studied to gain a new understanding of the changing reality, he says. Indigenous communities are very interested in university studies. "Young people are busy on the internet and have access to a lot of information." His task is to transform this massive information into knowledge or, in other words, guide students to make this information relevant to their lives. This way, they become experts who can help address the environmental crisis. Antezana's students come mostly from rural villages, with some from urban towns. The educational background in rural schools makes it challenging for students of rural origin to meet university curriculum requirements, says Antezana.

> The challenge is to enable all of them to flourish their knowledge because students of rural origin have less experience in reading and writing. However, they bring their own specific knowledge, especially regarding our relationship with nature. They have greater advantages in comparison to urban students. Some of their knowledge even constructively challenges our scientifically established knowledge. This is the moment of dialogue between different ways of knowing.

Antezana laments that most of his colleagues attempt to standardize knowledge in line with the existing curriculum instead of fostering students' ability to reflect. Even though academic freedom is a university standard, in practice, a professor is not allowed to change the curriculum, he complains.

Antezana doesn't like to distinguish between science and indigenous wisdom. "This distinction was made by the university because scientists were unable to recognize other forms of knowledge as equal." Even worse, says Antezana, many try to conceal indigenous knowledge, using it only to validate their own theoretical models and ideas.

Among the various strategies to deal with indigenous knowledge, as described by Sandra Harding (in Reiter, 2018: 39–62), one is to integrate endangered indigenous knowledge systems into "sciences originating in the global North" (ibid. 46). However, if this is not done in conjunction with efforts to safeguard traditional knowledge on the ground, the risk is that it ends up being done selectively, reinforcing the dominance of one authoritative knowledge system while ignoring the importance of preserving cultural differences (ibid. 47). The strategy of integrating Northern sciences into other sciences, she considers, has already been done selectively, but the issue is that "marginalized cultures are not strong enough" (ibid. 51) to make these selections themselves. Thus, she concludes in favor of a strategy of revaluing Southern scientific traditions and even modernizing Northern science by purging it of its Eurocentric elements (ibid. 55). While this may help better understand different perspectives and open the way for collaborative research "with both eyes open" (ibid. 97), it still assumes the possibility of a universally valid knowledge system and doesn't address the problem of performance within a hierarchically structured academic environment with significant inequalities of power and resources, even within the South.

Integration or appropriation and assimilation

Maria Emilia Montes Rodriguez, Ph.D., is a professor of linguistics at the Universidad Nacional de Colombia. She emphasizes the issue of persistent power imbalances, particularly in how to protect intellectual property rights. Western universities, guided by academic rules, often remain closed to many indigenous knowledge bearers, only welcoming their knowledge. She states, "It is imperative to consider how to better safeguard this knowledge from within academia to prevent what has happened to this knowledge before, as with the attempt to patent the formula of Yage (Ayahuasca)."

In her field of linguistics, she focuses on the protection of mythological narratives in the publication process. She explains:

> In 2016, we published a bilingual compilation of myths of the Tikuna, but we were uncertain about how to handle copyright. We struggled with attributing copyright to the ethnic group and the narrators, who were the sources of these tales. As scientists, it's challenging to properly manage ancestral knowledge. The rules are unclear. There's no issue with individual rights, but myths are part of the oral heritage of communities, and one must acknowledge that. However, when we publish the texts, the University distances itself, requiring us to sign a declaration confirming that copyright has been respected.

But she recognizes that in Colombia, government institutions and indigenous organizations are collaborating to create a system for protecting traditional knowledge.

This urgency arises because, for Bolivian Fernando Antezana, it is clear that

> every day we recognize more that indigenous knowledge provides answers to problems like environmental or development crises, which conventional science cannot solve. It doesn't make sense to debate which knowledge is superior. The boundaries between the two are not clear-cut. What could forestry experts do without their sense of smell? Dialogue should occur between equals, with mutual respect for each other's worldviews, to jointly solve problems.

So, even though Antezana acknowledges the value of indigenous knowledge, his rural students use the university to improve their status and gain recognition in the establishment. However, this done—within a Western supposed supremacy framework—could be a straightforward process of acculturation for students from rural backgrounds rather than the integration of indigenous knowledge into urban life. Antezana admits, "The majority will continue to live in urban areas, even if their work is in rural communities." Linhart (2008:29) describes them as a generation of "cultural irritation" or even intermediaries who may engage in indigenous activism, but where, even in indigenous academic environments, the quest for identity can be secondary to the goal of achieving status (ibid. 43).

Curriculum and established academic requests for eye-level dialogue

The limitations of curricula and academic requirements in universities were evident in the case of an Aymara University initiative in Tiwanaku. Milka Arteaga, a young professional who had also studied and worked at the state university San Andrés, valued the newly established "Universidad Indígena del Tawantinsuyo del Ajlla" of Tiwanaku courses because they strengthened her Aymara identity. Her teachers were Yatiris (healers and spiritual leaders) from different provinces in the Bolivian Altiplano. As they lacked academic titles, some academics followed their classes to sign and vouch for the formal recognition of the courses. However, this practical solution was ultimately rejected by the authorities. Thus, this university was closed, even in a country where the government officially proclaimed the decolonization of education (see also Arteaga, 2023).

The NGO Fundación Encuentro had a similar experience with postgraduate courses in intercultural therapy and community-based prevention of sexualized violence (Morales, 2020). First, the methodology and subsequently the academic offerings were developed by the director of the foundation, who had obtained her Ph.D. in Belgium, along with Aymara Yatiris from the Bolivian Altiplano who did not possess academic titles. They collaborated despite the fact that a private university, which initially accepted their curriculum and methodology, ultimately rejected the possibility of awarding academic recognition, in the form of diplomas, to the Yatiris. Only psychologists or social workers with university degrees had the opportunity to receive diplomas, even if they were students. The Yatiris, who were teachers, were excluded. In the end, the Foundation, in collaboration with the Yatiris, decided to establish equal conditions for all, forgoing academic recognition for a methodology that was certainly innovative and based on two systems of knowledge.

Maria Emilia Montes Rodriguez specializes in indigenous languages, where indigenous students from these cultures undoubtedly possess unique expertise. Montes Rodriguez shares the experience of Anastasia Candre, a poet and daughter of a wisdom bearer of the Huitoto people.

> She learned a lot from her father and, without having attended primary school or completed her bachelor's degree, she requested to attend Linguistics classes at the

> Leticia subsite of the National University. She learned and contributed her own insights. She was very active, and all her poems and chants were published. She is an intellectual who desired to receive her academic title, but she did not receive it because a bachelor's degree requires a high school diploma.

This differing treatment also applies to indigenous teachers at the National University of Colombia who teach without prior academic qualifications, as Montes Rodriguez informs us.

> In our faculty, since the mid-'80s, we managed to hire a permanent indigenous professor who spoke the Huitoto language. For 30 or 35 years, he taught his lectures (he passed away in 2022). He did not possess an academic degree and was hired under the title of "expert," which is not exclusive to indigenous professors. The issue is that he could never be promoted; he remained "stuck" in this position as an expert. To advance, he would have had to pursue academic training or enroll in regular courses to become a professor. Something similar happened at the Leticia headquarters in the Amazon, where anthropologists collaborated extensively with indigenous communities and indigenous wise individuals. It is not easy to hire a "sabio indígena" to teach a master's level class, even though students conduct their theses or master's degrees with them. They are hired as members of the research team or as speakers, tasked with translation and expert work, but there is no genuine parity. There always needs to be a fully employed professor who endorses everything.

Indigenous wisdom breaking into academic spaces

Times have changed, however, as the professors of Inga and Wayuunaiki at the National University now hold academic degrees. There are even indigenous professionals with master's or doctoral degrees. At Antioquia University in Medellín, the Kuna Abadio Green is a professor who teaches "pedagogics of mother earth" and has helped many young Kuna graduate and earn teaching or master's degrees. Thus, in practice, universities now prioritize indigenous professors with academic titles, Montes Rodriguez reports. But how can indigenous students earn academic degrees without forsaking their traditional ways of thinking and wisdom?

> It is not us [professors] leading the dialogue; it is them [students] when they enter academia. They adopt certain tools and begin to forge their own paths. For instance, Rafael Mercado was passionate about semiotics. Since his undergraduate studies, he began contemplating signs and symbols, the relationship between textiles and culture, and women's issues. It was challenging for him to complete his thesis because it lacked a conventional research structure. Instead, it was written continuously, resembling a weaving. In this work, he incorporated themes from semiotics, quoted numerous authors, and connected academic writings with his own ideas. To me, it was an excellent piece of writing, as it encompassed all the

> elements of linguistics, semiotic morphology, and offered a novel contribution from his own culture. Moreover, it displayed a high degree of reflection. A thesis need not adhere to the traditional research project structure. Some other students had already engaged in autoethnography. As a result, his degree was approved.

They later wished to publish his thesis, but the university press's editorial board rejected it, citing the absence of structure and academic style. Fortunately, another publishing house eventually published this work because "he is a very interesting indigenous intellectual, who speaks and writes exceptionally well," as Montes Rodriguez recalls. These challenges may stem from the fact that many indigenous cultures are primarily oral. If academic structures and regulations are overly rigid, it becomes crucial for individuals within the system to recognize and nurture potential. Montes Rodriguez concludes:

> I have supervised several indigenous thesis students. One can observe that they make fascinating contributions, providing facts, reflections, and highly interesting insights into bilingualism, dialectology, and how languages converge in their worldview. They excel in oral communication, possess a unique talent for discourse, and are adept speakers and skilled arguers. In a political context, they outshine everyone. They can also translate these ideas into writing. However, rural schools did not teach them enough how to translate their rich oral tradition into academic writing. Certain academic formats pose significant challenges for them, potentially causing paralysis. While adhering to an academic mold wouldn't necessarily alter the content, it would consume a considerable amount of time, which they may prefer to avoid. Therefore, an advisor's role should be to assist them by adapting the form, breaking it into chapters without disrupting its holistic nature. The substance remains intact; it merely assumes a more academically conventional shape. Each case is unique; it's not paternalism because what they bring is profoundly valuable. Instead, we seize the opportunity to enrich academia with these forms of knowledge.

Why university?

In his master's work on "Indigenous Universities," Jan Linhart (2008) analyzes the tension between different forms of wisdom and expression in the case of indigenous universities in Brazil, Bolivia, and Colombia. Linhart (ibid.: 20) states, "Meanwhile, models promoted by the state pursue the inclusion of indigenous people in the national educational system. Autonomous indigenous projects, on the other hand, start from an intercultural alternative that not only symbolically expresses self-determination but also serves as a model for a new concept of education." This new concept encompasses various forms of teaching and learning, closely related to indigenous knowledge systems (ibid. 52). Linhart further explores the question of why one should learn within a university context (ibid. 53).

Even in the prominent case of the university created by the Consejo Regional Indigena del Cauca (CRIC), they ultimately decided to adhere to the ordinary academic requirements to obtain state academic recognition, even if doing so required compromises (ibid. 44).

Montes Rodriguez refers also to the experience of the UAIIN, this Universidad Autónoma Indígena Nasa in the Colombian Cauca region:

> I was in a training program in 2007 for native language courses, and there was a strong debate about whether they should enter the accreditation system of Colombian universities or stay outside of it. If they stay outside, they can hire wise people and experts, and their degrees are recognized in their territory. But in the end, it seems they decided to join the system.
>
> The University of Cauca, a public university that works in the Nasa region, has also granted many indigenous people bachelor's and master's degrees in language revitalization. However, it was not easy for this institution to accept all the requests from indigenous individuals, such as hiring their traditional wise people.

Did those at UAIIN, in making the decision to enter the national system, lose their own way of being and sharing wisdom? Such as the rituals and conversations with nature, which, according to Western sciences, might not be considered a proper way of generating knowledge?

"The situation at UAIIN is the result of a challenging negotiation process between two poles," says Montes Rodriguez, but she also believes that things have changed. Western thinkers nowadays attempt to incorporate new perspectives that emerge from indigenous people regarding Mother Earth, holistic perspectives, the idea that we are all part of the same entity, ritualism, and criticism of development concepts. In the end, it's a form of Southern epistemology, which in Latin America is strongly influenced by Mexico, as well as by the Nasa people of Cauca and some Wayuu, according to Montes Rodriguez. She mentions Rafael Mercado, a linguist who graduated from Universidad Nacional. His books have been published and received awards, although their genre might be difficult to classify within conventional frameworks. Another example she mentions is Estercilia Simanca Pushaina, a Wayuu woman who wrote, "Manifiesta no saber firmar: Nacido 31 de diciembre." "It's a critical look at what happened to the Wayuu with the state personal registration system. The Wayuu didn't speak Spanish, and the officials didn't speak Wayuunaikki. So, they would put them absolutely mocking names on their ID card. Recently, there was an event asking for forgiveness and re-registering people, correcting their names." So indigenous scientists also have the potential to address discrimination that arises from the lack of recognition of indigenous knowledge and language, coupled with power imbalances. This is also exemplified by Chiquitano theologian Roberto Claudio Tomichá Charupa, Ph.D. Even though educational environments have not supported intercultural dialogue, and his own language was subject to discrimination in school, he had to navigate two parallel knowledge systems during his upbringing. However, he later succeeded in influencing even hierarchical structures such as the Roman Catholic Church and its doctrinal system. Since 2006, he has been a member of a movement of Christian indigenous theology, which was inspired by the Zapotec theologian Eleazar López and differs from traditional indigenous theologies.

We don't need to understand everything

> Every culture should learn in exchange with others, to improve. The inputs of western societies or Christianity are part of this, as was the case with me. My heritage is the Chiquitano culture, but I also learned from Europe, Africa, Asia, or other indigenous cultures of the Latin American region. This is the way I experienced it in my own family, which was open to other cultures.

Tomichá said this in an interview with the author (Tomichá, 2022). He is convinced that being open to everyone and avoiding conflicts was the key survival strategy, which he learned from his father. The rediscovery of his roots took place at the Pontifical Gregorian University, which is not necessarily a place where one would expect indigenous culture to become of importance. "I had different teachers," Tomichá said.

> I learned different things from everyone. Europe helped me to recognize the diversity in Christianity. In Rome, there were people from all other continents. Every person brought in their own perspective. And so I asked, what was my own contribution? A good example is the holy trinity. How can we understand this out of our indigenous culture? With the Chiquitano, the father and son flowed one into the other. Jesus is father and God. This is expressed in the terms they use. Indigenous people have little interest in a systematic analysis following western categories. They focus on life and the relations to other living beings and to nature, where all elements are living: Water, animals, plants, stones. . . How can we explain this to someone, which is not part of our culture? We need to know the cultural concepts of the other. So we use at least two languages: The language for our own people, and the other language for those who read our books. For instance, what does the holy ghost mean in an indigenous culture? First, we see that they exist in plural, and commonly they are called protectors. In the Andean culture, the term Ajayu [a kind of soul which gives force] comes very close. Is this expressing what the catholic doctrine means with the Holy Ghost? At the end every culture uses her own terms and significance, depending on, for instance, if they serve to accompany or to protect. In the old testament different terms are used, like "Ruah" which is wind or breath. In the New testament it is more a presence or fire. . . one can look at all this together. The point is to promote dialogue, and not to find a single definition. Even if with the concepts of Christ this is more difficult, because of so many existing images, which mark the concept, in the case of epiphany and creation there are more similarities between catholic dogma and indigenous religions. There is no consensus, but proximity.

I asked Tomichá how to communicate such diverse interpretations in a hierarchical institution as the Catholic Church. Tomichá considers that this is a question not only for indigenous but also for African or feminist theologies. Some Latin American bishops, he says, would have given support in the past, even if they didn't agree with all the contents.

> But this is not our aim at all: Women, Afro- or young people on the internet . . . all should elaborate on their own keys for interpretation and make their own experiences of mystics. Because without sentiment there is no theology. We are still at the beginning, looking for approximations and, in the case of indigenous cultures, for recuperation of our own linguistic expressions. But even if openness to new cultures and the interest to understand them are good things, we shall recognize that for many issues there will be no answer, or to phrase it better: There only will be diverse answers. From there, our interest arises to learn from each other. Even in many cultures, as in mine, you don't feel the need to understand everything.

References

Arteaga, Milka (2023), *Schwester P'itikilla*, January 26, 2023, https://blogs.taz.de/latinorama/schwester-ptikilla.

CEDI/PRATEC (eds.), (2007), *América Profunda*, Relatorías, conclusiones y acuerdos del coloquio y simposio realizado del 6 al 9 de diciembre 2003 en la ciudad de México, Lima. https://www.pratec.org/wpress/pdfs-pratec/america-profunda.pdf.

Escobar, Arturo (2018), "Transition discourses and the politics of relationality: toward designs for the pluriverse," in Reiter (2018), pp. 63–89.

Habermas, Jürgen (1981), *Theorie des kommunikativen Handelns*, Suhrkamp, Frankfurt

Harding, Sandra (2018), "One planet, many sciences," in Reiter (2018), pp. 39–62.

Linhart, Jan (2008), *Indigene Universitäten, Indigene akademische Bildung in Lateinamerika: Auf dem Weg zu einer interkulturellen Wissenschaft?*, München, M.A. Thesis, Ludwig Maximilians University. https://www.academia.edu/31373287/Indigene_Universitaeten_Indigene_akademische_Bildung_in_Lateinamerika_Auf_dem_Weg_zu_einer_interkulturellen_Wissenschaft?email_work_card=view-paper

Mansilla, Hugo Celso Felipe (2019), *Una visión crítica sobre la Filosofia Andina*, La Paz: Rincón Ediciones.

Morales, Katya (2020), "*Wir wollen die geraubte Zukunft zurückgewinnen*, Peter Strack interview with psychologist Katya Morales about intercultural therapy for girls survivors of sexual violence," in: ila 437, Bonn, pp. 19–21. https://www.ila-web.de/ausgaben/437/wir-wollen-die-geraubte-zukunft-zurueckgewinnen

Reiter, Bernd (ed.) (2018), *Pluriverse, The Geopolitics of Knowledge*, Duke University Press, Durham.

Schellhammer, Barbara (2019), "Truth and Reconciliation in Kanada, Zur Bedeutung indigener Philosophietraditionen für die Heilung des kulturellen Genozids an Ureinwohnern," in *polylog, Zeitschrift für interkulturelles Philosophieren*, No. 42, Wien, pp. 61–73.

Strack, Peter and Eckart Kühne (1989), *Los ritos religiosos de las reducciones jesuítas en conflicto por el control cultural*, in: Revista Paraguaya de Sociología, pp. 35–50.

Strack, Peter (1991), *Vor Gott, Gemeinschaft und den Gästen*, Verlag für Regionalgeschichte, Bielefeld

Strack, Peter (1994), "'Vill spannische bräuch und wunderliche Sitten.' Religiöse Festriten in Chiquitos im Blick der Europäer," in Eckart Kühne (ed.), *Martin Schmid 1694–1772, Missionar, Musiker, Architekt, Ausstellungskatalog*, Luzern, pp. 47–54

Tomichá Charupá, Roberto Claudio (2022), *Von Interkulturellem Dialog und dem Gesang der Vögel*, Peter Strack interview with Roberto Tomichá, October 19, 2022, https://blogs.taz.de/latinorama/von-interkulturellem-dialog-und-dem-gesang-der-voegel

5

Embodied Knowledge: Decolonizing Philosophy by Turning to the Nature that We Are

Barbara Schellhammer and Stan Wilson

Introduction

At the beginning of the twenty-first century, the process of decolonization has reached the hitherto unquestioned academic systems of knowledge in Europe and the US classics of the European philosophical tradition (e.g., Aristotle, Descartes, Kant, Hegel) and how they are taught at universities are now facing increasing criticism. This criticism is directed against a universalistic way of seeing the world which rests on Eurocentric and logocentric principles which seem to be taken for granted (Elberfeld, 2021; Lu-Adler, 2023).

As a consequence of a philosophy that understands itself as "abstract," "rational," and "detached," the so-called Western tradition has produced a huge gap between what it understands as "culture" or "civilization" and "nature" or "primitive instincts." This disconnect shows in how people exploit their own bodies as well as their natural environment. That is why Max Liboiron (2021) argues that *Pollution is Colonialism* and Annie Murphy Paul (2021) stresses the importance of thinking outside the brain, to tap into the vast knowledge system of our body, the physical spaces and the minds around us. Stan Wilson, a Cree Scholar, suggests that not only do we live in relationships but we *are* relational—which means that we "are sustained by our connection to the land" (Wilson, 2001: 92). We are interrelated, dependent, and thus also always accountable. Thinking relationally by turning to the nature that we are is one important yet often neglected way to decolonize philosophy and to face some of the most pressing questions of our times.

In our text, we start by outlining some significant traits of Western philosophical thought which created the aforementioned disconnect between body and thought as well as between nature and culture, and highlight some of its consequences. We then contrast this extremely influential way to think and see the world with the Indigenous concept of relationality stressing our natural interconnectedness. Here, we also refer to the very powerful awareness of a "sovereign erotics," which a number of Indigenous writers discover to overcome colonialism. Lastly, we point out how a genuine dialogue between European and Indigenous philosophies can help to heal the chasms between culture and nature, mainly by turning to the nature that we are. Writing this text as co-authors coming from different traditions, not only are we talking about "Western-Indigenous dialogues" and the potential of interwoven epistemologies but we continue to carry them out (Wilson and Schellhammer, 2021).

European philosophy: rational, detached, and male

One can of course read the history of Western philosophy from very different perspectives. However, one theme that seems to run through, is the tendency to describe man in his prominent position as a rational being who is able and supposed to control nature within himself and around him. Already Socrates aimed at ruling over one's willful desires and boisterous instincts—a longing that finds its most pungent expression in Plato's Chariot Allegory. The charioteer stands for intellect, the part of the soul whose task it is to guide the soul to truth. One of the horses represents positive traits and moral desires, the other irrational passions and lascivious appetites.

The urge to gain autonomy over everything that is wild, alien, and uncontrollable peaked in the age of the Enlightenment. There was of course, the experience of horrible wars that tempted philosophers like Thomas Hobbes to contain the animal in humans through social contracts. René Descartes sought for absolute truth based on reason that is free of doubt. Motivated by the French Revolution philosophers like Immanuel Kant demanded his readers to emerge from repression and their self-incurred immaturity using their own understanding without the guidance of another. With him the famous motto of the Enlightenment was born: *Sapere aude!* Have the courage to use your own reasoning!

Vast developments in the natural sciences helped to see nature not as an annihilating power, but to understand it in terms of laws. All of a sudden, people were no longer just exposed to nature; instead, they looked at it from a distance. Hans Blumenberg (2012: 66) uses the metaphor of a spectator who observes a shipwreck from the safe coast to explain: "Their distance from the monstrous of nature is not only that of the rocky shore, but that of self-consciousness." A new cosmology emerged that considered the universe as a mechanism explicable by discoverable principles, governed by our rational capacity.

These developments drew (safe) lines between everything that can rationally be captured and everything that resists our rational reach and therefore is diffuse, uncanny and threatening. Hartmut and Gernot Böhme (1996: 13) call the latter "the other side of reason": "from the reason's point of view it is the irrational, ontologically the unreal, morally the indecent, logically the alogical. The other side of reason, that is in content nature, the human body, fantasy, desire, feelings." This process of separating between what can be appropriated or assimilated and what needs to be suppressed and defeated created the modern definition of reality. In this definition, dominating the nature around us and the nature that we are went hand in hand. Or, as Gerhard Gamm (2009: 122) points out: "The mastery of nature rests on the suppression of self."

The perversion of the Enlightenment finds one of its most terrible expressions in the European imperialism subjugating everything that seems primitive, underdeveloped or uncultivated. This was driven and motivated by the "noble" attitude to bring "culture" and progress to all, to "help" others to become civilized beings. The conquering of supposedly "empty" lands was legitimized by universal principles "humanising" the non-European world. Enrique Dussel (1989: 21) got it right when he argued that the Cartesian "I think" downrightly lead to the unquestioned "I conquer" and "I enslave." Bernhard Waldenfels (1990: 62) talks about a "threefold monopoly of reason," or to be more precise, a certain understanding of reason:

> The adult is right over the child, the civilized over the so-called primitive, the healthy over the sick, not to mention animals or even plants without rights. In the case of children and primitives they are mere *pre*forms of reason, in the case of the sick they are *de*forms of reason.

Unfortunately, one must add another monopoly of reason here, which still shapes the mainstream understanding of philosophy: philosophizing is *(cis) male*. Lawrence Kohlberg's refuted notion that the highest level of moral development is based on abstract ethical principles tenaciously persists despite Carol Gilligan's well-known critique. It is symptomatic of a philosophy that regards spontaneity, contingency, emotionality, and corporeality as deficient—all of which are traits associated with women. The male dominance also shows in the arrogance of European philosophers, almost all of them men, who are so convinced of their philosophical claims "that any criticism is rejected with the demand that one must first reach the level of their philosophizing in order to even begin a philosophical conversation" (Elberfeld, 2021: 35).

However, in spite of these dire findings, it would be wrong to rebuke European philosophy with its focus on reason entirely—the question at stake is a *certain understanding* of "reason." Instrumental and thus often oppressive rationalism is an aberration of reason. It is a reason that has been amputated twice, in its cut from nature and thus from all of its relational references. Corine Pelluchon (2021: 288 f.) explains: "There is a destructive dialectic of modernity, an internal logic that explains the failure of rationalism and the emergence of phenomena that testify to extreme barbarism [. . .], but the cause of the evil is not reason itself." It is a reason that ego-logically ignores the fact that it always has to deal with an invincible demand of something other than self, something that is utterly alien and can never be captured or grasped. It is true, a reason that seeks to repress bodily desires, unconscious instincts or energies that resist our understanding leads to oppressive behavior and the addiction to control. Yet, especially after Nietzsche and Freud, we also know how important our reasoning is and that it is too short-sighted to just argue for the irrational. Falling from one extreme into the other has never been helpful. What we are after here, seeks to subvert the dichotomy between body and mind or nature and culture, it seeks to shed light on the interrelatedness of both. This brings us to the Cree concept of relationality: *wahkotowin*.

Wahkotowin (Cree)—self-*as*-relationship

The Indigenous protocol of situating oneself in a text suggests, to begin with, a greeting. It invites the reader into the text—and hopes to be greeted in return from the reader. It sets the tone for a relationship between the writer and the reader: *Taansii, watcheye* (Cree): "I greet you the reader with an open hand, symbolic of an open heart and mind." Protocols as such are important to make sure that the words that were traditionally conveyed through oral practices are still living, even in a written text. A text is seen as a living entity; it flows back and forth and develops its own spirit. Alice Keewatin (in Wilson and Schellhammer, 2021: 18) points out how living words impact our bodies, not just our minds: "When people read living words that have been placed on the page through protocols, their cellular knowledge and human DNA is activated." Since we are all connected through our basic natural makeup, a sacred space opens up which allows for knowledge transfer across cultures. Readers grasp foreign concepts and ideas on a deeper level or of higher consciousness with their bodies as they flow from the speaker (writer) to the listener (reader).

Another important reason for the protocol of situating oneself in his or her (con-) text is that the listeners, or readers, can place and understand the speaker's perspective. Part of this "situating" is to introduce yourself with your given name, in Cree *Nitisseeneekasown*—a term which also refers to *Nihtissi*, which translates to "belly button." This indicates a person's roots to the attachment at birth to their mother and to the mother's family line. It also refers to the ancestral ties to the land. The Cree ceremonial custom of taking the child's dried umbilical cord (*otissii*) and burying it in a special place near their home reflects the connection between the mother and child and symbolizes the connection to the earth and a particular piece of land. The word *otchiaan*, "I am from" in English, has the same root: *otissii*. The lineage is through the mother and her mother, her mother and so on—all are connected through the intergenerational umbilical cord. When people say where they are from, it can literally mean the female lineage as well as the place where the *otissii* is buried. In Cree cultures, women are the first teachers of the natural world and the Cree language which is rooted in the lands and waters they have occupied for centuries.

These protocols are reflections of the Cree concept *wahkotowin*, which translates to "relationality" in English (Wilson, 1999). Indigenous heritage languages were created and developed from observing natural forces of nature in order to live *within* it, not apart from it. They explain the binding relationship to the land that is held to be sacred because it is nature's way; it is therefore natural law. However, it is crucial to understand the meaning of *wahkotowin* from the original Cree language (*Nehynowaymowinik*) and how the words too are in relationship with each other. Understanding this relationship between and amongst the words of the language creates a deep understanding of the Cree worldview. That fact that everything is in relationship with each other is reflected in the language where even words live in families. Each of the Cree words has multiple connections to other words that are related. They form webs of interconnected words, which then become part of the consciousness of the wholeness, and inclusiveness of all things. It is not enough to learn just single words and translate them, this would just be a "thin" and technical understanding, mostly staying in the head. It would rationalize separate words and strip them off their deeper sense. Words carry the meaning of lived experiences—experiences that go back for many generations. Understanding word families means to also understand them with our bodies, to feel them, to live them. We don't own words as encapsulated containers to convey a message. If we engage with their living meaning, we experience our relationality. Here the phrase "embodied knowledge" becomes its concrete expression.

It is also reflected in the Indigenous notion of "self-*as*-relationship." Indigenous people's sense of self is planted and rooted in the land—it is more substantial than a propertied connection. It entails a deep sense of "relational accountability" (Wilson and Wilson, 1998), which means that each individual is responsible for their actions, but not in isolation.

> Individual responsibility for action must be in relation to all living organisms. It is this web of relationship with each individual in the center that stretches out in all directions. This is our understanding of how the universe is held together. We believe that the interconnection among all living organisms is essential for all life forms. The connections must be respected and honored.
>
> ibid.: 157

Stan Wilson describes an experience in which this natural connection or rootedness in a particular land showed itself in a striking manner. Together with his wife he went to a conference in Georgia where all of a sudden, he had an overwhelming feeling of being welcome while at the same time he felt a deep sense of sadness. There was a message coming from that place on the land he walked on: "Not in a human language, but it was nevertheless communication" (Wilson and Schellhammer, 2021: 75). Talking with Indigenous friends about his experience, he realized that he felt his bodily connection with his ancestors who once lived and died there. Their remains were buried in the ground and thus return into the circle of life. Our ancestral memories are in our blood, they are in our muscles, in our bones, in our hair. Lionel Kinuwa, one of Stan's friends, tells him:

> Many of us do not pay attention to these memories, because we are too busy paying attention to what's going on in the modern world. [. . .] But when someone starts to play the drum, our spirit is moved. This is because the vibrations of the drum are stirring up these old memories, our ancestral memories. These memories come out from the molecular structure of our being. This is also when you hear someone speaking your language, your molecular structure picks up those vibrations, because each language has its own peculiar patterns, and you feel good that somebody is speaking your language.
>
> ibid.

Decolonizing philosophy means to trust our body, to honor its ancient connection to the land, to learn from the natural wisdom we may not even be able to put in words. Sometimes research can be done in ceremonies; sometimes it even should be done in ceremonies, like playing a drum to engage in a rhythm that makes us feel our interrelatedness. This is what Shawn Wilson has in mind with his book *Research is Ceremony*. "Indigenous epistemology is all about ideas developing through the formation of relationships" (Wilson, 2008: 8), he states—and it is a ceremony that brings relationships together.

The liberating strength of "sovereign erotics"

Another very powerful way to decolonize philosophy is to turn to the nature that we *are*. A number of Indigenous writers, artists, and scholars have been discovering a liberating source within their bodies which they call "sovereign erotics" (e.g., Akiwenzie-Damm, 2003; Nelson, 2017; Rifkin, 2011; Bear, 2016). With that term mostly two-spirit[1]/queer people and women associate a creative force that helps them heal from patriarchal and heteropatriarchal gender regimes imposed on them through colonial oppression.

Based on what has been said before, it seems obvious that European traditions often have a hard time with the unruly power of unconscious desires and thus with everything that is related to erotic feelings. In her book *Sexkultur* ("Sex Culture"), Bettina Stangneth (2020: 10) writes:

> In the European tradition, sex is regarded as the antithesis of culture, and culture is the last bastion against the animal, from which we must distance ourselves at all costs. [. . .] sex is the epitome of our nature, and beyond that, nothing at all; first, it was considered the opposite of the divine, and much later on, the antithesis of reason. Nature is a constant obstacle, perhaps even an antagonist, an enemy.

Yet Eros plays an important role in European philosophy—the very name "Europe" comes from the Greek mythology of Zeus seducing the beautiful Phoenician princess Europe by transforming himself into a bull. Still, the uncontrollable life-force of passion and desire makes philosophers often feel uneasy. Waldenfels (1998: 151) talks of it as the "Achilles' heel of philosophical thought." Thinking about nature becomes tricky when thinking about a human being, because all of a sudden subject and object are the same. Essentially, it is thinking about oneself, which can be a daunting task—especially when you learned that mind and body are different, the body being inferior. "This turns sexual experience into self-awareness of the most complex kind imaginable. Nothing else brings us as close to the essence of humanity and to ourselves. Nevertheless, reflecting on sex is, oddly enough, not what thinking people—or even philosophers—like doing best" (Stangneth, 2020: 14).

In the context of imperialism, we see two contradictory reactions of colonizers stemming from this uneasiness or disregard of their bodily desires: admiring the

[1] The term "Two-Spirit" is an Indigenous-specific term explicitly used to differentiate Native concepts of gender and sexuality from those of non-Native queer people as well as from anthropological terms with clear colonial inclinations. Thus, the very practice of self-identification as "two-spirit" is a practice of decolonialization. Michelle Cameron (2005: 126) writes: "When Aboriginal queers decided to begin using the term two-spirited again, it was a sign of reclamation of the historical legacy that is unique to our First Nations. Two spirited identity can thus be viewed as a counterhegemonic identity, and as a term of resistance to colonization."

seemingly enormous potency of the "wild"[2] and suppressing cultures that openly express eroticism. What may be even more troublesome is to observe the dialectic dynamics between denying bodily desires in oneself on the one hand and abusing others on the other. Nora Samaran (2016) speaks about "rape culture" and explains that "violence is nurturance turned backwards." "[N]ormal intimacy needs, when distorted and denied, come out in distorted ways." Body, mind and spirit get out of balance and create chaos and pain by passing on this unbalanced self to the human, spiritual and natural world. Colonized sexualities internalize the crooked sexual norms of the dominant culture. Referring to Paulo Freire, Cameron (2005: 126) speaks of "cultural invasion" when intruders penetrate another culture's self-understanding enforcing the idea that sexuality and two-spirit practices are shameful, a sin and removed from any positive spiritual context. "When the colonizer or dominant society culturally invades Aboriginal society, it must be resisted; otherwise meanings are lost or distorted" (ibid.). Mark Rifkin's book *When Did Indians Become Straight?* has something similar in mind insisting "that heteronormativity functions as a key part of regulating native political self-representation and that queer analysis must grapple with the ways this ongoing history of *straightening* and *queering* powerfully shapes the possibilities for indigenous self-determination" (Rifkin, 2011: 316).

For Indigenous peoples, "sovereign erotics" means gaining back their bodies, in order to heal from the historical trauma, they continue to survive. It is a central mode of resistance regaining balance. For a number of Native writers this also entails returning to their stories, stories that were silenced by colonial powers. Kateri Akiwenzie-Damm (2003) started to collect erotic texts by Indigenous writers, because she realized, "[a] person could reach puberty, live her entire adult life, go through menopause and still not have stumbled across a single erotic poem or story by a First Nations writer." In her work *Power in my Blood*, Tracy Lee Bear (2016) explicitly raises the question: "If this is my body, where are my stories?" She noticed that a lot of the stories that she heard as a child were censored through a colonial lens stripping them off any erotic content.

> I wondered about the absence of the erotic in these stories. What forces were at play? In my family? In my community? If our stories teach us how to live well in all aspects of our lives, what happens when our erotic stories are erased? Is there a connection among the suppression of the erotic in our stories, colonialism, and endemic sexual violence against Indigenous women, girls, and genderful folk? Who defines my sexuality as an Indigenous woman?
>
> ibid.: 11

[2] What is striking here is that it seems particularly the case in the context of queer sexuality that non-Indigenous people claim to be Indigenous (a term used for people doing this is "pretneidan"—pretending to be Indian). However, they in turn seem to have an impact on Indigenous writers. Christos, for example, self-identifies as Menominee and two-spirit, but is not a member of the Menominee Tribe or any other Native American tribe. The same is true for Qwo-Li Driskill who was very influential encouraging Indigenous people to explore their "sovereign erotics." Michelle Cameron (2005) discusses these cases of harsh cultural appropriation referring to the meaning of "two spiritedness" which is deeply rooted in the history Indigenous cultures.

Stories refer to a subjective and experiential message, they convey a *relational truth* and thus offer an alternative episteme to Western knowledges which traditionally follow an objective and disembodied approach. It is striking that many of the texts on sovereign erotics are artistic; they are poems, fiction, and other kinds of creative prose. Hence, they offer an embodied method resisting a merely rational way to disseminate knowledge. They tap into a deeper knowledge system relating to a larger community in the natural and non-natural world. What seems to be so important about this way to write is, that it aims not only at the presentation of knowledge, rather it also seeks to transform peoples and societies.

Eroticism is enfolding its pervasive force in an ambiguous space on the edge of self and other, power and surrender, unease and fascination (Schellhammer, 2023: 98ff.). We have to let go and connect with nature or rather live the connection that has always been already there. In her powerful book *Split Tooth*, the Inuit artist Tanya Tagaq (2018: 62f.) writes, "We ARE the land, same molecules, and same atoms. The land is our salvation." And then once more, "The land is our salvation. Breathe, Fuck, Feel."

What Tagaq expresses here seems to be similar to what Melissa Nelson refers to with the term "eco-eroticism." Important ethical and political transformations emerge from a "contact zone" between our human bodies and other than human nature. This contact zone, she argues, "can provide critical eco-erotic experiences that are conducive to embodying an ethic of kinship so needed in the world today to address ecological and cultural challenges" (Nelson, 2017: 232). She, too, finds this place where we can "get dirty," because it is boundary-crossing, messy, nonlinear, relational, and often unsettling in stories rooted in "oral traditions about territorial attachment to ancestral places and beings" (ibid.). She demands us to "*re-Indigenize* our senses by relearning to listen, once again, to the languages of our four-legged, finned, and winged relatives, as well as those of our rooted and stationary kin: the plants and trees and stone grandfathers" (ibid.: 255).

Philosophizing with a tree or a dog? If we were able to allow us to "get dirty," to question our belief that we as humans are "outstanding" (in the truest sense of the word), we may understand on a different level. "Our bodies are filled with intelligences that are faster than and beyond the intelligences of our cognitive brains" (ibid.). This is what Annie Paul has in mind with her quest to explore ways to think outside the brain. If we were to reach outside of our brain, she writes, we are not bound by its limits, "[our thinking] can become as dynamic as our bodies, as airy as our spaces, as rich as our relationships" (Paul, 2021: xvii). Serious enlightenment (maybe a "new form of modernity"?), "should also demand from us to use our own sensuality without direction from another" (Stangneth, 2020: 30)—to rely on the sovereignty of our body.

A plea for "responsive reasoning" and "getting dirty"—concluding remarks

Colonialism is not over, it still sits in our minds and bodies and impacts the way we see ourselves and our natural environment. Philosophy plays a major role in this. It is entangled in a colonial history that continues to have an impact today in various forms of epistemic injustice and other forms of violence. At least three key properties often still set the tone for the way "proper" philosophy is done: rational, detached and male. When we finished our book *Indigeogy. Invitation to Learning in a Relational Way*, one of our first Indigenous readers said that the book has a "female spirit." Stan picked up on that right away, saying that it fits with the ritual of burying the umbilical cord of a newborn child in which the connection between the mother and child is symbolically reconnected to the earth. Thus, he undermined all three demands of what I (Barbara) learned about the academe, especially in philosophy: It is about a *ritual* or a *ceremony* (not rational in a narrow sense), it is *relational* as it reconnects a book to the earth (not detached), and it is "*motherly*" (not male).

If we engage in Western-Indigenous dialogues, we have a chance to uncover our deeply ingrained, often unconscious beliefs—a necessity to see blind spots and thus to avoid falling into reoccurring patterns. These patterns tend to come back in different guises. Max Liboiron (2021) sees one of these appearances in land pollution, which he reveals as a violent enactment of colonial land relations that claim access to Indigenous land. Similarly, Naomi Klein (2013) states in a conversation with Leanne Simpson, who was instrumental in the "Idle-No-More" movement for Indigenous sovereignty in Canada: "extraction isn't just about mining and drilling, it's a mindset—it's an approach to nature, to ideas, to people." Simpson agrees by saying,

> extraction and assimilation go together [. . .]. My land is seen as a resource. My relatives in the plant and animal worlds are seen as resources. [. . .] extracting is stealing—it is taking without consent, without thought, care or even knowledge of the impacts that extraction has on other living things [. . .]. Colonialism has always extracted the indigenous.
>
> ibid.

It is our human capacity to discern and reflect that can help us to see the dark, "extracting" side of the modern "enlightenment." However, it is obvious and yet often ignored that we can no longer practice this necessary critical reflection as a Western monologue. We will simply be stuck like Baron Munchausen trying to pull himself out of a mire by his own hair. What we have in mind here is a kind of *responsive reasoning* instead—a reasoning which "knows otherwise" (Shotwell, 2011). It abandons a manorial, individualistic, and rationalistic autonomy being aware of its manifold dependencies. This seems to be what Pelluchon (2021: 291) has in mind when she tries to argue for a new form of enlightenment that builds on a phenomenology of our life on earth, which illuminates our dependency from eco-systems, and other human and non-human living beings.

In a similar vein, Elberfeld (2021: 60) states that it is long time to let other philosophical traditions "proof-read" European philosophies—particularly those traditions who have become victims of a marginalization and annihilation through a self-righteous universalization of European values. This is where "responsive reasoning" unfolds its literal meaning. Leanne Simpson talks about "deep reciprocity" as the alternative to just extracting: "It's respect, it's relationship, it's responsibility" (in Klein, 2013). Living this responsive practice means to learn from the Cree notion *wahkotowin* including the deep sense for relational accountability. The Cree scholar Willie Ermine (2007) introduced the widely known concept of the "ethical space of engagement"—a space that "is formed when two societies with disparate worldviews are poised to engage each other," when both are open to reason responsively. He got his idea from Roger Poole's book *Towards Deep Subjectivity*. Ermine explains:

> The space is initially conceptualized by the unwavering construction of difference and diversity between human communities. These are the differences that highlight uniqueness because each entity is moulded from a distinct history, knowledge tradition, philosophy, and social and political reality. With the calculated disconnection through the contrasting of their identities, and the subsequent creation of two solitudes with each claiming their own distinct and autonomous view of the world, a theoretical space between them is opened.
>
> ibid.: 193[3]

The "ethical space of engagement" rests on least two ambivalent contradictions: First, distinct uniqueness and difference is needed for the intercultural engagement, and second, "objectivity" is born out of "deep subjectivity," however it is a special kind of *relational* objectivity. In this space, "interwoven epistemologies" develop their own weaving patterns in which each string brings in its own color or shape. This is also what Robin Kimmerer (2013) has in mind with her metaphor of *Braiding Sweetgrass* when she weaves Indigenous wisdom, scientific knowledge and the teaching of plants together. Every attempt to assimilate, or to exoticize, to "rescue" language in a mere museal manner without seeing its deeper meaning, kills the ethical "space of engagement"—moreover, it is prone to again kill Indigeneity in philosophy and in our societies.

Granted, this "braiding" may be difficult to do without "getting dirty," without listening to the messy language of our body and our lands. However, it is an important key to decolonize philosophy as it opposes some of the main missions of the European expansion that were legitimized and driven by philosophical arguments, all of which aimed at purity and perfection. Some of the major strategies used in that vein were to Christianize, to civilize and to modernize (Elberfeld, 2021: 63). How attractive and yet ethically questionable (to say the least) the quest for a "civilized purity" is, shows Alexis Shotwell (2016) in her book *Against Purity*. Part of being "pure" is to be a self-contained being. "Purity practices—in ideology, in theory, and in practice—work to delineate an inside and an outside; they are practices of defining a 'we'" (ibid.: 13). The Indigenous concept of relationality stands in sharp contrast to this possessive individualism. It resists "the core idea that our selves are owned by us [which] functions as a categorical move to lay out a map of who can own others" (ibid.: 12). Purity is not just an illusion;

[3] An interesting variation of Ermine's concept can be found in Morgensen's (2011) *Spaces between Us. Queer Settler Colonialism and Indigenous Decolonization*. Exploring how differences between "Indigenous" and "settlers" determine the status of being "queer," he suggests that modern queer identities came out in Native as well as non-Native contexts by engaging profound distinctions Indigeneity makes within settler societies.

it is dangerous, as it is a rhetoric which delimits the world into something separable and homogenous. It divides and racializes. Yet "[w]e are all already polluted. We have more micro-organisms in our guts than we have cells in our bodies [. . .]. We are, in other words, continuous with everything" (Biss, in ibid.: 10). Instead of searching for self-contained purity, we should honor the fact that we are always already related.

Honoring our embodied relational knowledge decolonizes philosophy. This is what we can learn from Western-Indigenous dialogues. Since philosophy is not just happening in the academe, but has an impact, it has the chance to transform our way of living. Indigenous chiefs in the Yukon invited youth to form what they call a *First Nations Climate Action Fellowship*. In their action plan, the young fellows state: "We are living in an era of Disconnection. The imbalance in our Lands [. . .] is a reflection of the imbalance within ourselves [. . .]" (RVAP). Their approach, which they call "Reconnection is Climate Action," is based on this diagnosis. An approach that solely relies on scientific premises falls short, because it is deaf to the voice of creation—a voice we will never be able to fully understand. Therefore, we have to learn to "listen otherwise" (Lipari, 2009) and to practice forms of "responsive reasoning" that emerge out of an ethical space of engagement.

References

Akiwenzie-Damm, K. (ed.) (2003), *Without Reservation: Indigenous Erotica*, Cape Croker Reserve: Kegedonce Press.

Bear, T. L. (2016), *Power in My Blood: Corporeal Sovereignty through the Praxis of an Indigenous Eroticanalysis*, https://doi.org/10.7939/R3348GS66 (accessed May 30, 2023).

Blumenberg, H. (2012[5]), *Schiffbruch mit Zuschauer*, Frankfurt a. M.: Suhrkamp.

Böhme, H. and G. Böhme (1996[3]), *Das Andere der Vernunft. Zur Entwicklung von Rationalitätsstrukturen am Beispiel Kants*, Frankfurt a. M.: Suhrkamp.

Cameron, M. (2005), "Two-spirited Aboriginal people: continuing cultural appropriation by non-Aboriginal society," *Canadian Women Studies*, 24(2/3): 123–7.

Driskill, Q-L., D. H. Justice, D. Miranda, and L. Tatonetti (eds.) (2011), *Sovereign Erotics: A Collection of Two-Spirit Literature*, Tucson: The University of Arizona Press.

Dussel, E. (1989), *Philosophie der Befreiung*, Hamburg: Argument.

Elberfeld, R. (2021), *Dekoloniales Philosophieren*, Hildesheim: Georg Olms.

Ermine, Willie (2007), "The ethical space of engagement," *Indigenous Law Journal*, 6(1): 193–203.

Gamm, G. (2009), *Philosophie im Zeitalter der Extreme. Eine Geschichte philosophischen Denkens im 20. Jahrhundert*. Darmstadt: wbg.

Kimmerer, Robin Wall (2013), *Braiding Sweetgrass. Indigenous Wisdom. Scientific Knowledge, and the Teachings of Plants*, Minneapolis: Milkweed.

Klein N. (2013), "Dancing the world into being: a conversation with Idle No More's Leanne Simpson," in *yes!* March 6, 2013, www.yesmagazine.org/social-justice/2013/03/06/dancing-the-world-into-being-a-conversation-with-idle-no-more-leanne-simpson (accessed May 29, 2023).

Liboiron, M. (2021), *Pollution is Colonialism*, Durham: Duke University Press.

Lipari, Lisbeth (2009), "Listening otherwise. the voice of ethics," *The International Journal of Listening*, 23, 44–59.

Lu-Adler, H. (2023), *Kant. Race, and Racism. Views from Somewhere*, Oxford: Oxford University Press.
Morgensen, S. L. (2011), *Spaces Between Us. Queer Settler Colonialism and Indigenous Decolonization*, Minneapolis: University of Minnesota Press.
Nelson, M. (2017), "Getting dirty: the eco-eroticism of women in Indigenous oral literatures," in H. Barker (ed.), *Critically Sovereign: Indigenous Gender, Sexuality, and Feminist Studies*, New York: Duke University Press, 229–260.
Paul, A. M. (2021), *The Extended Mind*, Boston and New York: Mariner.
Pelluchon, C. (2021), *Das Zeitalter des Lebendigen. Eine neue Philosophie der Aufklärung*. Darmstadt: wbg.
Rifkin, M. (2011), *When Did Indians Become Straight? Kinship, the History of Sexuality, and Native Sovereignty*. Oxford: Oxford University Press.
Samaran, Nora (2016), *The Opposite of Rape Culture is Nurturance Culture* (posted February 11, 2016), https://norasamaran.com/2016/02/11/the-opposite-of-rape-culture-is-nurturance-culture-2 (accessed August 30, 2023).
Schellhammer, B. (2023), "Reading split tooth lessons of the sovereign erotic I take away and still grapple with," *ZKS*, 43(73): 94–107.
Shotwell, A. (2016), *Against Purity. Living Ethically in Compromised Times*, Minneapolis: University of Minnesota Press.
Shotwell, A. (2011), *Knowing Otherwise. Race, Gender, and Implicit Understanding*, Pennsylvania University: Pennsylvania University Press.
Stangneth, B. (2020[2]), *Sexkultur*, Hamburg: Rowohlt.
Tagaq, T. (2018), *Split Tooth*, Toronto: Penguin.
The Yukon First Nations Reconnection Vision + Action Plan (RVAP). *Draft Summary of Engagement*, www.yfnclimate.ca/yfnrvap (accessed February 4, 2023).
Waldenfels, B. (1990), *Der Stachel des Fremden*, Frankfurt a. M.: Suhrkamp.
Waldenfels, B. (1998), *Grenzen der Normalisierung*, Frankfurt a. M.: Suhrkamp.
Wilson, S. and B. Schellhammer (2021), *Indigegogy. Invitation to Learning in a Relational Way*, Darmstadt: wbg Academic.
Wilson, S. and P. Wilson (1998), "Relational accountability to all our relations," *Canadian Journal of Native Education*, 22(2): 155–8.
Wilson, S. (1999), "Honoring our relations. Aboriginal spirituality as comprehensive relational accountability," *The Canadian Social Studies Journal*, 23(3): 76.
Wilson, S. (2001), "Self-as-relationship in Indigenous research," *Canadian Journal of Native Education*, *25*(2), 91–92.
Wilson, S. (2008), *Research is Ceremony. Indigenous Research Methods*, Halifax & Winnipeg: Fernwood.

6

Perhaps We Didn't Need a Bridge: In Dialogue with Indigenous Mathematics

Edward Doolittle and Amber Hughes

The missing bridge

The Königsberg bridges problem is a classic problem in nearly universal, conventional mathematics, sometimes (inappropriately, we believe) called "Western mathematics"; we suggest "global mathematics," in contrast with Indigenous mathematics, as alternative terminology. The story of the Königsberg bridges problem is foundational in the history of graph theory in mathematics and is often invoked in the early stages of education in graph theory. The problem, dating from the early 1700s, is to find a walking path that crosses each bridge exactly once. The reader is invited to try the problem; a well-known map of the region as it was in the early 1700s is in Figure 1. (Since that time, changes in the geography of the space have changed the problem.)

Our intentions

Over the past months, we (Edward and Amber) have engaged in a discursive, collaborative process of generating new understandings pertinent to the Indigenous mathematics education field. We have been thinking about the methodological implications of our experience and process, and how this resonates with what we know about not only existing educational theories, but, more importantly, Indigenous epistemologies, or ways of knowing. To do so, we have examined some aspects of our own positioning in the space, as well as key moments (or narratives shared within the dialogue) which highlight among many things the centrality of experiential knowledge, spiritual knowledge, dialogical ways of knowing and the creation of ethical space for discourse in giving life to new and shared meaning. We see these aspects of our process as examples of ways in which we can draw strength from, and contribute to the continuance of, Indigenous traditions of knowledge creation within mathematics.

Through centering Indigenous ways of knowing in our processes of coming-to-know, we consider new possibilities for creating a shared space of mathematical

Figure 1 Königsberg Bridges map view, Zeiller, M. (1652), Topographia Prussiae et Pomerelliae, Merian.

understanding which is the culmination of more than global perspectives of what the discipline entails. More specifically, we seek to understand how our very ways of knowing, mathematically, can be formed and shaped through ongoing Indigenous contribution. Hence, this work is about redefining the boundaries completely of how we think about and engage with mathematical processes through shared, ethical spaces of dialogue (Ermine 2007) and through holistic perspectives of what the discipline encompasses. We see our contribution as part of a growing foundation for further projects related to expanding on understandings of mathematics: what it is, what it entails, and what it could do for future generations.

Giving importance to experiential knowing

As we are reflecting on the dialogue that has unfolded between us so far, or what is recorded of it and what we can recall, the prominence of experience in giving shape to our unfolding foci within the discussion becomes quite clear, and therefore how it is we have constructed our own shared meanings from the interaction. These topic choices are not so much "choices" as they are organic extensions of the conversation and have been very much informed by our own personal and professional experiences. Perhaps there is not so much a direct, one-to-one correspondence nature of connection between dialogue and immediate experience, but rather a more mercurial, roundabout way that the topics to which we have chosen to give our attention are in many ways connected to multiple levels of experiential, intuitive, perhaps even spiritual ways of knowing, that we were not consciously aware of in those moments (Botha 2011).

To give an example of this, there are multiple inclusions in the threads of conversation from the very beginnings of our collaborative journey, of the importance of ancestral ways of knowing, which give rise to the ways we have envisaged "ways forward" in this discipline and space: more specifically, the importance of Indigenous and non-Indigenous peoples coming to an understanding of their own connection to "spiritual knowledges," which connect us all to our own distinct pasts, places and people in ways that many of us have completely forgotten, or have never been taught.

Amber has held a particular interest in ancestral ways of knowing, and reconnecting to her pre-Christian, Celtic "roots," for some time; becoming deeply motivated by this in her own personal journey. Looking back, Amber might discern that such a focus links back to her upbringing with a mother who has always been deeply conscious of the stories of her ancestors and mapping out her family tree. As Amber moves forward now in various directions of her work and comes to her own understandings as a non-Indigenous researcher in the spaces of Indigenous mathematics education, she can clearly see how this experiential knowing infiltrates her current ways of being and doing, and how this is evidenced in the tone and direction of some of the discussions that have unfolded between Edward and herself. In those moments of conversing, it wasn't as apparent. However now, looking back over the unfolding discourse, she can see the patterns emerging and the gravitas this concept has had in her overall schema, or worldview, of understanding in this field.

In many ways, the experiential also links to the importance of understanding our own positionality, or "roots" in this space; and further still, how it is that our ancestral ways of knowing might have implications for our ways of knowing and creating meaning, now. Interestingly, philosopher Ludwig Wittgenstein showed through his development of a theory on language that our interpretations of concepts and knowledge, and the ways we develop "meanings" are the "result of historical, inherited backgrounds upon which we learn to distinguish between what is 'true', or meaningful, and what is not" (Wittgenstein 1969; cited in Lerman 2001). Note the emphasis on "inherited" here, and "historical," as though there is a part of our knowing that both precedes and precludes us, in our own bodily experience. More recent understandings in work on intergenerational trauma add further weight to this concept, and the view that it is the experiences of trauma that could not be processed by our predecessors, that come through us in various ways in the present time, often in embodied ways, through time immemorial. The suggestion is made that our present and lived realities are an alchemy of the past/historical; and so in ways, our "knowing" is the ongoing result and impact of what has already happened in the past; both to ourselves, and to our ancestors, and this is what we have to learn how to navigate as a collective, and for the betterment of humanity (Borden and Wiseman 2016; Simon 2023).

We may return to the idea of "betterment" at another time, but for now, these observations circle around the main points being offered, that our ways of knowing are constantly being informed by our experiences, from all dimensions of the present, past and ancestral, and in ways that might outwardly seem quite indiscernible. Cultural/discursive psychology theories in mathematics tell us that meanings are not singular, or static, and are experienced and internalized from a range of discursive situations (Lerman 2001). While we acknowledge these contributions as being relevant and established parts of the discourse within educational contexts, we continue to expand on them to consider potentially deeper, philosophical questions about the historical/ancestral and collective experience (plural), and their bearing on theories of consciousness. This kind of excavation exceeds the parameters of this contribution, however, and so we acknowledge that in any "holistic" representation of our understandings, these questions must be in some way reckoned with, as they are also a part of our experience, one way or another.

Further to the point of the significance of experience to our ways of knowing, Edward shared a story about his approach to the Königsberg bridges problem in the light of his own personal experience, and how when he reimagined different ways of approaching the scenario drawing from Indigenous, spiritual ways of knowing, he imagined a completely different solution to the problem than what was being sought in conventional mathematics (Doolittle 2018).

First, let us consider the experience of most when engaging with the Königsberg bridges problem. In the conventional presentation, the experience is generic, the same for everyone, literally boxed-in by the diagram provided and the context in which the experience is delivered (usually a book or a lecture hall, both rectangular, bound spaces).

The usual solution to the Königsberg bridges problem repeats endlessly the analysis of mathematician Leonhard Euler who concluded that the tour of the city crossing

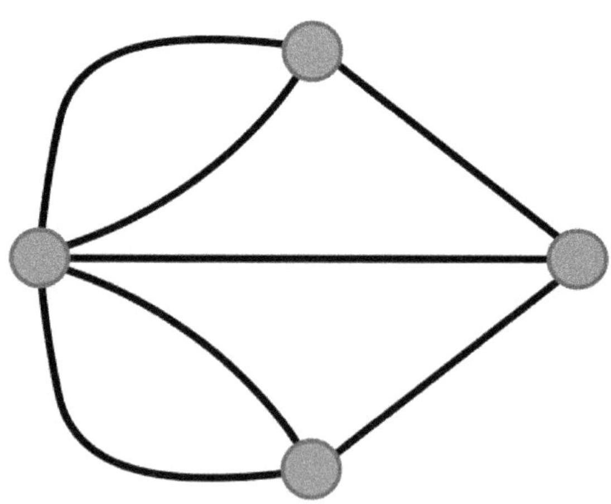

Figure 2 Euler's map of Königsberg.

each bridge exactly once was not possible (unless another bridge were built) because one must transition into or out of at least two of the four regions of the city (left bank, right bank, and the two islands) an even number of times, but each region only had an odd number of bridges. Euler's analysis has been codified in the graph theory that he founded, and his story is invoked by the names given to aspects of the story: Eulerian circuit, Euler's Theorem. One problem with Euler's approach is believing that the simple abstraction, the graph of four points and seven lines, completely captures the situation.

On the other hand, Edward connected the problem to wider varieties of experience: dialogue with his next-door neighbor, Ken Wilson, who engaged in a spiritual journey along the Grand River, traditional homeland of the Kanyen'kehake and the others of the Six Nations in Ontario, from its source to its mouth (Wilson 2016; Wilson and Anderson 2021); contemplation of the spiritual requirement to give thanks (perhaps even at its source) to the Pregel River which forms the necessary but generally neglected "being" at the heart of the Königsberg bridges problem; and the spiritual experience of praying for guidance on the matter. Those different experiences—Indigenous, experiential, dialogical and holistic—led to a different understanding of the problem. Contrary to the conventional analysis and answer, it is possible to walk over each of the bridges exactly once, if we (for example) start on one of the islands, cross to a bank, head to the boundary of the Pregel River drainage basin, walk around that boundary to the "other side" of the river (topologically there is only one side unless we artificially limit ourselves to the city), and then complete our travels within the city.

While we will reflect on the spiritual implications of this scenario in subsequent sections, let us draw attention in this moment to the role of experiential knowledge, and the story sharing that had allowed a particular understanding to unfold, in a completely new and seemingly "unconnected" experience. We consider the ramifications processes like these might have for the mathematical problem-solving activity in the first instance, and how the drawing together of experiential and even spiritual ways of knowing might prove more "reasonable" in the navigation of realities, mathematically.

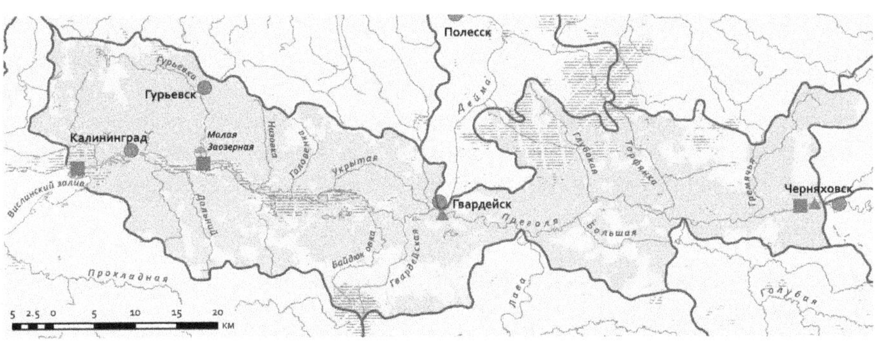

Figure 3 Pregel River drainage basin.

Implications

The points offered thus far are to highlight further or position some assessments on ways of thinking about what it is to "know," and how it is that knowing, i.e., the process of acquiring meaning, is facilitated in the first place. If we can accept that historically, and culturally, different ways of thinking have existed around how knowledge is formed and shaped, and constitute multiple ways of knowing in many cases (i.e., intellectual, experiential, embodied, spiritual, intuitive, emotional, or sensory—to name a few), then we can accept that whatever we bring to bear on an understanding in the present moment is always the result of not only our multiple subjectivities, but our cumulative experiences and "knowings" from other times (both of, and not of, this lifetime/experience) (Drawson, Toombs and Mushquash 2017; Gutiérrez 2022; Lavallée 2009; Simon 2023).

We might also reason that within the sciences and mathematics, typically only "some" of these ways of knowing have been given space in the discourses around what constitutes valuable knowledge (Moreton-Robinson 2016; Stavrou and Murphy 2022). In recent decades, and particularly within the discourses of critical Indigenous studies, attention has been drawn toward unravelling how particular forms of knowing have been silenced in the business of what "counts" (Hokowhitu 2016) and in terms of educational foci. As such, implications for Indigenous peoples globally have been felt largely in the silencing of traditional and cultural knowledges in any "meaningful" curricula in schools, vis-à-vis mathematics. We accept that while there may exist very different interpretations of what valuable knowledge is, and should be, and according to whom; the implications for those this privileging excludes are felt perhaps most intensely in mathematics (Parkinson and Jones 2019). Hence, more needs to be "done" to understand how any kind of discourse might be thought about between the various positions on the fundamental act of making meaning.

Where, for example, do experiential knowledge or embodied ways of knowing come into prominence within mathematical processes and discourse? If we can accept that such ways of knowing play a significant role in our forming of understandings, cognitively, then where is this recognition extending beyond the role of providing greater access, or "scaffolding," to the dominant ways of thinking about what mathematics is and does (Hughes 2022; Stavrou and Murphy 2022)?

Let's go back to the story shared by Edward, around how he was able to imagine a completely alternative solution to a mathematical problem through an integration of his experiential and spiritual ways of knowing, i.e., Indigenous ways of knowing. In that specific instance, a new resolution was found that completely precluded the need for any in-depth mathematical problem solving or abstraction. A new way of making meaning from that context was brought to life. We might be able to propose, therefore, that the mathematics was simply not "needed"—or was it?

We could pose the argument here that through the inclusion of experiential and spiritual ways of knowing, which we may otherwise refer to collectively as "Indigenous" ways of knowing, we may not need to include any global mathematics in our solving of real-world problems, and therefore, such contributions relegate global mathematics to a state of redundancy. At various times we have even considered the questions around what mathematics might be needed in times of drastic global shifts to ways of life brought about by climate change, for instance.

However, we refrain from making any gross generalizations around the potential irrelativity of global mathematics, as we know it, and rather, offer a moment for contemplation on the potential ways of creating space for dialogue between different ways of knowing, different ways of creating meaning. Between global, non-Indigenous systems of thought (the ones that are mandated in schools and universities) and Indigenous ways of knowing, which until relatively recently, have been denied any space within the academe, or curricula (Nakata 2007). And so, we consider, can the "two co-exist in dialectic with each other, much like thinking and speech?" (Lerman 2001: 7).

This point brings us onto the next, that of dialogue, and what it means to engage in dialogical ways of knowing, which are very much akin to Indigenous epistemological perspectives if we consider our collective knowledge on the subject.

Dialogical reasoning as more sophisticated iterations of "truth"

We have reflected on the observation that in First Nations oral tradition, we are discouraged from sharing or explaining our interpretation of the meaning, or "moral" of the story if you will. We talked about why that is, and how, in the case of knowledge transmission approaches, which much of our formal education systems are built on, the act of "instructing" the recipient (of knowledge) on how it ought to be understood removes the agency of the recipient in forming and shaping their own understandings, drawing from their own experiences, and what it therefore means to "them." This indeed was a criticism leveraged against some of Vygotsky's earlier work on moving from the "abstract" to the concrete in his models of knowledge acquisition. His take on the matter was that it was only after a concept was named, in the abstract, that understanding of said concept could start to unveil in the mind of the recipient, or "knower," in their own unique ways. Many at the time, however, saw this type of modelling as reductive and "transmissive" in approach, relegating the role of the knower to that of a passive recipient. We draw from these types of arguments to propose that the removal of this agency results in static, one-way, directive forms of transmission which do not allow for any other input, or creativity, much less the creation of "new" types of understanding.

Edward reflected on how dialogical forms of knowing, in which active participants co-create new understandings through shared space and discourse, produce incrementally more sophisticated understandings, which consider the perspective and contribution of the "other" in the exchange, but in ways that are unique to them. Previously, we elaborated on the significance of experiential knowledge, and how it forms and shapes our understanding of current realities in ways of which we are often not consciously aware at the time. Dialogical knowing goes further to suggest that not only might a multitude of experiential knowings be impacting on us in any one moment, and from across a great expanse of time, but also the act of sharing this process with others produces increasingly more complex forms of knowing that any "singular" knower would not possibly have access to on their own. It is not always the objective of knowledge creation through dialogue to find therefore "common ground," or "sameness," among and between the various constituents, but to share with them in a process in which all those involved have equal contribution to the creation of more complex, more sophisticated understandings, which are informed by a collective of voices.

This might perhaps be a way we can think about the process of coming-to-know "with Place," or "with Country," which is a defining feature of Indigenous epistemologies globally, and that has not been as robustly understood among the mathematics and science communities thus far for its lack of anything quantifiable or measurable. There are of course, some exceptions to this assessment. Cajete, for instance, unpacks what he defines as "7 orientations" toward the redevelopment of Indigenous science education, which does specify in point 3 "An epistemology of Indigenous science" (Cajete 2008). Collective work by Lloyd et al. with Bawaka Country explored Indigenous ontology in relation to mathematics, through the concept of "Ganma," "where the salt water and the fresh water meet and mingle" (Lloyd et al. 2016). Recent work by Nicol and others (Nicol et al. 2023) has also more deeply accounted for the ways in which "Place" might be implicated in ways of knowing in STEM education, in which the utility of the "Sciences" and their applications is driven toward the creation of more conscious, purposeful activities that nurture humanity's relationship with Place.

What we must grapple with epistemologically, however, is that according to Indigenous ways of knowing, Place also has voice in what there is to know, and so learning how to know with Place is a part of what is considered valuable knowledge-producing activity. Moreton-Robinson tackles this assertion in her case for an Indigenous Women's standpoint theory, citing "Place" in ways of knowing as a distinguishing factor between Indigenous, and white-feminist standpoint perspectives, among others (Moreton-Robinson 2013). For now, let us return our thoughts to the importance of dialogue in coming-to-know through Indigenous ways of knowing.

We considered how in global mathematics, the notion that there are "correct" ways of doing and being, and that something must be "proven" objectively to be seen as valid sense-making, prevent any notions of discourse or dialogical understanding that we might find in Indigenous epistemological perspectives. Once a theory of understanding has been "proven" mathematically, it is very difficult to overturn or counter such positions through any other means. As Gutstein contends, the power of number in dispelling any other claims to the contrary is one of the reasons why Mathematics for Social Justice was seen as instrumental in social activist developments (Gutstein 2006). If you can make a case for why a given social phenomena is unjust, or disproportionately represents certain populations, and you can prove it using mathematical skills and processes, you are a force to be reckoned with in larger political spheres.

The point we would like to emphasize with respect to dialogue however, is that the universal nature of mathematics that has been a hallmark of the discipline in modernity excludes the possibility of discourse and therefore the openness to new ways of thinking within the discipline, which may enable collective thinkers to progress to incrementally more sophisticated understandings. Conversely, Indigenous epistemologies seem to embed this understanding in their very structure; that is, there is never "one way" of knowing something, and often "many ways" of seeing are accepted, if not encouraged, as appropriate ways of coming-to-know.

In the next section, we consider our use of mapping methodology as an approach toward building on our experiential, and dialogical processes of coming-to-know, and as an outward expression of some of the more fundamental components of Indigenous approaches to knowledge creation.

Discussion of our mapping methodology

Conversations/dialogue

We reflect that the conversation that emerged over time, and between us two, evolved organically, and with no apparent structure or focus other than to share experiences, observations and stories around our ways of knowing in mathematics, and Indigenous (education) contexts. We might liken this to Yarning methodology in Australia (Bessarab and Ng'andu 2010), or Sharing Circles commonly used in Indigenous research methods in North America, as the conversations encompassed many aspects and layers of our individual experiences. There was a degree of vulnerability in this process, as we each offered very authentic contributions to the dialogue, through reflection and sharing of our lived, felt and known realities. The exchange integrated deep listening, a term Edward uses within the context of First Nations oral traditions, where the listener provides space for the person speaking in an authentic way, without the need to interject or "control" the conversation and speaks in response to what has been offered in the dialogue in a reciprocal way. In this way, all parties must engage in deep listening to further the dialogue, as they must be actively processing the new information and offering their own understandings in ways that can be deemed "responsive" to the discourse.

Shortly after many of the conversations, a time was set aside for reflection in which Amber would write down any interesting or salient parts of the dialogue in a journal. This journal was then shared with Edward in an ongoing and reflective way, to ensure that the points being made were considered by both parties, and that any additional information felt to be significant was also appended and included, although that was rarely needed. Sometimes the discourse would also continue via email, in a continuous exchange of reflection and sharing, and so was not always limited to the moments where we were actively involved in dialogue.

Coding and graphing

After a substantive volume of written reflections had been documented, we began the process of collaboratively identifying concepts and larger ideas from the discourse, in more of an inductive approach. This would sometimes involve paraphrasing or condensing larger ideas down to the "main" ideas being discussed, determining any existing relationships between the various concepts inherent in the discourse. We noted that the variety of relations that exist between two connected, or disconnected concepts, was quite immense. For example, we could pose that some concepts seemed to be connected to each other in a "is the same as" way; others "is opposite to"; while others might be thought of as "is a result of," or "is associated with." Ways of relating concepts to each other were flexible and varied, and we did not try to put any stipulations on how this could occur. If there was a connection between concepts of any kind, it was documented.

These concepts, and connections between concepts, could then be systematically documented and mapped, with recursions of the same conceptual links permitted to define more "strongly connected" concepts within the data set. Concepts mentioned together more frequently, therefore, would have a stronger degree of connection or proximity within the visual map output, and would be less likely to be moved further away from each other as new information was added to the map. Concepts mentioned together by association or proximity were linked only two concepts at a time, corresponding to the idea of graph in mathematics, the same idea as in Euler's solution to the Königsberg bridges problem. An initial attempt at depicting the graph of connections was prepared with the graph visualization tool "neato."

Clustering and mapping

The graph as it stood was difficult to interpret, so we further processed it with additional tools from the Graphviz software suite: "sfdp" (scalable force-directed placement) was used to place related concepts as close as possible to one another, and "gvmap" was used to identify clusters of related concepts and use them to produce a map of regions of connected concepts. The resulting map appears in Figure 4. In the map, each shade represents a different cluster of concepts. (Further development of this map, and a more detailed discussion of the technicalities involved, will appear elsewhere.)

How to read the map

Let us return to the points made earlier about the plurality of knowledge and understanding within many Indigenous knowledge systems, in which no "one" objective truth exists, but rather, multiple iterations of truth that are grounded in multiple subjectivities, inclusive of subjectivities of Place.

We also acknowledge that this representation is the culmination of our (two) collaborative inputs, and so represents a particular "space" in which dialogue has been shared. This map, we intend, would take on a variety of visual-spatial outputs, or versions, depending on the persons of input to the discourse, and so determining this as an overall or exhaustive representation of what the discipline entails should not be the objective of this exchange.

Conclusions from the map

While we cannot make generalizations from our mapped representation of discourse and cannot claim that the "findings" somehow implicate broader themes about Indigenous mathematics in a variety of contexts, we can consider how the map might be useful for furthering our understandings of this field.

We would like to consider the possibility that since the map emerges from dialogue, then as other conversations take place with other participants, the map will grow organically, as an extension of the Indigenous mathematics (education) dialogue. The map and process behind it is never static, as incrementally more complex ways of understanding are arrived at through continued discourse, and with others. We draw these points back to those offered previously around some of the foundational aspects of Indigenous epistemologies, in which knowledge making processes are never static and "one way," but dynamic, collaborative, and responsive.

We also consider that "clusters," or groups of concepts within the map which seem to be less populated, might be indicative of where further understanding is required. For instance, the fact that some of the regions of the map contain fewer concepts and connections than other regions, may suggest that our understanding of ways of creating shared space between different ways of knowing is more limited than our understanding of Indigenous knowledge and belief systems, or global mathematics and theories of education.

Potential uses of the map that are less understood currently might relate to how we can bring others into this shared discourse and extend the process beyond us two. We do not intend that some larger, generalizable output might be the result of including more people, or other perspectives, in this process. We simply wish to understand how the map might change and transform, and therefore can be seen as dynamic and ever-changing, in particular contexts.

Edward's interests in the technical aspects of capturing these kinds of transformations via "version control" methodology (used extensively in software development), together with our collective appreciation of the need for greater representation of kinds of educational goals, objectives, and discourses that better reflect the realities for Indigenous peoples, and, in vastly different contexts, add momentum to our pursuit of furthering these processes and understandings.

Final observations and conclusions

It is interesting to note that the same mathematics commonly used to analyze the Königsberg bridges problem, graph theory, is also used (but in a far more sophisticated way) in the mapping methodology. Developments in mathematics and computing, and the resources available in modern times, make it possible to use technology, mathematics, and science in ways that more clearly reflect our ways of knowing and being. It is often suggested that Indigenous people need to be educated "better," but we contend that it is technology, mathematics, and science that needs to be improved so that they achieve human goals rather than forcing humans into compatibility. For nearly 300 years, we have been boxed into conformity with Euler's vision of the original problem and its essentially simple solution. Now with the aid of modern technology, we can begin to form graphs to suit our true needs and values.

However, there is still and always will be a danger that we do not understand the full extent to which global technology, mathematics and science are forming us, molding us, and leading us away from our Indigenous heritage. We still, and always, should continue to ask, "What do we lose when we study mathematics?" not just "What do we gain?" (Doolittle 2007). And we should continue to resist the loss of culture we would

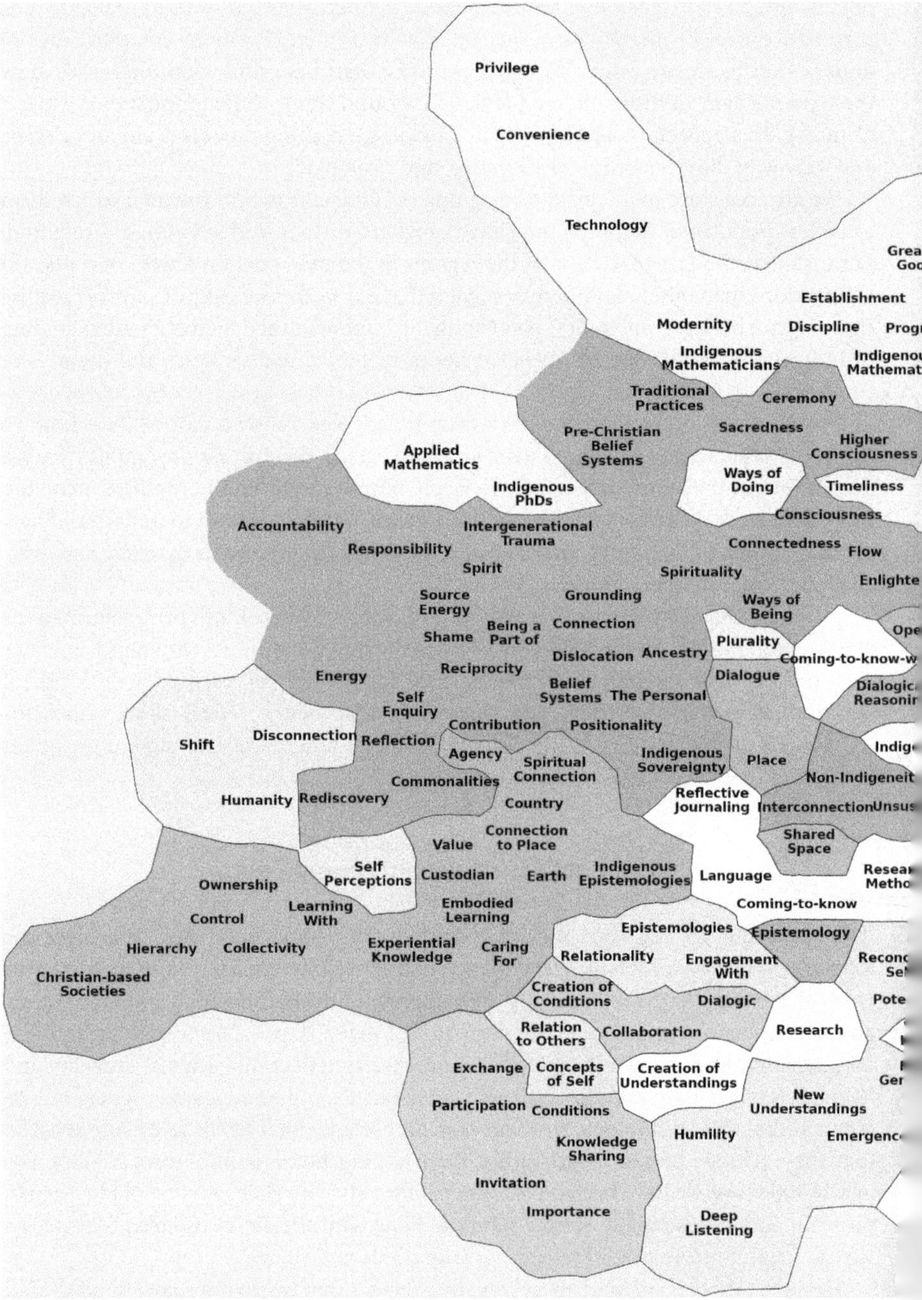

Figure 4 Concept map of our conversations, with clustering and mapping.

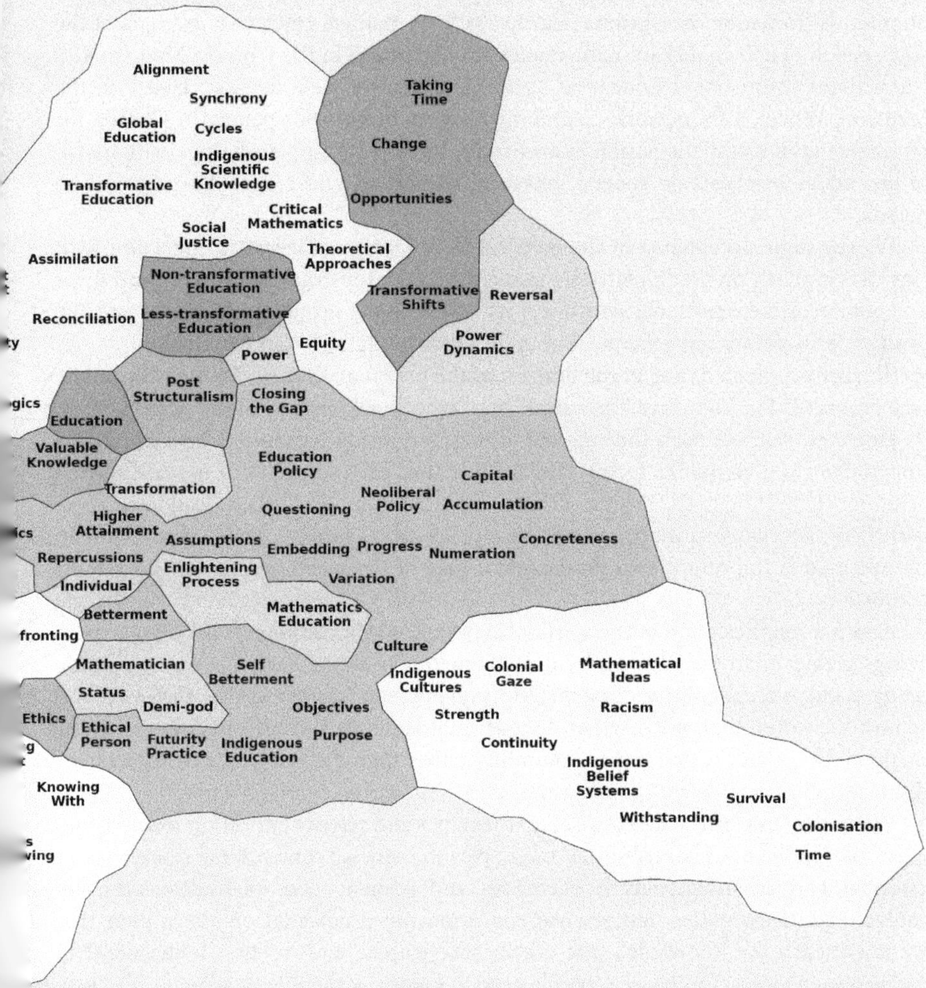
Alignment
Synchrony
Global Education
Cycles
Indigenous Scientific Knowledge
Transformative Education
Critical Mathematics
Social Justice
Assimilation
Taking Time
Change
Opportunities
Theoretical Approaches
Non-transformative Education
Less-transformative Education
Reconciliation
Transformative Shifts
Reversal
Power Dynamics
Equity
Power
Post Structuralism
Closing the Gap
Education
Valuable Knowledge
Transformation
Education Policy
Capital
Neoliberal Policy
Accumulation
Questioning
Higher Attainment
Assumptions
Embedding
Progress
Numeration
Concreteness
Repercussions
Enlightening Process
Variation
Individual
Betterment
Mathematics Education
Culture
Mathematician
Self Betterment
Indigenous Cultures
Colonial Gaze
Mathematical Ideas
Status
Demi-god
Objectives
Strength
Racism
Ethics
Ethical Person
Futurity Practice
Indigenous Education
Purpose
Continuity
Indigenous Belief Systems
Knowing With
Survival
Withstanding
Colonisation
Time

experience if we blindly followed the global way. We caution that the "transformative" power of education, generally viewed in a positive light, can mean transformation away from Indigenous ways of knowing and ways of being. We wonder what "non-transformative" or "less-transformative" education might look like.

An example of what we might lose in the study of mathematics is spirituality, which is generally viewed as incompatible with "rigorous" global mathematics, with a small number of "romantic" exceptions (Hardy 2012: 30; Kanigel 1992). An example of the suppression of spirituality in mathematics may be found in the Ontario Mathematics Curriculum (Ministry of Education 2020: 7), where we see a diagram based on the Medicine Wheel, a framework commonly used by Indigenous people in Canada for representing "holism," the notion that we must develop as complete beings, with more-or-less equal emphasis on mental, physical, emotional, and spiritual aspects of the person.

We see some acceptance of the physical in modern mathematics education with (say) its emphasis on manipulatives and the concrete to abstract continuum, and some acceptance of the emotional with (say) its emphasis on group work, affect, and the positive feelings that can generate. But in the Ontario Mathematics Curriculum, we see spirituality suppressed: it is in the domain of the personal, the self, without the public face permitted for "cognitive," "physical," and "emotional" aspects of the person.

However, we have seen through our Königsberg bridges example, how a spiritual imperative (to give thanks to the Pregel River that gave so much to us) and related ceremony (traveling to the source of the river to make an offering) can lead to creative, surprising, alternative understandings. In any case, even without obvious direct benefit, the spiritual is important as a fundamental part of the whole person in Indigenous contexts.

Another manifestation of the spiritual is prayer, which could involve dialogue with beings greater than us. We wonder if mathematics could be such a being, a gigantic being as old as creation itself (first there was no one, then there was one, then two, and so on). We would then not learn or discover mathematics, but instead come-to-know mathematics, a task which requires humility rather than the arrogance of knowing or discovering.

When contemplating the future of mathematics and science education research and what role "Indigenous contribution" might play therein, we contend that research is an essential part of Indigenous mathematics and science. Too often discussions of Indigenous mathematics and science education are predicated on the notion that we must teach the knowledge that others have gained, and for the "betterment" of Indigenous peoples. Indigenous people need to take back the role of generators of new ways of knowing, as they have always been, until recently. Every understanding they had, every means and mode of survival which carried them from the time of creation to the time of contact, was theirs, and their survival, growth, and flourishing from time immemorial to today are testament to the power of their "research methods." In contrast, the more prominent global culture, including global technology, mathematics, and science, is only a few hundred years old, and is already at risk of destroying itself.

We contend that research is a necessary part of the whole cycle of teaching and learning: we cannot survive if we do not adapt to a changing world, and we cannot

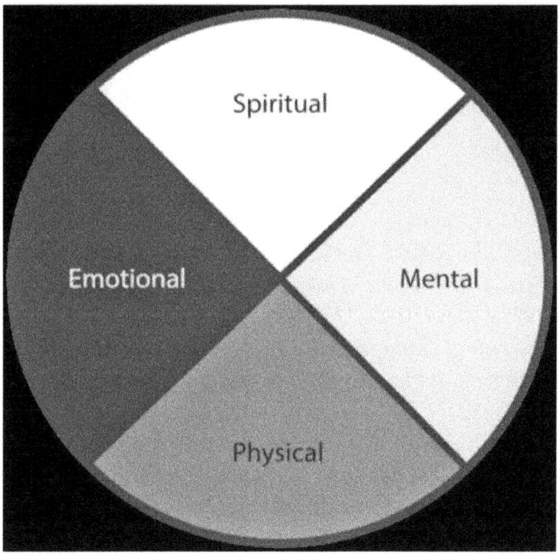

Figure 5 A holistic medicine wheel.

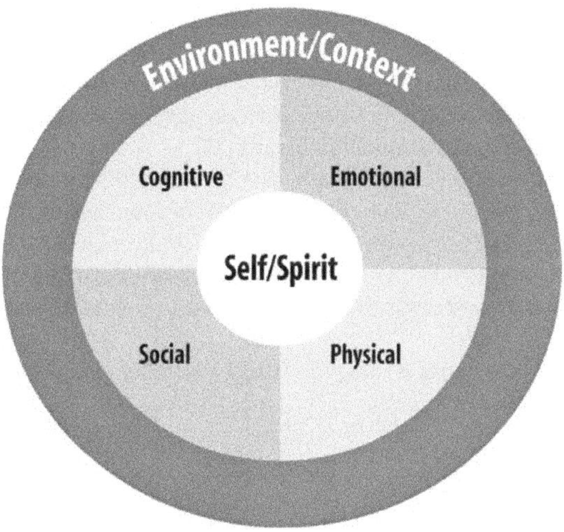

Figure 6 The modified medicine wheel in the Ontario Mathematics Curriculum, with spirituality pushed to the personal domain and out of the public domain.

survive as independent, Indigenous nations if our only adaptations are those provided to us by others. We envisage a form of mathematics to evolve that considers the active creation of new understandings, new meaning and knowing, through shared, dialogical processes in which all, inclusive of Place and disparate subjectivities, have a say in not only "what" there is to know, but "how," and, perhaps most importantly, "why."

We hope we have drawn attention to some possibilities, founded on sharing, the creation of shared spaces, collaboration and dialogue, for regenerating and recovering our ancient traditions of coming-to-know, so that we may come-to-know-again.[1]

References

Bessarab, D. and B. Ng'andu (2010), "Yarning about yarning as a legitimate method in Indigenous research," *International Journal of Critical Indigenous Studies*, 3(1): 37–50.

Borden, L. L. and D. Wiseman (2016), "Considerations from places where Indigenous and Western ways of knowing, being, and doing circulate together: STEM as artifact of teaching and learning," *Canadian Journal of Science, Mathematics and Technology Education*, 16(2): 140–52.

Botha, L. (2011), "Mixing methods as a process towards Indigenous methodologies," *International Journal of Social Research Methodology*, 14(4): 313–25.

Cajete, G. A. (2008), "Seven orientations for the development of Indigenous science education," *Handbook of Critical and Indigenous Methodologies*, 487–96, SAGE Publications, Inc.

Doolittle, E. (2007), "Mathematics as medicine," Proceedings of the 2006 Annual Meeting of the Canadian Mathematics Education Study Group = Actes de La Rencontre Annuelle 2006 Du Groupe Canadien d'Etude En Didactique Des Mathematiques (30th, Calgary, Alberta, Canada, June 3–7, 2006), Canadian Mathematics Education Study Group.

Doolittle, E. (2018), "Off the grid," in S. Gerofsky (ed.), *Contemporary Environmental and Mathematics Education Modelling Using New Geometric Approaches*, 101–21, Switzerland: Springer International Publishing AG.

Drawson, A., E. Toombs, and C. Mushquash (2017), "Indigenous research methods: a systematic review," *The International Indigenous Policy Journal*, 8(2).

Ermine, W. (2007), "The ethical space of engagement," *Indigenous Law Journal*, 6(1).

Gutiérrez, R. (2022), "A spiritual turn: toward desire-based research and Indigenous futurity in mathematics education," *Journal for Research in Mathematics Education*, 53(5): 379–88.

Gutstein, E. (2006), *Reading and Writing the World with Mathematics: Toward a Pedagogy for Social Justice*, Taylor & Francis.

[1] Edward is Kanyen'kehake (Mohawk) from Six Nations in southern Ontario. He has a PhD in pure mathematics (partial differential equations) from the University of Toronto. He is currently Associate Professor of Mathematics in the Department of Indigenous Knowledge and Science at First Nations University of Canada in Regina, Saskatchewan, Canada.

First Nations University occupies the atim kâ-mihkosit (Red Dog) Urban Reserve which is situated on Treaty 4 Territory, the original lands of the Cree, Saulteaux, Dakota, Lakota, Nakoda, and homeland of the Métis Nation.

Amber is an Australian, non-Indigenous woman. Her family are of Anglo-Celtic and other European descent, and have resided in Awabakal Country (otherwise known as Newcastle, NSW) for many generations. Amber is currently nearing completion of her PhD on Indigenous Mathematics Education in Australia in the School of Education, University of Newcastle.

Amber would like to pay her respects to the traditional owners of the lands she calls home: the Awabakal and Worimi peoples. She would also like to extend her gratitude and respect to Elders past and present, for their ongoing care and custodianship of the Awabakal and Worimi regions.

Hardy, G. H. (2012), *A Mathematician's Apology*, Canto Classics, Cambridge: Cambridge University Press.

Hokowhitu, B. (2016), "Monster: post-Indigenous studies," in A. Moreton-Robinson (ed.), *Critical Indigenous Studies: Engagements in First World Locations*, 83–101, The University of Arizona Press.

Hughes, A. (2022), "Equitable outcomes and Indigenous learners of mathematics: deconstructing Australian education policy," *International Journal of Inclusive Education*, 1–17.

Kanigel, R. (1992), *The Man Who Knew Infinity: A Life of the Genius Ramanujan* (reprint), New York: Washington Square Press.

Lavallée, L. F. (2009), "Practical application of an Indigenous research framework and two qualitative Indigenous research methods: sharing circles and Anishnaabe symbol-based reflection," *International Journal of Qualitative Methods*, 8(1): 21–40.

Lerman, S. (2001), "A cultural/discursive psychology for mathematics teaching and learning," *Sociocultural Research on Mathematics Education*, 3–17, Routledge.

Lloyd, B., S. Suchet-Pearson, S. Wright, L. Burarrwanga, R. Ganambarr, M. Ganambarr-Stubbs, B. Ganambarr, and D. Maymuru (2016), "Morrku Mangawu—knowledge on the land: mobilising Yolŋu mathematics from Bawaka, North East Arnhem Land, to reveal the situatedness of all knowledges," *Humanities* (Basel), 5(3): 61.

Ministry of Education (2020), "Mathematics 2020: The Ontario Curriculum Grades 1–8," Queen's Printer for Ontario.

Moreton-Robinson, A. (2013), "Towards an Australian Indigenous women's standpoint theory," *Australian Feminist Studies*, 28(78): 331–47.

Moreton-Robinson, A. (2016), "Race and cultural entrapment: critical Indigenous studies in the twenty-first century," in A. Moreton-Robinson (ed.), *Critical Indigenous Studies: Engagements in First World Locations*, United States of America: The University of Arizona Press, pp. 102–15.

Nakata, M. (2007), *Disciplining the Savages: Savaging the Disciplines*, Canberra, ACT: Aboriginal Studies Press.

Nicol, C., J. S. Thom, E. Doolittle, F. Glanfield and E. Ghostkeeper (2023), "Mathematics education for STEM as place," *ZDM—Mathematics Education*.

Parkinson, C. and T. Jones (2019), "Aboriginal people's aspirations and the Australian curriculum: a critical analysis," *Educational Research for Policy and Practice*, 18(1): 75–97.

Simon, T. (2023), "Thomas Hübl: alchemizing individual, ancestral, and collective trauma," *Insights at the Edge*.

Stavrou, S. G. and M. S. Murphy (2022), "Failures indigenising school mathematics: a narrative inquiry," *The Australian Journal of Indigenous Education*, 51(2).

Wilson, K. (2016), "Dunnville to Port Maitland: the end of the road, but not the journey," *Muscle and Bone: Walking the Haldimand Tract*.

Wilson, K. and M. Anderson (2021), "The promise and peril of walking Indigenous territorial recognitions carried out by settlers," *International Journal of Religious Tourism and Pilgrimage*, 9(2): 7.

Wittgenstein, L. (1969), *On Certainty*, eds. G. E. M. Anscombe and G. H. von Wright, tran. D. Paul, Oxford: Blackwell.

7

Western and Indigenous Knowledge Systems: Attainment of the Sustainable Development Goals and the Umunthu Developmental Approach

Chance Chagunda

Introduction

Contemporary sustainability literature offers various approaches and theories on how to achieve the Sustainable Development Goals (SDGs) by 2030. However, current literature related to the SDGs and how to attain them appears to be biased toward the Western Knowledge System (WKS) and its critical thought and ideas. For example, the Human Development Approach (HDA) aims to expand opportunities for people, which can lead to a good standard of living, a long healthy life, and living in harmony with each other and nature (Stewart et al., 2018). The centrality of human development focuses on breaking the cycle of poverty and other obstacles to sustainable development (Chagunda, 2014; Noyoo, 2019). Burkey (1996) defines human development as a process by which an individual develops self-worth and becomes more self-assured, self-sufficient, cooperative, and tolerant of others through becoming conscious of his/her deficiencies as well as his/her potential for positive change. HDA involves the broadening of choices of individual human beings to achieve the desired well-being, including enjoying long and healthy lives (UNDP, 1990). This is why the United Nations Development Programme (2000) defines the HDA as the method of enlarging people's capabilities and choices. This approach advocates for an individual's human dignity and freedom whereby an individual human being has the right to education, a long and healthy life, improved access to resources, and self-worth, just to mention a few (UNDP, 2016).

Another approach to achieving development, including the SDGs, is the capability approach. However, this approach puts emphasis on individuals as the locus of development, following the Western knowledge of individualism. Sen (2000) united the concepts of freedom and functionings to deliberate on capability that is understood as the freedom of a person to choose between different ways of living. Gutierrez (1973), Bonino (1975), and Nissiotis (1997) question this type of development as it focuses on an individual's liberation from all that limits or keeps an individual from self-fulfillment. Unlike capability and HDA, which consider development as freedom whereby the outcome for development is enabling human beings to realize their rights and live in freedom, the umunthu developmental approach (UDA) focuses on the fundamental notions of reciprocity, interconnection, and service, and the developmental outcome is the community's well-being.

It can be argued that most African countries do not incorporate the African Indigenous Knowledge System (AIKS), such as the UDA, into their national strategies and approaches to achieve the SDGs. Therefore, it is not surprising that the African Development Bank (2022) argues that Africa's progress toward achieving the SDGs and targets needs strategic acceleration. The author argues that while all African governments have made significant efforts toward achieving the SDGs, incorporating the SDGs and Africa's Agenda 2063 goals into national strategies and development plans alone may not be enough. Africa Agenda 2063 refers to a joint framework for inclusive growth and sustainable development for Africa to be achieved by 2063. Actually, Agenda 2063 is an extension of the pan-African drive over centuries, and, among other things, it is intended to eliminate poverty, provide opportunities for all Africans to have decent and affordable housing, and industrialize economies, and achieve gender parity (Brickstone, 2023). Without integrating African Indigenous Knowledge Systems, such as the UDA, achieving SDGs will remain a significant challenge in Africa.

The United Nations (2022) contends that the SDGs represent a global commitment to address pressing social, economic, and environmental challenges facing humanity. The 17 SDGs are interlinked; among other things, they aim to eradicate poverty, promote equity, and ensure environmental sustainability by 2030 (African Development Bank, 2022; United Nations, 2022). The author argues that achieving SDGs requires a holistic approach to development that integrates diverse knowledge systems (a comprehensive and multifaceted approach) that harness the knowledge, wisdom, and experiences of diverse cultures and societies.

According to the Economic Commission for Africa (ECA) (African Development Bank, 2022), the SDGs are grouped into five pillars: people, prosperity, planet, peace, and partnerships. The ECA (African Development Bank, 2022) states that most African member states (especially those south of the Sahara) are not on track to meet the intended SDGs and targets. In terms of the People Pillar, nearly 40 percent of all Africans still live in extreme poverty. As for the Prosperity Pillar, economic growth has lagged behind real income growth in every sub-region of Africa between 1980 and 2019 (African Development Bank, 2022). Even periods of modest growth have not translated into higher standards of living for most Africans, resulting in the paradox of greater income inequality. To address this poor performance, homegrown developmental interventions in the form of AIKS and robust community-driven efforts need to be integrated with mostly Western-crafted approaches for Africa. The author argues that heavy reliance on Western nations for funding or loans to implement SDGs can have unintended negative consequences.

This chapter argues that the UDA can be integrated with the WKS to achieve the SDGs in Africa. The author asserts that a multidimensional approach integrating various knowledge systems, such as WKS and AIKS, can lead to attaining SDGs. According to the United Nations (2022), society should approach SDGs, thus requiring individual and community participation. Therefore, the chosen developmental approaches should encompass collective hard work and participatory methods at both central and peripheral levels. Despite its significant potential, the UDA does not receive the credit it deserves.

This community-focused approach is capable of achieving most of the SDGs due to its interconnectedness and collective characteristics toward sustainable development. It is important that African countries implement their own indigenous developmental approaches to boost development. Among other things, corruption and self-enrichment must be arrested so that the government revenues can be used to implement SDGs (Chagunda, 2021). Corruption deprives Africa of developing meaningful livelihoods. African leaders in 2018 pledged to work together to defeat this evil (Tankou 2018). However, corruption has continued to stifle development and people's well-being across the continent; even with the COVID-19 pandemic funding, some African leaders and government officials enriched themselves personally through corruption. Some of these leaders are from Malawi, Nigeria, Zimbabwe, Somalia, and South Africa (Fagbadebo and Dorasamy 2021). Corruption will make it difficult to achieve the SDGs and promote people's well-being.

Integrating Western and AIKS provides a promising approach to addressing complex development issues holistically (Kugara et al., 2021). Therefore, this article explores the potential synergy between Western and AIKS in the pursuit of sustainable development and the attainment of the SDGs. Specifically, it examines the UDA as a case study to demonstrate how African Indigenous values can complement Western paradigms in the pursuit of sustainable development. Through a comparative analysis, this article aims to establish the synergies between Western and African Indigenous knowledge and their collective capacity to guide effective and inclusive strategies for achieving the SDGs.

Western and indigenous knowledge systems: a comparative overview

Comparative analysis of Western and AIKS reveals synergies that can enhance the attainment of SDGs. For a long time, the colonial system reinforced the notion that African countries were incapable of development and technical contribution. Due to notions like these, Western Knowledge Systems are perceived as providing technical expertise and innovation, while Indigenous Knowledge Systems are considered to provide a holistic perspective and respect for the interconnectedness of all life (Mbiti, 1970; Kugara et al., 2021). Integrating these knowledge systems can lead to more comprehensive, culturally sensitive, and effective solutions for sustainable development.

Western knowledge systems

The WKS, among others, refers to the values and cultures of Western civilizations, including social norms, ethical values, and technologies, collectively found within the

Western sphere of influence (Kugara, Mdhluli, Daswa, Matshidze, and Ramavhunga, 2021). This knowledge system is characterized by empirical analysis, scientific inquiry, and a reductionist approach. Rooted in Enlightenment principles, it values objectivity, quantification, and the separation of subject and object (Akena, 2012). Proponents of either the WKS or AIKS are always motivated by their relationship with the society in which they reside (Akena, 2012).

The author argues that the WKS has led to remarkable technological advancements and improvements in various sectors, shaping modern societies and economies in some parts of the world. However, in Africa, not much progress has been achieved. Schools in Africa are where most people first encounter and learn about the WKS (Grange, 2007). Gough (1998) suggests that advocates for the WKS often portray it as superior and universal, making it difficult to integrate it with the AIKS. It is imperative that learners in school curriculums learn that knowledge systems can coexist and be integrated to achieve a common goal of SDGs and people's well-being (Akena, 2012).

Development should not overlook economic, social, cultural, and political processes that aim to improve continuously the well-being of the population (UN, 1986; Peet and Hartwick, 2009). The capability approach is rooted in the individualistic perspective of development. According to Nussbaum (2000), the capabilities approach involves enabling individuals to have and do everything that life demands, thereby allowing for a dignified existence. Sen (2001) recognizes obstacles that hinder people's development and advocates for developmental freedoms. These include political freedoms, economic facilities, social opportunities, transparency guarantees, and protective freedoms that can enhance a person's overall capability. From a Western standpoint, advancement under the capability approach refers to valuable actions and circumstances that contribute to an individual's well-being (Sen, 1999). The value of functioning, as proposed by Sen, is both personal and subjective and varies across time and individuals. Limited or no functioning can prevent people from living their desired life, leading to their exclusion from social relationships (Sen, 2000). Therefore, capabilities can be perceived as an individual's freedom to choose between different ways of living and to utilize various skills to generate income or fulfill desired aspirations. Similarly, when Sen (1999) addresses agency, he refers to an individual's capacity to perform actions that they value and have reason to value.

Western knowledge system contributions to the SDGs

In 2015, the United Nations established the 17 SDGs as a continuation of the Millennium Development Goals (MDGs). The SDGs cover all possible social, economic, and natural aspects (UN, 2022). Achieving the SDGs is a complex process that requires diverse understandings from political and economic discussions and must therefore be addressed in all areas of society (UN, 2022). The WKS has played a crucial role in advancing several SDGs. Advancements in medicine, technology, and infrastructure have contributed to improvements in health (SDG 3) and access to clean water and sanitation (SDG 6). Economic growth driven by industrialization and innovation has led to poverty reduction (SDG 1) and enhanced livelihoods (SDG 8).

Scientific research and technological innovation have informed climate action (SDG 13) and environmental conservation (SDG 15).

Western Knowledge Systems through HDA aim at attaining SDGs. HDA can be perceived as the process of widening the choices of human beings to attain their desired well-being (UNDP, 1990). It is argued that this approach is a people-focused approach to advance the needs of human beings, as exemplified by the World Bank's approach in its World Development Reports (Alkire and Deneulin, 2009). It can be argued further that the purpose of development is to widen people's choices as individuals, allowing them to enjoy long and healthy lives. For this chapter, the HDA sets the basis for discussion in the sense that it advocates that human beings can attain development individually.

The capability approach, which is part of WKS, is an evaluative approach advocated by Amartya Sen, who proposed that social arrangements should be understood from the freedom people have to achieve the instrumental freedoms they value (Alkire, 2005). For Sen (1999), capability is understood as the freedom of an individual person to choose between different ways of living. When it comes to functionings, Sen states that the instrumental freedoms one may have in one's life are political freedom, economic facilities, social opportunities, transparency guarantees, and protective security (Sen, 2000). It must be emphasized that the value of functionings is personal and subjective; as such, they vary across time and from an individual to an individual. Sen (2000) understands that deprivation of capability prevents people from developing.

Indigenous knowledge systems

African Indigenous Knowledge Systems are deeply rooted in the traditions, cultures, and ecosystems of Indigenous communities worldwide (Noyoo, 2015). These knowledge systems are holistic, relational, and context-specific, emphasizing interconnectedness with nature, spirituality, and community. They encompass a wide range of expertise, including ecological stewardship, traditional medicine, and cultural practices (Treblanche-Greef, 2019; Chagunda, 2021). AIKS offers insights into sustainable resource management, resilience, and community well-being that can enrich development approaches. At the center of AIKS lies Ubuntu or Umunthu,[1] a Bantu word derived from languages spoken in Malawi, Zambia, South Africa, Eswatini, and other African countries. It means "humanness" and refers to an individual who is wrapped up in the community and people's relationships with others (Etieyibo, 2017; Chagunda, 2021).

It is everyone's responsibility to clear obstacles to development so that the community has enough food, builds solid houses, dresses well, and constructs roads and bridges not just for individuals but for the community at large. At the center of

[1] Umunthu derives from the Nguni languages and is used to mean humanness. This concept states that a person is a person through other people. The term incorporates principles of community solidarity and collective responsibility for the good of all those who are members of the community. Umunthu places emphasis on the importance of social responsiveness and cohesion.

being human lies community solidarity and collective responsibility for the good of all those who are members of the community (Mbiti, 1970; Etieyibo, 2017; Chagunda, 2021). This understanding is key to empowering each other.

The AIKS views people from a more holistic and collective perspective, where community development is valued (Treblanche-Greef, 2019). In contrast, the WKS emphasizes a more individualistic orientation toward life, which is linked to individual development (Schiele, 1994). After independence, some African countries included indigenous knowledge systems in their national plans and developmental strategies. For example, Tanzania followed the African indigenous approach of Ujamaa, Zambia applied humanism, and Malawi adopted umunthuism, as explained earlier (Chagunda, 2023). Scarritt (1971) explains that the government built schools and clinics and provided services that nurtured a cooperative relationship with the economy. In indigenous knowledge, African communities generally understand that development is a communal affair and not just an individual issue (Nussbaum, 2003; Chagunda, 2021).

Umunthu, rooted in the AIKS, emphasizes the interconnectedness of all living beings and the importance of community and compassion (Mbiti, 1970). It emphasizes collective well-being, cooperation, and the idea that an individual's humanity is defined through relationships with others (Chagunda, 2021). Umunthu offers a perspective that resonates with several SDGs, such as no poverty (SDG 1), zero hunger (SDG 2), and good health and well-being (SDG 3), by highlighting the significance of shared responsibility and social support systems (UN, 2022). Treblanche-Greef (2019) argues that umunthu, a developmental approach, is an African indigenous knowledge that is not fully represented in academic discourse or in global sustainable development literature. Furthermore, Treblanche-Greef (2019) explains that this indigenous knowledge system provides an alternative to the WKS, which advocates for altruism, cohesion, and harmony between all humans and nature. There is an excessive imposition of WKS in literature.

This can be attributed to Western academics having more resources at their disposal, foreign experts being forced on African institutions to do work because the funding is from a particular donor country, or fear that the donor money will end up being corruptly used by government leadership, among other reasons mentioned. According to Chagunda (2021), Umunthu means community solidarity, collective responsibility, and developmental work for the good of all those who are members of the community. In Malawi, during the tenure of the first president of the country, Kamuzu Banda, umunthu was expressed through collective development and responsibilities for looking after those who could not support themselves (Chagunda, 2021). For instance, during the Kamuzu Banda regime, farmers were asked to group themselves in what was known as the "Farmers Club." Such clubs were required to register with the relevant authority, in this case the Ministry of Agriculture. Farmers Clubs were eligible for agricultural loans to boost their agricultural production. Once the harvest was sold, they had to pay back the original loan and use the profit for various things such as paying for school fees, buying school-related items, building or maintaining their homes, or buying livestock, to mention a few. To make sure that the club did not default on their payment, when any member of the club was unable to cultivate the field due to sickness or maternity, for example, the club members cultivated that land on behalf of the specific individual. This requirement helped to boost collective development, as no one was left behind due to natural ills.

African indigenous knowledge contributions to the SDGs

The UDA, also referred to as the Umunthu approach, has been utilized in achieving developmental goals in Africa since the pre-colonial period. According to Chagunda (2021), UDA is appropriate for the African continent as it emphasizes collective development, and it entails transformative characteristics such as mutual caring for the well-being of others, supportiveness, cooperation, companionship, reciprocity, and interdependence with others.

The UDA provides unique insights into achieving the SDGs from a different perspective. For instance, traditional ecological knowledge can inform sustainable land use (SDG 15) and biodiversity conservation (SDG 14), while Indigenous practices of communal resource management align with goals of responsible consumption and production (SDG 12). Furthermore, Indigenous approaches to governance and decision-making can promote social justice (SDG 10) and gender equality (SDG 5) by valuing inclusivity, participation, and cultural diversity.

To achieve SDG 3, which aims to ensure healthy lives and promote well-being for all, a holistic developmental approach is necessary, as it includes the promotion of interconnectedness and collective well-being (Mbiti, 1970; Koster, 1996). For instance, SDG 4 seeks to ensure inclusive and equitable quality education and promote lifelong opportunities for all, which is closely linked to the UDA. Before the colonial period, indigenous communities tried their level best to live a balanced life for the sake of sustainable development (Noyoo, 2015). Since the pre-colonial, colonial, and post-independence periods, a lot has happened to the Umunthu way of life. Some various political parties in Malawi, in order to fight the British colonial government more easily, united under one umbrella body so that they would have a unified voice in fighting for liberation and better socio-economic conditions for Black people (Windel, 2009; Briggs, 2013).

Umunthu Developmental Approach: a case study

It is important to note that the UDA, which is part of African Indigenous Knowledge Systems, focuses on communities rather than individuals, as is the case in WKS, to promote development. Pre-colonial indigenous institutions such as families, villages, and traditional authorities were vital in providing numerous forms of social welfare systems and development. This form of social arrangement is popularly referred to as Umunthu. This word is derived from the Bantu languages, and it means humanness. At the hub of being human lies community solidarity and collective responsibility for the good of all the members of the community (Chirwa and Naidoo, 2014). Umunthu has been passed orally from generation to generation, and as such, Umunthu is not knowledge that an individual is born with, but what an individual acquires throughout one's life; therefore, oral tradition plays a critical role in transferring the African philosophy of life (Letseka, 2000). Umunthu is intended to promote the common good of family and community (Le Roux, 2000).

The African government leadership should revisit indigenous knowledge systems, such as the UDA, to draw lessons on how development can be truly sustainable and how self-reliance can be reignited in citizens. It is suggested that the UDA can positively contribute to Africa's development and should be integrated to achieve the SDGs. This approach aligns well with the African understanding of "development" as the well-being of the community, not just the individual.

SDG 15 aims to protect, restore, and promote the sustainable use of terrestrial ecosystems while sustainably managing forests, combating desertification, and halting and reversing land degradation and biodiversity loss. This goal can be best achieved through the integration of UDAS and Western approaches. Development should focus on the community, not just the individual. When everyone is expected to play a pivotal role, humanity has arrived at the Umunthu way of life, where the community's well-being is more important than that of an individual. Ignoring indigenous developmental interventions will make it difficult to achieve SDGs by 2030.

In order to achieve the SDGs and improve people's living standards, UDAs must be integrated into government mechanisms. For instance, the South African government in the Constitution of the Republic of South Africa (No. 108 of 1996) and various children-related legislations are influenced by Ubuntu (Umunthu) and as a state, it recognizes that it is responsible for providing leadership in various ways for the benefit of all citizens (National Treasury, 2016, 2023).

Natarajan (2002) warns governments around the world not to undermine indigenous development interventions that have been used by indigenous communities to produce or manage food, medicines, and shelter, as this knowledge could be lost forever. Sneed (2019) argues that the world should acknowledge that African indigenous development approaches at the local community level play a substantial role in achieving developmental outcomes that benefit both current and future generations.

The integration of Western and AIKS presents challenges, including differences in epistemology, communication barriers, and power dynamics. Western paradigms may dismiss Indigenous knowledge as anecdotal or unverifiable, perpetuating colonial legacies. Bridging these gaps requires mutual respect, co-learning, and recognizing the validity of multiple knowledge systems.

Conclusion

Achieving the SDGs requires a multidimensional approach that incorporates diverse knowledge systems. Both Western and AIKS offer valuable insights and tools for addressing complex development challenges. The UDA is a compelling example of how African Indigenous philosophies can complement Western paradigms to achieve sustainable development. To realize fully the potential of Western–African Indigenous synergies, open dialogue, cultural sensitivity, and equitable collaboration are necessary.

Acknowledging the interconnectedness of all knowledge systems enables societies to leverage the strengths of both Western and African Indigenous perspectives, fostering a holistic and inclusive approach to sustainable development. Embracing diverse knowledge systems is essential for creating a more just, equitable, and sustainable future for all as the global community strives toward the SDGs.

In conclusion, this paper argues that relying solely on WKS for implementing developmental projects, including SDGs, without incorporating African Indigenous knowledge systems is problematic. The UDA offers an alternative to the WKS by emphasizing altruism, cohesion, and harmony between all humans and nature. This approach has transformative characteristics, such as mutual caring for the well-being of others, supportiveness, cooperation, companionship, reciprocity, and interdependence with others. As such, the integration of WKS and AIKS can be a life-changing intervention for SDGs.

Despite not receiving the credit it deserves, the UDA is a community-focused kit to the extent that even if implemented on its own, it is capable of achieving some of the SDGs due to its interconnectedness and collective characteristics toward development. African countries should intensify the fight against corruption and self-enrichment to integrate or implement their own indigenous developmental approaches so that the revenues can be used to implement SDGs. The paper suggests integrating the UDA with the WKS to achieve SDGs, as this aligns well with the African understanding of "sustainable development" as the well-being of the community, not just the individual. It is important to note that the SDGs are interlinked, and achieving one goal can aid progress toward others. The UDA is a holistic approach that promotes interconnectedness and collective well-being, making it a valuable tool for integration with the WKS in the pursuit of achieving SDGs.

References

African Development Bank (2022), *Africa Sustainable Development Report 2022.* U/UNECA//AFDB/UNDP. Available at www.afdb.org/en/documents/2022-africa-sustainable-development-report

Akena, F. A. (2012), "Critical analysis of the production of Western knowledge and its implications for Indigenous knowledge and decolonization," *Journal of Black Studies*, 43(6) (September 2012): 599–619.

Alkire, S. (2005), "Why the capability approach?," *Journal of Human Development*, 6(1): 115–35. Available at https://ssrn.com/abstract=2118610.

Alkire, S. and S. Deneulin (2009), "Introducing the human development and capability approach," in S. Deneulin (ed.), *An Introduction to the Human Development and Capability Approach*, London: Routledge, pp. 19–45.

Bonino, J. M. (1975), *Doing Theology in a Revolutionary Situation*, Philadelphia: Fortress Press.

Brickstone, (2023), "8 Facts about the African Agenda 2063," available at https://brickstone.africa/african-agenda-2063

Briggs, P. (2013), *Malawi: The Bradt Travel Guide*, Lilongwe: Ulendo.

Burkey, S. (1996), *People First: A Guide to Self-Reliant, Participatory Rural Development.* London and New Jersey: Zed Books.

Chagunda, C. (2023), "The policy challenges of Africa's changing demography and social structures," in D. Besharov, N. Gilbert, and K. J. Baehler (eds.), *The Oxford Handbook of Governance and Management for Social Policy*, Oxford: Oxford Academic.
Chagunda, C. (2022), "Interrogating the Structural Adjustment Programme (SAP) in Malawi: impacts on the livelihoods of the working class and poor people," *Brazilian Journal of Social and Labour Economics*, 4(e022016), https://doi.org/10.5260/zbjvse.v4i0.22016
Chagunda, C. (2021), "Development aid, democracy and sustainable development in Malawi—1964 to date," *Cadernos de Estudos Africanos*, 41. https://doi.org/10.4000/cea.6446
Chagunda, C. (2014), *South Africa's Social Assistance Intervention as a Building Block of a Developmental State*, doctoral dissertation, University of Cape Town.
Chirwa, G. and Naidoo, D. (2014), "Curriculum change and development in Malawi: a historical overview," *Mediterranean Journal of Social Sciences*, 5(16): 336–45.
Conservation International (2023), "Mapping natural capital: Where are the ecosystems that people rely on?," (report). https://www.conservation.org/projects/mapping-natural-capital
Etieyibo, E. (2017), "Ubuntu and the Environment," in A. Afolayan and T. Falola, *The Palgrave Handbook of African Philosophy*, New York: Palgrave Macmillan, available at https://doi.org/10.1057/978-1-137-59291-0_41
Fagbadebo, O. and N. Dorasamy (2021), "Political leadership, corruption, and the crisis of governance in Africa: a discourse." *African Renaissance*, 18(1), (March): 27–52.
Gough, N. (1998), "All around the world: science education, constructivism, and globalisation," *Education Policy*, 12(5): 507–24.
Grange, L. L. (2007), "Integrating Western and Indigenous knowledge systems: the basis for effective science education in South Africa?," *International Review of Education*, 53(5–6): 577–591.
Gutierrez, G. (1973), *A Theology of Liberation*, London: SCM Press.
Kugara, S. L. T. D. Mdhluli, T. J. Daswa, P. E. Matshidze, and N. E. Ramavhunga (2021), "The significance of multicultural methodologies on African indigenous knowledge," in N. E. Ramavhunga (ed.), *Ethical Research Approaches to Indigenous Knowledge*, Cape Town: ABC Publishers, pp. 1–20.
Le Roux, J. (2000), "The concept of '*Umunthu*': Africa's most important contribution to multicultural education?," *Multicultural Teaching*, 18(2): 43–6.
Letseka, M. (2000), "African philosophy and educational discourse," in P. Higgs, N. C. G. Vakalisa, T. V. Mda, and N. T. Assie-Lumumba (eds.), *African Voices in Education*, Juta: Lansdowne.
Mbiti, S. J. (1970), *African Religions and Philosophies*, New York: Doubleday and Company.
Natarajan, B. (2002), "Conservation of biological diversity and indigenous traditional knowledge as exemplified by the cases of Asia and Africa," available at https://journals.co.za/doi/10.10520/EJC61376?utm_source=TrendMD&utm_medium=cpc&utm_campaign=Indilinga_African_Journal_of_Indigenous_Knowledge_Systems_TrendMD_0
National Treasury (2016), *Budget Review 2016*, Pretoria: Government Printers.
National Treasury (2023), *Estimates of National Expenditure 2023*, Pretoria: Government Printers.
Nissiotis, N. A. (1975), "Introduction to a Christological phenomenology of development," in A. Kee, *A Reader in Political Theology*, Philadelphia: Westminster Press.
Noyoo, N. (2019), *Social Policy in Post-Apartheid South Africa: Social Re-Engineering for Inclusive Development* (1st ed.), London: Routledge. Available at https://doi.org/10.4324/9780429273605.

Noyoo, N. (2015), *Social Policy and Human Development in Zambia*, Pretoria: Kwarts Publishers.

Nussbaum, M. (2000), *Woman and Human Development, the Capabilities Approach*, Cambridge: University Press.

Peet, R. and E. Hartwick (2009), *Theories of Development: Contentions, Arguments, Alternatives* (2nd ed.), London: The Guilford Press.

Sen, A. (1999), *Development as Freedom*, Oxford University Press: New York.

Sen, A. (2000), *Social Exclusion: Concept, Application, and Scrutiny. Social Development Papers no. 1.: 1–60*. Available at https://www.think- asia.org/bitstream/handle /11540/2339/social-exclusion.pdf

Sen, A. (2001), *Development as Freedom*, Oxford: Oxford University Press.

Scarritt, R. J. (1971), "Elite values, ideology, and power in post-independence Zambia," *African Studies Journal*, 14(1): 31–54.

Sneed, A. (2019), "What conservation efforts can learn from Indigenous communities," available at https://www.scientificamerican.com/article/what-conservation-efforts-can-learn-from-indigenous-communities

Stewart, F., G. Ranis, and E. Samman (2018), *Advancing Human Development: Theory and Practice*, Oxford: Oxford University Press.

Schiele, H, J. (1994), "Afrocentricity: Implications for higher education," *Journal of Black Studies*, 25(2): 150–69.

Tankou, E. A. (2018), "President Buhari of Nigeria launches the AU theme of the Year 2018 on fighting corruption in the continent," available at https://au.int/en/pressreleases/20180129/president buhari-nigeria-launches-au-theme-y

Treblanche-Greef, A. C. (2019), "Ubuntu and environmental ethics: the West can learn from Africa when faced with climate change," in M. Chemhuru, *African Environmental Ethics*, The International Library of Environmental, Agricultural and Food Ethics, vol. 29, Springer: Cham.

United Nations (UN) (1986) *General Assembly: Declaration on the Right to Development*, available at http://www.un.org/documents/ga/res/41/a41r128.htm.

United Nations (2016), *The SDGs Report 2016*, available at https://unstats.un.org/SDGs/report/2016/The%20Sustainable%20Development%20Goals%20Report%202016.pdf

United Nations (2022), *SDGs.*, available at https://www.un.org/sustainabledevelopment/climate-change

United Nations Development Programme (UNDP) (1990), *Human Development Report 1990: Concept and Measurement of Human Development*, New York: United Nations Development Programme.

United Nations Development Programme (2000), *South Africa: Transformation for Human Development*, Pretoria: UNDP.

United Nations Development Programme (2016), *Regional Human Development Report 2016 Africa:* Accelerating Gender Equality and Women's Empowerment. New York: United Nations Development Programme.

Windel, A. (2009), *British Colonial Education in Africa: Policy and Practice in the Era of Trusteeship*, Bowdoin College. Available at http://onlinelibrary.wiley.com/doi/10.1111/j.1478-0542.2008.00560.x/abstract

8

Kitheka ki matu: A Forest Has Ears

Mark Lawrence

Introduction

Not for the first time, the so-called "Western" world finds itself disenchanted with its prospects. Wrestling with worsening ecological crises threatening survival, the sociotechnical imaginary of the Anthropocene hopes to implement epic visions of the future (Jasanoff and Kim, 2015). But this dreamscape is more nightmarish than fanciful; every step we take to combat undesirable change only reveals how much more we need to understand about our relationships with the world around us and with one another (Grear, 2015). If there is really any such thing as a "Western" or "modern" perspective (Latour, 1993), increasingly it seems the formerly healthy skepticism upon which it was built has now become cancerous. Scientific engagement with the world based on a staunch refusal to accept received wisdom without rigorous testing of assumptions is now treated cynically by the society it helped create. A short leap to pessimism has many resolutely ignoring mounting evidence of ecological catastrophe on the assumption that with time running short, anything goes.

Thus, we have again granted ourselves the privilege of looking for alternative dreamscapes imagined by so-called "indigenous," ostensibly non-"Western" cultures (Michel and Gayton, 2002; Bala and Gheverghese, 2007; Drouin-Gagné, 2014; Gorelick, 2014; Kealiikanakaoleohaililani and Giardina, 2016; etc.). But via exposure to the worldview of the Akamba of Kenya's Kibwezi Division, I intend to argue that searching for an "indigenous" science considering the abundant failures of "Western" approaches reproduces their central mistakes while hobbling the power of alternatives being sought to provide responsible, effective ways of engaging with the world around us. First, it assumes we can collapse a myriad of practices of problem-solving human interactions with nature into the single term *an* indigenous science. Second, it assumes an identifiably "indigenous" science, when in fact indigeneity has a complicated career both as an intellectual object and as a highly mutable set of political ecological relations. Third, even sincere hopes for a dialogue between "Western" and "indigenous" worldviews overlook the reality of what dialogue entails. Fourth, there is a need for a broader concept of "dialogue" that emphasizes nonhuman agency.

Confusing science with political economy

"Western" science is a slippery concept. Distinguished from astronomical and mathematical efforts in antiquity, it has a relatively short history, and concern has been expressed about reliance on unobservable data as science has engaged with more theoretically abstract propositions from the late nineteenth century (Snively and Corsiglia, 1998). Paralleling abstractions of use and exchange value in industrializing economies and individual rights and responsibilities in societies democratizing in the same period, science is clearly a subculture of a larger "Western" project (Aikenhead 1996, 1997). As such, "Western" science proceeds from a philosophical distinction of Being and Material that has had dramatic negative consequences for the ecosphere, focusing on the latter rather than the former.

Yet calls for a dialogue with ostensibly "indigenous" sciences just as dangerously reproduce old mistakes of assuming that "indigenous" is readily distinguishable from "Western." Part of the difficulty is that fundamental questions about ontology, epistemology, and methodology have yet to be decolonized. Within the domain of Science Education, value judgments still assume "indigenous" perspectives must meet supposedly higher standards of "Western" science (Good, 1995). Obvious bias immediately disrupts any objective effort to compare different cultural approaches to science by ontologically assuming a single, universal standard for assessing the legitimacy of whatever's presented as "real." For example, faced with the question "Do gods, spirits, etc. exist?" when searching for explanations of how natural processes function, "Western" science at best dismisses the inquiry as irrelevant for lack of evidence and sufficient experimental methodology to supply that deficiency. By contrast, like many other "indigenous" cultures, that of East Africa's Akamba people takes for granted that ignoring the presence of unseen *aimu* spirits weakens understanding of how the world functions, thereby putting ourselves at risk.

This reveals another important difference between "indigenous" and "Western" worldviews during the Anthropocene. Environmental risk has become a principal product of late modernity, not merely its consequence, with the ability to mitigate exposure now supplanting wealth as the centrally defining characteristic of class inequality (Beck, 1992). Always exposed to the threat of natural disasters, today's manufactured risks (Giddens, 1999) without historical precedent drive a process of reflexive modernization (Beck et al., 2003), a generalized, increasingly transnational critique of industrial capitalism in the face of developments the long-term consequences of which we cannot adequately assess. "Western" science once provided ways to overcome material scarcity as history's defining characteristic, but it is unable to offer any response to today's scarcity of confidence.

Risk and anxiety are central to Akamba cosmology, too, but in significantly different ways. Instead of emphasizing uncertain futures, what is at stake is a destabilized sense of the past. The relation of "sustainability" to an uncharted prospect is far less meaningful than its role regarding continuity of culturally accepted interactions of human and nonhuman worlds for "indigenous" populations. Indeed, there are many cultural variations in the definition of risk domains, explanations of what risks entail, and strategies for coping with risk (Göbel, 2001).

"Western" science stems from the so-called Enlightenment tradition, which both atomized parts of nature to be examined separately from the whole to which they belong and assigned greater significance to individuals than to larger networks of culture, society, and political economy with which they are entangled. Enormously beneficial for enabling a more fulsome set of problem-solving discussions, emphasis on *quantity* of perspectives is very different from "indigenous" approaches, which typically begin with an emphasis not on the individual but on the group, thereby implying prioritizing *quality* of inquiry and explanation. Among the Akamba, access to wood and water resources traditionally was regulated by reference to gender, age, the purposes to which resources would be put, etc. Epistemologically, therefore, what counts as knowledge for the Akamba is not inherently universal (as "Western" science assumes its propositions are until contradicted by further evidence) but instead relative to circumstances.

Likewise, Akamba resources are not considered in isolation from a shifting set of relationships to other lifeforms (spiritual, animal, mineral, etc.) and processes (rainfall, wind patterns, etc.). "Western" science isn't incapable of synthesizing disparate sets of conclusions into systemic explanations, but even ecology, as a complex example of a system-oriented perspective, first assumes compartmentalizing knowledge of hydrology, chemistry, botany, biology, climatology, etc. Nature as a whole emerges as an assemblage only in the final instance of inquiry, while "indigenes" typically begin with that sense of the oneness of things, which thereafter is maintained throughout examination of its constituent elements.

This is not to ignore the reams of data that "Western" science generates but draws attention to how the particularity of circumstance quickly gets subsumed within categorical concepts such as nature, society, economy, etc. (Latour, 2005). The crucial difference of "indigenous" perspectives is that innumerable associations that can be made between things in time and space are always present, while "Western" disenchantment in the Anthropocene insists on working through proxies.

So, caution is warranted if dialogue between "indigenous" and "Western" science is undertaken based on confusing the latter for the political economy of which it's a part. For instance, while the African Research Survey of the 1930s was meant to further the aims of British and other colonialists to categorize, universalize, and thereby generally simplify the political project of directing development of the continent toward metropolitan interests, the scientists involved kept emphasizing the heterogeneity of local, vernacular conditions (Tilley, 2011). The Survey was meant to deploy "Western" science to overcome stereotyped African barbarism defiantly resistant to the use of colonized territories, resources, and peoples as part of the imperial project. But science proved as subversive as domineering, not least of all as Africans themselves gradually shifted from being objects of study to becoming partners enlisted as informants, translators, and assistants. An effort to standardize administrative control was significantly reoriented, particularly considering the Survey's (and therefore Africans') roles in contributing to the emergence of such sciences as ecology and social anthropology. Caricatures of "Western" science as productive only of negative results and somehow distinct from "indigenous" input prove unhelpful in any search for "dialogue."

Problems with indigeneity and indigenization

In 2023, three Canadian women were criminally charged with posing as Inuit to gain access to federal subsidies. Believed to be the first prosecution of this kind, the case reveals another problematic issue; namely, who is an "indigene"?

Icelanders are not generally considered "indigenous" despite a history not especially different from that of, say, Australia's Aborigines. Likewise, Greenland's Inuit *are* considered "indigenous" despite arriving later than the territory's Danish population (and likely displacing earlier Arctic residents). Or what about the struggles of the Labrador Métis to be recognized as "indigenous" without being aboriginal, for instance (Plaice, 1990)? We need to address satisfactorily social injustices rather than indigeneity as such. After all, an insistence (as in Australia (Povinelli, 1998)) that those claiming to be indigenes conform to archaic lifestyles in order to secure legal recognition of claims only exacerbates the precarity of many attempting to make headway in the "modern" world (Barcham, 1998).

If indigeneity amounts to nothing more than claims of original occupancy of location, it has little to offer and only presents barriers of distinction (i.e., "this is ours because we were here first" instead of maintaining contact with the shared planetary reality of crisis). Equally problematic are deeply entrenched stereotypes of indigeneity defined in terms of spatial and sociocultural isolation, today amply contradicted by self-images of ostensibly "indigenous" peoples well connected to global networks of transformative political and economic processes (Cadena and Starn, 2007; Forte, 2010; Clifford, 2013; Gerharz et al., 2017). Even applied historically, notions of isolation impose racist characterizations of indigenes as "backward" (Béteille, 1998) due to supposed nonparticipation in circuits of long-term economic and cultural exchange that rarely was the case.

Narrowly defining indigeneity denies realities of struggles for recognition, not least of all in Africa, where such populations have histories of long-distance transhumance (Hodgson, 2011). Despite recognition as "indigenous" according to the 2007 UN Declaration on the Rights of Indigenous Peoples, Mbororo in Cameroon are nonetheless locally considered migrant "strangers" (Pelican, 2009). The notion of Akamba occupying an ethnic region (Ukambani) halfway between Nairobi and Mombasa has more to do with British colonial constructions of territory now taken for granted as "Kenya" than it does with a history dating to the fifteenth century of repeated movement back and forth from north-central Tanzania[1]. Similarly, while indigeneity is often defined with respect to histories of loss due to European settler colonialism, populations in parts of the world such as Southeast Asia, which did not go through this experience, nonetheless lay claim to the "indigenous" label (Baird et al., 2017; Baird, 2020). Such analysis applies to much of Africa, too (Igor, 2006; Balaton-Chrime, 2013; etc.).

[1] Indeed, a complicated history of land squatting, eviction, and mass resettlement schemes since at least the 1930s has continually altered the boundaries of what constitutes Kibwezi Division, the primary area of focus in this chapter, though Kibwezi Town has almost always been considered its economic and administrative center.

Talk about dialogue with "indigenous" perspectives grows more common but continues to suffer persistently from colonial bias (Tsang and Yan, 2001). Calls for "indigenous Social Work" really amount to grafting "Western" models of professional practice onto native lifeways made subservient to them. Sooner or later, recognition of how imported standards only poorly fit local conditions must drive a need for "authentization" involving local practitioners and strategies (Walton and Abo El Nasr, 1988).

The "indigenous" has always been a non-discrete set of objects and subjectivities regarding them; culture is always adoptive as a matter of sustainability. Therefore, positionality becomes crucial when judging "indigenous" as offering much-needed alternative perspectives to "Western." What gets talked about and how it is talked about depend on who values doing so in which ways and with which means to act. Certainly, the system that appropriated "indigenous" resources first essentialized the supposed difference of indigenous peoples (i.e., insisting they're incapable of managing themselves in the modern world), thereafter elaborating a politics that criticizes attention to difference as undemocratic and intolerant to disrupt efforts of peoples so essentialized to claim rights of access to what has been appropriated. Social Work has yet to face adequately the cruel paradox of striving to find common ground for international efforts but being discouraged by the particularities of local culture (Park, 2005; Lawrence, 2013). In addition, what are we to make of "indigenes" living as labor migrants, political refugees, and internally displaced persons (Keating, 2016), or of peoples tagged as "indigenous" by external interests who refuse the appellation (Castaneda, 2004)?

Not just the humans, but also the trees, waters, and animals of Kibwezi

Beyond thorny issues of what we mean by "science" and "indigenous," we must recognize the roles of *non*-human agency in any debate about how to proceed in the Anthropocene. Plants and animals are not consulted about their attitudes toward environmental policymaking; nonetheless, they have political agency beyond their simple materiality as resources (Mateer, 2018). Trees tend not to grow at the speed of market exchange, for instance, thereby requiring a negotiated relationship with timber interests. Likewise, many animal species displaced by human development have adapted by colonizing urban spaces, challenging our own sense of dwelling there. "Affective logics" influencing political decisions about conservation reflect historically changeable interactions of human and nonhuman bodies of all sorts (Wilcox, 2017). Certainly, situationally reflexive Akamba ways of acknowledging such agency are utilized in response to challenged wood and water sustainability given climate change and a globalizing political ecology abetted by "Western" science and its policy outcomes.

This overlaps "rights of nature" (or RoN) legalization of nonhuman natural entities (Stone, 1972, etc.). In 2018, for instance, to protect wild rice as a dietary staple and centerpiece of culturally significant indigenous traditions, the White Earth Band of Ojibwe in northern Minnesota granted the plant legal protection against industrial water pollution and the introduction of genetically modified organisms into the ecosystem (LaDuke, 2019). RoN initiatives seek to move beyond narrowly defined conversations about the scope of anthropocentric entitlements to nature. In 2008, Ecuador became the first country to constitutionally enshrine juridically enforceable ecocentric rights (Kotzé and Villavicencio Calzadilla, 2017), with no fewer than 409 RoN initiatives in 39 countries today (Putzer et al., 2022). The International Union for Conservation of Nature and the United Nations have pursued similar efforts for more than a decade, and some would extend new standards of socio-ecological justice not only to natural but even artefactual nonhumans such as artificial intelligence (Donhauser, 2019; Dauvergne, 2020; Gellers, 2021).

Yet the natural world is not composed merely of individual entities but of aggregates, including species and ecosystems (Guim and Livermore, 2021: 1350). Impacting any judgment of the level of legal protection from harm under particular circumstances, how should such aggregates be defined? Besides, using anthropocentric legal standards to formalize static ecocentric rights misses the point, which is about dynamic practices of human-nonhuman interaction.

Importantly, much of what counts as knowledge is non-verbal; narratives of science and politics are not representative of the whole range of possible expression (Bloch, 1998). What matters more than the indigeneity of populations involved are material aspects of local landscapes through which people rework their collective past, present, and future (Walley, 2004; Shetler, 2007). Versus "Western" science restricting dialogue by imposing conditions of "accuracy," what matters is how those conditions are defined and redefined as contexts in which memory is deployed change. For instance, the placename "Kibwezi" means "to stand akimbo," Akamba farmers originally disturbed by the noise of the British colonial railway infamously remembered as the Lunatic Line for its history of mismanaged construction (including costing the lives of some of its workers). Locals were seen walking away from their farms (*shambas*) to stand by the railway, hands on hips and elbows turned outward, frowning in disapproval.

Significantly, this history has reemerged almost as a kind of right to stand askance about disruptions imposed by the arrival of the new Standard Gauge Railway (SGR) built for Kenya by China starting in 2011 (see Figure 7). Disruptions have been many, not limited to drastic reduction of water supply, interruption of *shamba* access, and alteration of migration patterns leading to increased human-wildlife conflict. Most visible has been substantial damage to Kibwezi Forest. Housing a neglected section of the Kenya Forestry Research Institute (KEFRI), it was a source of household fuelwood, a habitat for resident monkeys and baboons, and included *mathembo* shrines. Rather than hiring large numbers of workers required to cut SGR's path with chainsaws, the Chinese used bulldozers to push down trees. Some were giant baobabs easily hundreds of years old; other species were also of great cultural value to the local people and scientific community alike. Most downed timber was simply dragged to the side of the construction route and left there; locals were denied permission to take advantage of this unhappily supplied source of fuelwood without paying considerable fees.

Troubles with the Chinese overlap a much longer history of grievances about the 22,475-acre, British-owned sisal plantation called the Dwa Estate on Kibwezi Town's

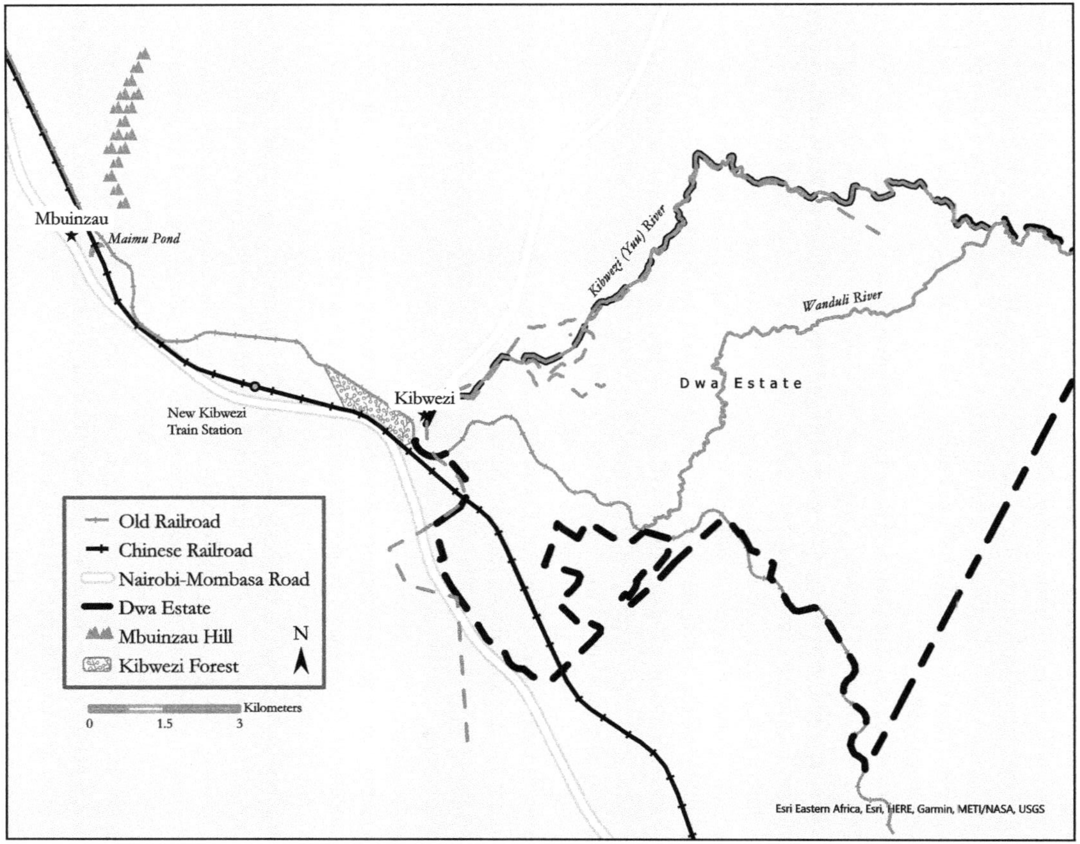

Figure 7 Old and new railways with respect to indigenous socioecologies of Kibwezi Forest and Maimu Pond.

eastern edge (see Figure 8). One of the region's original colonial landowners, since 1909 Dwa has alienated indigenous populations of at least thirteen so-called "lost" villages. Violent encounters between seven newer villages largely made up of descendants of evictees and the Estate's guards (*askari*) have followed, especially along Dwa's sprawling western border where it follows the Kibwezi River. A small area squatted just south of Kivungoni Dam is most hotly contested, the site nicknamed Baghdad at the height of American difficulties in Iraq, though the area has an older connection to a village named Kilui, supposedly drowned by vengeful *aimu* in the 1920s for the sin of hubris and indifference to neighbors at a time British rubber planters were beginning to disrupt life in Syetali, Kya Mathiti, and another "lost" village adjacent to the original railway, Kakangani.

Askaris have assaulted men and boys herding animals, stripped and raped women, bulldozed and torched *shamba* crops and buildings, installed concrete boundary "beacons" at night, and once even sprayed high concentrations of herbicide to clear vegetation, including crops. These so-called "Dwa Lines," established in 1962, 1981, 1995, 2003, and 2013, still scar the landscape. Only a third of the Estate is used, so when talk of constitutional reform in 2003 defined the ownership of land as "productive" use, Dwa bulldozed the remaining area of Syetali to plant fresh sisal. In 2015, the same landscape was modified again when SGR got government permission to bulldoze 200 acres. Outraged that not they, but the Estate was compensated for this land clearance, in 2016 local people tried to blockade the Syetali construction site, insisting that if changing land use was an option, the government should recognize still-remembered histories of indigenous settlement. Dwa's 947-year lease of land is invalid, they argue, because it was negotiated with the British, not the Kenyan government, which shouldn't compensate the Estate for lost land before adjudicating village claims that would vacate the plantation's expectation of being taken seriously by the Chinese.

Appropriately, the search for dialogue between "Western" and "indigenous" approaches to the Anthropocene's crises becomes grounded in site-specific circumstances. Grievance already "authorized" by the advent of the first railway, the town's name has developed by usage a meaning not directly deducible by the strict denotation of the word Kibwezi. Sentimentally idiomatic, generative of diffuse meanings across both space and time, emphasis on the unspoken reveals that before and beyond the situating of locality within any discursive context, there are nonrepresentational actions and interactions of "taking-place" (Anderson and Harrison, 2010). Instead of abstract conceptualization of matters where we might hope to easily discuss the relative merits of one or another approach to problem-solving, meanings belong to the many, both to what their originators intend and to what others encountering them interpret them as being.

This brings us into the orbit of affective theory in geography (Thrift, 2007; Pile, 2009), focusing on enactment of place, which "valorises those processes that operate before . . . conscious, reflective thought" (McCormack, 2005: 122). Acknowledging the active agency of nonhuman landscape elements, sensation becomes as fundamental as signification, with affect encompassing more than emotion (Massumi, 2002). Complaints are not only about quantified volumes of water now missing in the Kibwezi River or about particular fig trees cut down to make room for another multistory

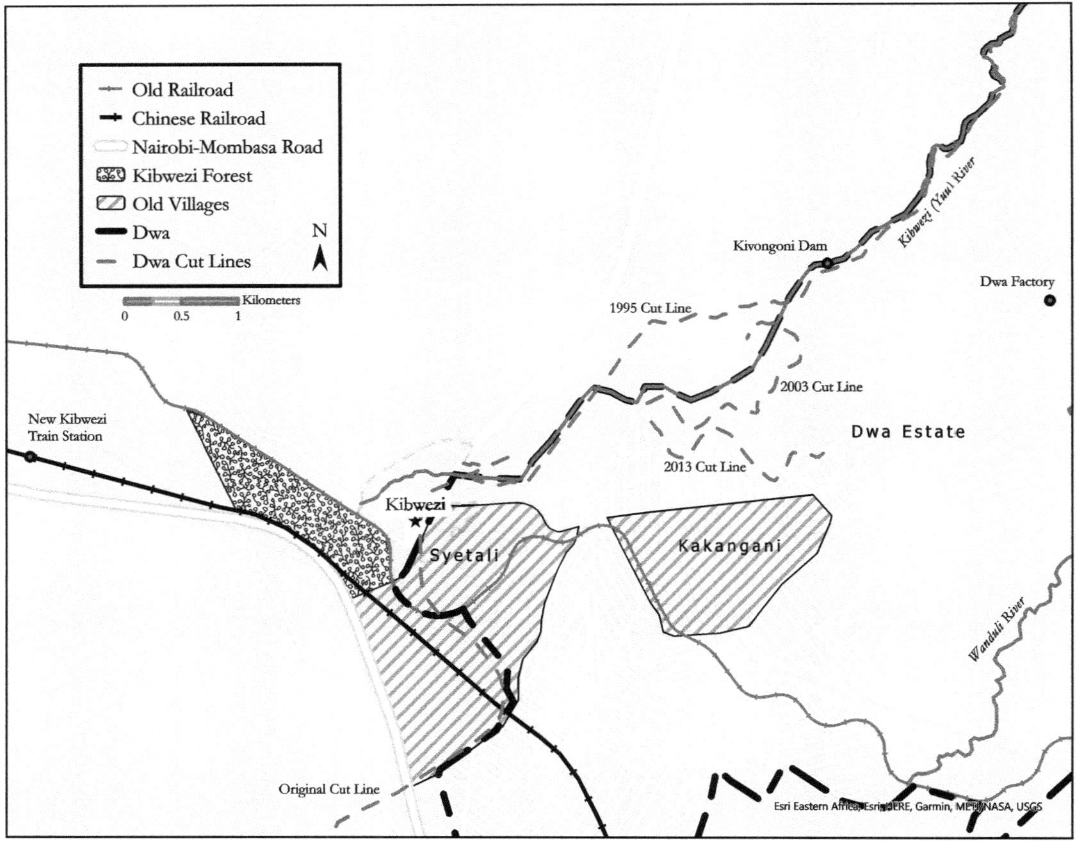

Figure 8 Contested indigenous spaces within the sisal plantation of the Dwa Estate.

building erected by "strangers" but also about changing relationships with forests and fish across long generations of time having much to do with defining the community's identity. Likewise, even though specific landscape markers of former villages bulldozed by British colonists or today's Chinese railway workers can only rarely be pointed to, the sense of place people have about them is hardly less substantial.

So, place should be understood as a field of intensities constantly changing relations with each other. Representational efforts such as maps, photographs, scientific reports, and government policies and plans all render the world intelligible only by reducing lived experiences to second-order phenomena. By contrast, affect has both the seeming insubstantiality and unavoidable presence of an "inventory of shimmers" (Seigworth and Gregg, 2010), though we shouldn't revert to anachronistic habits of romanticizing superhuman indigenous connection with nature. Not "always already sutured into a progressive or liberatory politics" (Seigworth and Gregg, 2010: 10), affect's stubborn neutrality should not be mistaken for simple indifference to existing conditions. Instead, "elud[ing] easy polarities and contradictions" as neither an event fixed in time, a location in space, nor a body positioned at a particular juncture of the two, affect is an interval composed of intensities of interaction "that pass body to body (human, nonhuman, part-body, and otherwise)" (Seigworth and Gregg, 2010: 1). Or again, "Every body and every thing . . . is a mediator, part social, part natural, lacking discrete boundaries and internalizing the multiple contradictory relations that redefine and rework them" (Swyngedouw, 2004: 18).

Nonrepresentationally, we are emphasizing not the environment or society as categorical entities but how these are always performatively emerging into view. The Akamba universe is saturated with such energies never entirely containable by a single body, human or nonhuman. Beyond "Western" distinctions of good or evil, Akamba concepts of *uwe* and *uoi* reflect intensities and trajectories toward maintenance of order or its disruption, respectively, including relations between the visible world of the living and that of *aimu* spirits of the dead still actively occupying the same landscape. A *mu'unde mu'uwe* is able to manipulate this energy "positively," but he or she necessarily understands how to work *uoi* and therefore is susceptible to becoming *mu'unde mu'uoi*. Such ability often involves deployment of a kit of supplemental means of exerting it called *kyondo* (typically kept in a small bag). Some individuals are even said to be *mganga*, the Ki-Swahili term for shapechangers.

Challenging our understanding of risk in the Anthropocene, while *mu'unde mu'uoi* work in ways we might read as "bad," there is more to their roles in Akamba landscapes than that. From Mbuinzau Hill northwest of Kibwezi comes the story of Mr. Mathu, who, as a young man, went to work in the tea plantations in the Rift and Kenya's West Country after he had finished at the mission school in Kibwezi. He came back with a well-practiced English accent and a Brooke Bond Kenya bush hat and other mannerisms intended to affect his sense of assimilated self-importance compared to those who had stayed on Mbuinzau. But then he built his house in the colonial rather than traditional style without getting permission from *aimu* and his elders. A much-feared *mu'unde mu'uoi* who owned a bar on the highway already had used his power to kill a man coveting his youngest wife by binding his victim's fate to the survival of a tree soon uprooted by a government road-widening project. One night, the *mu'unde mu'uoi* encountered *aimu* upset with Mathu's new cinder-block house. A *mganga*, the sorcerer changed himself into a whirlwind and blew down the house, with Mathu inside.

Whether such tales are true is beside the point. "Modern" housing is now common throughout Kibwezi, with the current fashion to build sprawling "Nigeria houses" in the style of mansions seen on West African soap operas broadcast on Kenyan television. At issue is not change per se but a "right" sort of relationship to place, appropriately affective change. Mr. Mathu soon rebuilt his house in nontraditional materials again, but only after appeasing the supposedly aggravated *aimu*.

Especially intense affective Akamba geographies are associated with sacred *mathembo* forests and springs, typically with *aimu* present. Again emphasizing a dynamic rather than static definition, an *ithembo* is much more than just a location; the stem verb –*thembo* means "offerings." Therefore, a place defined by action, more than a location in the present, it is a zone of historical and also potential happenings instead, always connected to particular types of flora and fauna. Some are named for specific human beings, others for nonhuman animals (a leopard, snakes, etc.) or types of trees, and still others for stories about specific relationships (e.g., between an ostrich and a young girl). Reflecting ecological relationships, many *mathembo* are hierarchically interrelated, especially when several are part of the same hydrology. The positioning of sites within this hierarchy influences how and why people react as they do to the impacts of environmental change (including those caused by Dwa and the SGR).

Mathembo still exert gravitational force for social action, despite the colonial reorientation of local political economies. Christian groups often target the felling of individual trees associated with *ithembo* ritual, especially since churches gained Black leadership in the late 1990s. Yet even nominally Christian Akamba engage a *mu'unde mu'uwe's* services at desiccated *mathembo* springs in hopes of convincing *aimu* to provide relief during unexpectedly long dry seasons. Likewise, *mathembo* frequently serve as important rhetorical devices for commenting on "modern" water development projects. The National Council of Churches of Kenya constructed a wellhead around *Ithembo ya Kwakyongo* on Mbuinzau Hill in 1985. Dry by 2011, local people are convinced Christianity disturbed the site's energy. Similarly, locals often claim water "doesn't like" concrete commonly used in the construction of modern wells and dams.

Such cultural ecology serves as a basis for considering the possibility of dialogue with arguably similar "Western" concerns about sustainability and human impacts on nature. But dialogue alone is fruitless if it fails to generate action to mitigate such impacts. Around Kibwezi, it is instructive to observe *Ngolano*, a women's group that traditionally maintained rules of access to *mathembo* and was empowered to mete out social punishments on those who disrespect local sacred geographies, especially since its *Kingo'ole* male counterpart infamously was banned by the British colonial administration ostensibly for abusing capital punishment powers (Luongo, 2011). In 1988, *Ngolano* famously mobilized to chastise a man for shooting a 17-foot python in an *ithembo* on the Kangesu River near Kibwezi Town. Despite the dangers it presented, the presence of a creature that big was taken as a sign of a healthy ecosystem. Consequently, killing it in an *ithembo* disrupted the balance between *uoi* and *uwe*; locals were convinced the failure of a borehole drilled with modern equipment to provide irrigation water was punishment delivered by *aimu* with which the *Ngolano* women were in contact. Moreover, persistent water shortages throughout the Kangesu area since then also symbolize the connection of the individual to the community, with nature's response to one person's misbehavior becoming a pervasive, collective punishment.

Ngolano's gendered character is also worth noting given the significant number of female-headed households in Kibwezi Division. When SGR cut through Maimu pond at the foot of Mbuinzau Hill in 2015 (disrupting hydrological connections between several *mathembo*), *Ngolano* interrupted Chinese work crews by occupying the site, but this time, with no successful outcome. While male labor migration to Nairobi and Mombasa has been a common feature of life in the region for more than a century, Mbuinzau people attribute *Ngolano*'s failure to the fact that some of the women's husbands, sons, and brothers had taken jobs with the railway or associated enterprises.

This powerfully echoes Whatmore's call for "an upheaval in the binary terms in which the question of Nature has been posed and a recognition of the intimate, sensible, and *hectic* bonds through which people, organisms, machines, and elements make and hold their shape in relation to each other in the business of everyday living" (1999: 26, emphasis added). The simplest example is that, like hopes for productive dialogue of any sort, the SGR is about making connections—a train track seems an unproblematic route between two points, origin and destination, with comings and goings functionally equivalent. But deeply held Akamba tradition insists the same path should be avoided whenever possible to thwart *uoi* finding its way across the landscape of human interactions. This is not far removed from pleas to "Western" consumers to rethink wasteful habits of resource use. Here, too, nonhuman agencies play an important role. Since comings and goings are deeply meaningful to the Akamba, when important sacrifices are made at Kwa Muthita *ithembo* (close to the source of a main river), a spotless black bull must be walked completely around the foot of Mbuinzau Hill. For similar reasons, Dwa's managers have assisted in bull sacrifices at Kivungoni Dam almost every July since violent clashes with squatters in the 1990s and early 2000s (in a "Western" inflection, trying to atone for repeatedly following the same wrong path vis-à-vis community relations).

Intimate bonds of human and nonhuman forces extend beyond animal presences, colors vitalizing microspatializations for the Akamba also. Traditionally, red, black, and white cloths hang in trees marking *mathembo* entrances. White is peaceful and cleansing; black is associated with watchfulness and an appeal to *aimu* for goodwill. Houses constructed of modern materials and outfitted with such typical consumer goods as televisions and refrigerators nonetheless sometimes are painted with a broad black band just above the foundation. Red is associated with judgment, danger, and related connotations, including cursing. In 2015, small acts of sabotage involved relocating red SGR construction flags or planting decoys embodied with carefully prepared curses against work crews and their equipment. In these ways, the SGR's technological bond with local geography repeatedly is made "hectic" as the railway tracks' linearity is complicated by different sorts of localization and movement. Red cloths often appear hanging from the sharp barbs of sisal planted along the Dwa Lines, too, silent but visible protests about the land dispute there.

Together with animals and colors, plants also exert multiple forms of agency in local confrontations with the Anthropocene's impacts. Affective geographies ensure

Kibwezi Forest's trees are not merely trees, for instance (Rocheleau and Ross, 1995). They simultaneously mark vestiges of displaced people's ancestral villages, Dwa Estate's expectation of compensation for resources bulldozed by the Chinese, the temporal edge of sisal soon to transform from a low-profit fiber crop into lucrative biofuel, and in the KEFRI arboretum, parts of a revitalized scientific discourse about species richness now lost to the SGR. Just as familiar landscapes have been rendered unfamiliar in disfigurement, the trees still standing and the trees pulled down are no longer recognizable as what they once were presumed to be. Nor are any of these versions of Kibwezi's trees superior, categorically speaking, with different interests vying for their version to have the greatest affective impact on anybody who is paying attention.

The limit—and opportunity—of dialogue

Importantly, reports of situational religiosity, of the hierarchical importance of certain *mathembo*, and of the fuzzy distinction between *uoi* and *uwe* all reveal that even what appears to be a moral economy of local resistance to change (Scott, 1976) is only ever a temporary agglomeration of affective intensities full of internal contradictions. Together with postcolonial struggles over the power of discourse, this suggests that possibilities of dialogue between "indigenous" and "Western" approaches to environmental crisis cannot at the outset be presumed to have an objective basis.

Whether any of this conforms to "Western" bias about what constitutes evidence and methodologies for assembling and testing hypotheses is beside the point. As noted: (1) "Western" science itself needs to be properly contextualized; (2) indigeneity is a political object, not ontologically real; and (3) while the agency of "indigenous" peoples has suffered significant erasure, that of nonhuman life has also via imposition of an even more inert status. Given this much of a burden placed on the possibility of engaging in dialogue regarding matters of pressing importance in the Anthropocene, it becomes vital that we appreciate what "dialogue" really is all about, and what it isn't.

Bakhtin's (1981) distinction of the "epic" and the "novel" is helpful here. The former aims to have its audience assimilate a hegemonic discourse, while the latter is never a finished composition but endlessly disturbed by so-called "hybrid utterances." No matter how deliberately either party to a dialogue might try to craft only one frame of interpretation, the process of interaction vitalizes an endless referencing of indefinite chains of sources. The selective assimilation and/or contestation of such permanently disturbed narratives by many different constituencies simultaneously is what Bakhtin refers to as heteroglossia. The heteroglossic bears some relationship to Kristeva's "semiotic" (1980) insofar as parties involved in any dialogue are never themselves defined by identities fixed entirely in language; instead, an unspoken, visceral, affective component of who we each are and how in particular contexts we relate to the world around us matters just as much.

So, meaningful engagement with indigenes will never amount to a simple sorting procedure of picking and choosing from among approaches to nature-society relations those preferred for purpose. Otherwise, contending forces can combine and recombine; efforts to restore traditionally rural Hawaiian indigeneity have been pushed by mostly urbanized activists, for example (Linnekin, 1983). Similarly, some people in Kibwezi insist *aimu* are not necessarily exclusively Akamba ancestral spirits. Reflecting historical changes in Ukambani's wider political ecology, they may be Maasai, Gikuyu, or even European, all associated with specific geographies and/or family lineages. At least one *ithembo* is named for an infamous British-born Dwa General Manager (GM).

Whatmore's talk of "hectic" bonds is echoed by Luongo's talk of "fraught" Akamba experiences trying to manage the balancing act of living between *uoi* and *uwe* (2011: 7). Near "Baghdad," where Dwa *askari* guard Kivungoni Dam, is a dense stand of "bleeding" bamboo that is said to be the reincarnated spirits of the drowned villagers of Kilui. Dwa's GM ordered it removed despite warnings this would disturb Kilui's anguished *aimu*. Predictably, the bamboo bled at the first cut of the *askaris*' machetes. When workers fearfully ran away, the manager took up the task and died in mysterious circumstances the next day. The truth of such stories matters less than the way the guilty of Kilui were narratively deployed to punish Dwa's guilt, and how such heteroglossic reframing of human relations with nonhuman nature implies possibilities of renewed examination, including concerning the coming of SGR to Kibwezi. The allure of dialogue needs to be tempered by the realization that it may not lead us to a conclusive understanding of things.

Bakhtinian heteroglossia is particularly emphasized in the reinvention of Kliundo Meli, a powerful *mu'unde mu'uoi* who strongly influenced the whole region at the time the British were constructing the Lunatic Line. Following the railway, missionaries from the Presbyterian Church of East Africa (PCEA) claimed land stretching from Kibwezi Town all the way to the crest of Mbuinzau Hill after negotiating with Kilundo. But when he and his neighbors were pushed off their lands, Kilundo supposedly used his power to kill three missionaries without touching them and then disguised their burial sites. Even today, PCEA's school is visited by missionaries from abroad who never travel in Kibwezi Town alone and always stay inside a walled compound otherwise. Like any *mu'unde mu'uoi*, Kilundo's reputed power was as much technical as embodied, and locals believe he often buried his *kyondo* beneath an *ithembo* tree to recharge the spiritual energy of his accoutrements.

Kilundo has lately been reinvented as a chief to rally resistance to government collaboration with the Chinese. Yet the precolonial Akamba did not have chiefs; communities were organized along clan lines by gender and age grades—even today, *Ngolano* groups have no leaders except to the degree an organizational structure has been borrowed from Western cultures. It was the British colonial authority that imposed the Chiefs Authority Act in 1937, creating salaried civil service positions for administrators of Locations (assistant chiefs for Sublocations, only subordinate to Village Elders elected by local people). The Act was maintained with only minor revisions after independence and subject to only partial repeal in 1997.

Especially on Mbuinzau Hill, there are those who believe *mathembo* forests and springs fared better when chiefs wielded unchecked power. During the droughts and famines of the 1980s, Chief Munandu Mwololo protected critical resource areas through the use of corporal punishment and intimidation. But because of how hated the Act was generally, it is fascinating that Kilundo's often referred to as a chief given such authority (promulgated decades after his death in 1901) derives from the government now blamed for allowing SGR to destroy local landscapes. The reinvention of Kilundo also suggests PCEA and Dwa, which followed its arrival, somehow swindled the chief. Yet this, too, assumes today's definitions of property ownership and market exchange can be applied to the situation existing more than a century ago when Kilundo was in no position to have negotiated with PCEA as it claims he did.

There is also Kilundo's *kyondo* and the *ithembo* tree under which he supposedly buried it. An *ithembo* is not personal property, and the *mu'unde mu'uoi*'s disruptive actions do not easily align with *mathembo* rituals' socially reinforcing purposes. While some insist that Kilundo's *kyondo* was destroyed shortly after his death, today, the chief's *kyondo* lives on, many of those outraged by the wanton destruction of Kibwezi Forest insisting Kilundo's tree could not be removed, the *mu'unde mu'uoi*'s power still invested there actively working to punish supporters of SGR. It does not matter which tree is supposed to be Kilundo's, especially given that *mathembo* known to be in the forest are associated with the Kibwezi River's water, now diverted for SGR's construction. Instead, *mathembo* are sites for affective commitments to environmental and sociocultural conservation and preservation in the face of development pressures.

Dialogue is not likely to be straightforward since affective geographies are energized by precisely this condition of changing local conditions paradoxically exploited to reaffirm a sense of place (Kottak, 1980). This helpfully echoes "new" ecology's (Scoones, 1999) "postequilibrial" emphasis on adaptive disturbance. Indeed, as with indigeneity (Ellen, 2007), it is important to wonder when and why equilibrium is emphasized (Gillson et al., 2003). Wrongly presuming copses, springs, and other *mathembo* landscapes in tiny enclaves on Mbuinzau Hill or along the Dwa Lines are sustainable and might be expanded amounts to an unjustified "forest fundamentalism" (Buttel, 1992: 19). But protests highlighting devastation wreaked by the Chinese reveal much about how interests might be organized to promote alternative visions of socioecological development locally.

From a "postequilibrial" perspective, many supposedly "traditional" and local *mathembo* have come into being because of, not in resistance to, trends generally collectively considered "modern" and global. Dialogue thus involves selective adoption of the terms of debate used by one party and their repurposing by the other. Hirtz (2003: 887) emphasizes that today, "being recognized as traditional or indigenous requires the employment of modern means" (see also Plaice, 2010; Barnard, 2019, etc.). Hope of dialogue between "Western" science and "indigenous" approaches thus requires realizing the latter have now taken on the character of Deleuze and Guattari's "metaphor of the rhizome" (Ferreira and Devine, 2012; Deleuze and Guattari, 2013). In botany, a rhizome is a plant stem sending out roots in multiple directions, growth not coming from a centralized point of origin. It is similarly appropriate to speak today of "planetary indigeneity" (Sidaway et al., 2014) in terms of how "indigenous" peoples seek novel ways and spaces in which to overcome being defined only by political oppression and also how such peoples are defined by translocal actors (including, of course, academics writing about indigeneity).

Rhizomatically, an *ithembo* may be claimed as ancient and intensely indigenous even as it's often a material expression of how local people have adopted, modified, and resisted Westernization. For example, different interests repurposing the figure of Kilundo Meli claim different "lost" villages as his home. Those who say the "Chief" came from Syetali do so in part because it is the only village ever to manage any kind of collective access to Kenya's courts, even if no judgment in favor of ancestral land claims has been issued. Some Syetali descendants also lay claim to the residual affective power of Kilundo's name because of, not despite, the fact the community's early Christian converts supposedly burned his *kyondo* after his death. Meanwhile, those claiming Kilundo belonged to Kakangani say Syetali suffered evictions after his *kyondo* was destroyed, not before, and note Kakangani was the village where Dwa GMs built their railway station and first boundary gate. There is supposed to be an *Ithembo ya Kakangani* at the symbolic center of this capitalist resource exploitation, revealed to local people only after the arrival of the plantations.

At the end of the day, while "Western" science certainly has advanced our *knowledge* of environmental processes, what Basso (1996) appreciated is that for his Western Apache informants, "*wisdom* sits in places" (emphasis added). Arguably a common feature of "indigenous" worldviews, this difference in consideration of the relationships of humans to nonhumans may be vital to our navigating the challenges of the Anthropocene. Like the Western Apache, the Akamba move through their world via narratives about particular locations, and these stories "make you live right [and] make you replace yourself" (Basso, 1996: 38). An important question for any possible dialogue, therefore, is what sorts of stories can we tell one another about how humans and nonhumans should "dwell together" today (Heidegger, 1977: 323)? Beyond simply sharing information, what kind of wisdom is needed to overcome unhelpfully persistent divisions of "science" and "non-science," "indigenous" and "non-indigenous," nature and society, etc. (Green, 2012)? Finally, what are the limits of talk, writing, or other familiar forms of expression; that is, what might be gained by close examination of more affective kinds of interactions with one another, nonhuman experiences, and the geographies with which they're dynamically related?

References

Aikenhead, G. (1996), "Science Education: Border crossing into the subculture of science, *Studies in Science Education*, 27: 1–52.

Aikenhead, G. (1997), "Towards a First Nations cross-cultural science and technology curriculum," *Science Education*, 81: 217–38.

Anderson, B. and Harrison, P. (2010), *Taking-Place: Non-Representational Theories and Geography*, Farnham, UK: Ashgate.

Baird, I. (2020), "Thinking about indigeneity with respect to time and space: reflections from Southeast Asia," *Espace populations sociétés/Space populations societies* 2020/1–2, online at https://journals.openedition.org/eps/9628.

Baird, I. G., P. Leepreecha, and U. Yangcheepsujarit (2017), "Who should be considered 'Indigenous'? A survey of ethnic groups in northern Thailand," *Asian Ethnicity*, 18(4): 543–62.

Bakhtin, M. M. (1981), *The Dialogic Imagination: Four Essays*, M. Holquist (ed.), Austin: University of Texas Press.

Bala, A. and J. G. Gheverghese (2007), "Indigenous knowledge and Western science: the possibility of dialogue," *Race & Class*, 49(1): 39–61.

Balaton-Chrime, S. (2013), "Indigeneity and Kenya's Nubians: seeking equality in difference or sameness?," *Journal of Modern African Studies*, 51(2): 331–54.

Barcham M. (1998), "The challenge of urban Maori: Reconciling conceptions of indigeneity and social change," *Asia Pacific Viewpoint*, 39(3): 303–14.

Barnard, A. (2019), "The politics of indigeneity," in A. Barnard, *Bushmen: Kalahari Hunter-Gatherers and Their Descendants*, Cambridge: Cambridge University Press, pp. 18–37.

Basso, K. H. (1996), *Wisdom Sits in Places: Landscape and Language among the Western Apache*, Albuquerque: The University of New Mexico Press.

Beck, U. (1992), *Risk Society: Towards a New Modernity*, London: Sage Publications.

Beck, U., W. Bonss, and C. Lau (2003), "The theory of reflexive modernization: problematic, hypotheses and research programme," *Theory, Culture & Society*, 20(2): 1–33.

Béteille, A. (1998), "The idea of indigenous people," *Current Anthropology*, 39(2): 187–92.

Bloch, M. (1998), *How We Think They Think: Anthropological Approaches to Cognition, Memory, and Literacy*, Oxford: Westview Press.

Buttel, F. H. (1992), "Environmentalization: Origins, processes, and implications for rural social change," *Rural Sociology*, 57(1): 1–27.

Cadena, M. and O. Starn (eds.) (2007), *Indigenous Experience Today*, New York: Berg.

Castaneda, Q. E. (2004), "'We are not indigenous!' An introduction to the Maya identity in Yucatan," *The Journal of Latin American Anthropology*, 9(1): 36–63.

Clifford, J. (2013), *Returns: Becoming Indigenous in the Twenty-First Century*, Cambridge: Harvard University Press.

Dauvergne, P. (2020), *AI in the Wild: Sustainability in the Age of Artificial Intelligence*, Cambridge, MA: MIT Press.

Deleuze, G. and Guattari, F. (2013), *A Thousand Plateaus: Capitalism and Schizophrenia*. Minneapolis: University of Minnesota Press.

Donhauser, J. (2019), "Environmental robot virtues and ecological justice," *Journal of Human Rights and the Environment*, 10(2): 176–92.

Drouin-Gagné, M. E. (2014), "Western and Indigenous sciences: colonial heritage, epistemological status, and contribution of a cross-cultural dialogue," *Ideas in Ecology and Evolution*, 7(1): 56–61.

Ellen, R. F. (ed.) (2007), *Modern Crises and Traditional Strategies: Local Ecological Knowledge in Island Southeast Asia*, New York: Berghahn Books.

Ferreira, M. L. and D. Devine (2012), "Theater of the oppressed as a rhizome acting for the rights of indigenous peoples today," *Latin American Perspectives*, 39(2): 11–26.

Forte, M. C. (2010), *Indigenous Cosmopolitans: Transnational and Transcultural Indigeneity in the Twenty-First Century*, New York: Peter Lang.

Gellers, J. C. (2021), "Earth system law and the legal status of non-humans in the Anthropocene," *Earth System Governance*, 7: 100083. https://doi.org/10.1016/j.esg.2020.100083.

Gerharz, E. et al. (eds.) (2017), *Indigeneity on the Move: Varying Manifestations of a Contested Concept*, New York: Berghahn.

Giddens, A. (1999), "Risk and responsibility," *Modern Law Review* 62(1): 1–10.

Gillson, L., M. Sheridan, and D. Brockington (2003), "Representing environments in flux: case studies from East Africa," *Area*, 34(4): 371–89.

Göbel, B. (2001), "Risk and culture in the Andes: differences between Indigenous and Western developmental perspectives," in G. Böm, J. Nerb, T. McDaniels, and H. Spada (eds.), *Environmental Risks: Perception, Evaluation and Management (Research in Social Problems and Public Policy, Vol. 9)*, Bingley: Emerald Group Publishing Limited, pp. 191–220.
Good, R. (1995), "Comments on multicultural science education," *Science Education*, 79: 335–6.
Gorelick, R. (2014), "Indigenous sciences are not pseudosciences," *Ideas in Ecology and Evolution*, 7: 43–55.
Grear, A. (2015), "Deconstructing anthropos: a critical legal reflection on 'Anthropocentric' law and Anthropocene 'humanity,'" *Law and Critique*, 26(3): 225–49. https://doi.org/10.1007/s10978-015-9161-0aSterli.
Green, L. J. F. (2012), "Beyond South Africa's 'indigenous knowledge—science' wars," *South African Journal of Science*, 108(7): 1–10.
Guim, M. and M. A. Livermore (2021), "Where Nature's rights go wrong," *Virginia Law Review*, 107(7): 1347–419.
Heidegger, M. (1977), "Building dwelling thinking," in D. Krell (ed.), *Martin Heidegger: Basic Writings*, New York: Harper & Row, pp. 319–39.
Hirtz, F. (2003), "It takes modern means to be traditional: on recognizing indigenous cultural communities in the Philippines," *Development and Change*, 34(5): 887–914.
Hodgson, D. (2011), *Being Maasai, Becoming Indigenous: Postcolonial Politics in a Neoliberal World*, Bloomington: Indiana University Press.
Igor, J. (2006), "Becoming indigenous peoples: difference, inequality, and the globalization of East African identity politics," *African Affairs*, 105(420): 399–420.
Jasanoff, S. and S. Kim (eds.) (2015), *Dreamscapes of Modernity: Sociotechnical Imaginaries and the Fabrication of Power*, Chicago: University of Chicago University Press.
Kealiikanakaoleohaililani, K. and C. P. Giardina (2016), "Embracing the sacred: an indigenous framework for tomorrow's sustainability science," *Sustainability Science*, 11: 57–67.
Keating, N. B. (2016), "Kites in the highlands: articulating Bunong indigeneity in Cambodia, Vietnam, and abroad," *Asian Ethnicity*, 16(4): 566–79.
Kottak, C. P. (1980), *The Past in the Present: History, Ecology, and Cultural Variation in Highland Madagascar*, Ann Arbor: University of Michigan Press.
Kotzé, L. and Villavicencio Calzadilla, P. (2017), "Somewhere between rhetoric and reality: Environmental Constitutionalism and the rights of nature in Ecuador," *Transnational Environmental Law*, 6(3): 401–33.
Kristeva, J. (1980), *Desire in Language: A Semiotic Approach to Literature in Art*, New York: Columbia University Press.
LaDuke, W. (2019), "The White Earth band of Ojibwe legally recognized the rights of wild rice. Here's why," *Yes! Journalism for People Building a Better World.*
Latour, B. (1993), *We Have Never Been Modern*, trans. Porter, C.; Cambridge: Harvard University Press.
Latour, B. (2005), *Reassembling the Social: An introduction to Actor-Network-Theory*, Oxford: Oxford University Press.
Lawrence, M. (2013), "The moon at the edge of words: problems of spatiality and diversity in the global agenda for social work and social development," in R. Lutz (ed.), *Diversity and Respect: Problems of Perception in the Global Agenda for Social Work* (Internationale Sozialarbeit, Band 9), Oldenburg, Germany: Paulo Freire Verlag, pp. 21–50.

Linnekin, J. S. (1983), "Defining tradition: variations on the Hawaiian identity," *American Ethnologist*, 10(2): 241–52.
Luongo, K. (2011), *Witchcraft and Colonial Rule in Kenya, 1900—1955*, Cambridge: Cambridge University Press.
Massumi, B. (2002), *Parables for the Virtual: Movement, Affect, Sensation*, Durham, NC: Duke University Press.
Mateer, J. (2018), "Rebel elephants: resistance through human–elephant partnerships," in S. Wilcox and S. Rutherford (eds.), *Historical Animal Geographies*, London: Routledge, pp. 121–31.
McCormack, D. (2005), "Diagramming practice and performance," *Environment and Planning D: Society and Planning*, 23: 119–47.
Michel, H. and Gayton, D. (2002), "Linking Indigenous peoples' knowledge and Western science in natural resource management: adialogue," *Journal of Ecosystems and Management* 2(2): 1–12.
Park, H. (2005), "Culture as deficit: a critical discourse analysis of the concept of culture in contemporary social work discourse," *Journal of Sociology and Social Welfare*, 32(3): 11–33.
Pelican, M. (2009), "Complexities of indigeneity and autochthony: an African example," *American Ethnologist*, 36(1): 149–62.
Pile, S. (2009), "Emotion and affect in recent human geography," *Transactions of the Institute of British Geographers*, 35(1): 5–20.
Plaice, E. (1990), *The Native Game: Settler Perception of Indian/Settler Relations in Central Labrador*, St. John's: Institute of Social and Economic Research, Memorial University of Newfoundland.
Plaice, E. (2010), '"Making Indians': debating indigeneity in Canada and South Africa," in D. James, E. Plaice, and C. Toren (eds.), *Culture Wars: Context, Models, and Anthropologists' Accounts*, New York: Berghahn Books, pp. 45–56.
Povinelli, E. A. (1998), "The state of shame: Australian multiculturalism and the crisis of indigenous citizenship," *Critical Inquiry*, 24: 575–610.
Putzer, A., T. Lambooy, R. Jeurissen, and E. Kim (2022), "Putting the rights of nature on the map: a quantitative analysis of rights of nature initiatives across the world," *Journal of Maps*, 18(1): 89–96.
Rocheleau, D. and L. Ross (1995), "Trees as tools, trees as text: struggles over resources in Zambrana-Chaucey, Dominican Republic," *Antipode*, 27: 407–28.
Scoones, Y. (1999), "New ecology and the social sciences: what prospects for a fruitful engagement?," *Annual Review of Anthropology*, 28: 479–507.
Scott, J. (1976), *The Moral Economy of the Peasant: Rebellion and Subsistence in Southeast Asia*, New Haven: Yale University Press.
Seigworth, G. J. and M. Gregg (2010), "An inventory of shimmers," in M. Gregg and G. J. Seigworth (eds.), *The Affect Theory Reader*, Durham and London: Duke University Press, pp. 1–28.
Shetler, J. B. (2007), *Imagining Serengeti: A History of Landscape Memory in Tanzania from Earliest Times to the Present*, Athens: Ohio University Press.
Sidaway, J. D., C. Y. Woon, and J. M. Jacobs (2014), "Planetary postcolonialism," *Singapore Journal of Tropical Geography*, 35(1): 4–21.
Snively, G. and J. Corsiglia (1998), "Discovering indigenous science: Implications for Science Education," paper presented at the Annual Meeting of the National Association for Research in Science Teaching (71st, San Diego, CA, April 19-22).
Stone, C. D. (1972), "Should trees have standing—toward legal rights for natural objects," *Southern California Law Review*, 45: 450.

Swyngedouw, E. (2004), "Modernity and hybridity: nature, *regeneracionismo*, and the production of the Spanish waterscape, 1890–1930," in T. J. Barnes et al. (eds.), *Reading Economic Geography*, Malden, MA: Blackwell, pp. 189–204.
Thrift, N. (2007), *Non-Representational Theory: Space, Politics, Affect*, London: Routledge.
Tilley, H. (2011), *Africa as a Living Laboratory: Empire, Development, and the Problem of Scientific Knowledge, 1870–1950*, Chicago: University of Chicago Press.
Tsang, A. K. T. and M.-C. Yan (2001), "Chinese corpus, Western application: the Chinese strategy of engagement with Western social work discourse," *International Social Work*, 44(4): 433–54.
Walley, C. J. (2004), *Rough Waters: Nature and Development in an East African Marine Park*, Princeton: Princeton University Press.
Walton, R. G. and M. M. Abo El Nasr (1988), "Indigenization and authentization in terms of social work in Egypt," *International Social Work*, 31(1): 135–44.
Whatmore, S. (1999), "Hybrid geographies: rethinking the 'human' in human geography," in D. Massey et al. (eds.), *Human Heography today*, Cambridge: Polity Press, pp. 22–39.
Wilcox, S. (2017), "Savage jaguars, king cats, and ghostly *tigres*: affective logics and predatory natures in twentieth-century American nature writing," *The Professional Geographer*, 69(4): 531–8.

Part Two

Challenges

9

Conceded Land Rights, Indigenous Territoriality, and Cooperation: Contradictions and Challenges

Volker von Bremen

Intercultural cooperation

For many indigenous peoples, the framework of International Cooperation[1] is a field of interaction in their relationship to the dominant society of their surroundings. Although there is a broad variety of interpretation, the term "project" is automatically associated with cooperation offered by local, national, and international organizations, either by state and government institutions, or by private and civil society organizations with the declared purpose to help, to support, or to strengthen indigenous peoples, their communities and their members.

In general, the different actors cooperating in those contexts come from fairly different cultural backgrounds that condition their relationship and practice. Team members, coordinators, and directors of local, national, and international organizations are formed and socialized within their specific cultural context of family and society in which they grew up. In the practical work context, it might make a considerable difference whether a person grew up in rural or urban surroundings, in a more traditional or modern context, conditioned by specific intellectual, artistic/aesthetic, as well as practical settings and experiences influenced by certain (religious) beliefs, ethical values, practical experiences, etc. Indigenous actors come from their own, and generally quite different cultural and historical backgrounds determined, among others, by their own languages, cosmovisions and spirituality, relationship to environment, their proper ways of organizing their communities, of making their living, as well as their specific and very diverse experiences with colonization and its impacts on the process of becoming part of national societies.

In the context of those settings that determine intercultural cooperation, the different actors—being of indigenous as well as non-indigenous background—often

[1] In the context of this article, the term "international cooperation" refers to the established fields of international cooperation for development where governmental as well as non-governmental institutions and organizations are involved.

consider certain basic principles they had grown up with as normal, obvious, and self-evident ("natural"). They are not questioned but are rather taken for granted. In general, they are not considered as topics to be mentioned and reflected on in the dialogues and negotiation processes within the context of cooperation. With the effort to build a good and constructive relationship as a foundation to achieve a successful cooperation, there is the attempt to identify commonalities between cooperating partners, looking for harmonic feelings, sympathy, and a state of mutual understanding. In order to establish ways of cooperation in a constructive manner, however, it is important to be conscious about one's own principles as well as those used by the other actors within cooperation—the so-called "partners"—especially those principles that are not directly apparent as well as those that easily seem to be shared by everybody.

Based on broader experiences of cooperation with indigenous peoples, here are some examples of topics where different principles of understanding and interpretation might be taken for granted and applied as "obvious," each one of them with a huge variety of aspects that need to be analyzed and considered in the practical context of cooperation in order to be able to enter into a mutual relationship with sufficient sensitivity:

- dynamics and organization of economic action and behavior;
- concepts and perception of "nature," "environment," and human-nature-relationship;
- concepts of space and time;
- principles of decision-making;
- principles of socio-political organization and representation;
- role and status of human rights.

The broad diversity of indigenous peoples in present times derives from many different factors related not only to their cultural history and development but also to the very specific experiences with colonialism they had been exposed to throughout history up to the present. Wherever cooperation with indigenous peoples is pretended to strengthen their lives in terms of sustainability, there is the need and necessity to consider the specific situation of the corresponding indigenous group, community, and people, as well as their cultural and historical background, in order to overcome preconceptions and prejudice.

One of the most important and justified fields of cooperation has been and continues to be the struggle of indigenous peoples to have and to secure the access to their land and territories. Within political, socio-economic, as well as legal and judicial contexts, the recognition of land rights within established national and international jurisdictions is considered to be the major topic of quarrel, conflict, and negotiation. In legal terms and within ruling judicial systems and concepts, there has been quite some progress within the last decades. In Latin America, almost every country has ratified Convention No. 169 of the International Labour Organization (ILO).[2] All of them

[2] See ILO (1989).

accepted the United Nations Declaration of Rights of Indigenous Peoples (UNDRIP)[3] in 2007. Indigenous rights are recognized as constitutional rights including their rights to land. In addition, according to specific national legislation, a considerable number of indigenous communities have received either land titles or the recognition of "indigenous land" with exclusive rights to live on (like in Brazil). However, there is a great number of communities confronted with land conflicts and threats. Cases of ongoing administrative quarrels for decades, unresolved judicial cases over unbelievably long periods of time, as well as the existence of communities and groups without any kind of protection and legal recognition and support still exist in almost all countries that recognized well-defined indigenous rights.

And even those communities with land titles and judicial security of their land rights and access to land are losing this security in many cases because of present political and economic tendencies and their impacts on local levels. Because of those tendencies, indigenous land rights are violated and ignored, endangering survival in their territories.[4]

Due to those facts that, unfortunately, have to be observed as a general tendency, the dedication to secure and apply for land rights continues to be one of the most central issues, also in the challenges of cooperation.

Another area of cooperation related to land and land rights is dedicated to the support of indigenous peoples in their possibilities and capacities to organize and make their living on their land under present-day conditions with a perspective of sustainability in ecological, social, and economic terms. There is a huge variety and diversity of initiatives dealing with questions related to this field of cooperation in terms of methodological approaches, production, distribution and marketing, land use, sustainable management of natural resources, gender, organizational processes and development, security and defense, just to name some examples.

Considering the basic goal of cooperation to support initiatives in order to create and have a sustainable impact on possibilities of indigenous life in their territories,[5] there is the need to get to know and understand the peculiar concept of territoriality used and applied by the specific indigenous peoples living within the corresponding territory. This is only a first step to create sensitivity toward indigenous approaches that are the basis for their further organizing and motivating initiatives.

Indigenous land use strategies

Most communities continue disposing of knowledge about a differentiated sustainable life with their territories and the use of their resources based on their cultural history and development. However, according to the degree and growth of dependence upon market society and externally produced consumer goods, there is a tendency that their

3 See United Nations (2007).
4 Among many others, see Bremen (2018: 16, 30) and IWGIA (2024).
5 In different Latin American countries, the concept of "gestión territorial" is used to identify this field of action.

corresponding knowledge mentioned above and its application in practical life-shaping contexts might diminish. This is very much related to the specific experiences with colonialism[6] that conditioned and changed their way of life with deep impacts, the integration of elements derived from colonial influences, and the degree of dependence on dominant regional, national, and international actors, their values, and principles. As living cultures and societies, all indigenous communities and groups that live in regular or permanent contact with surrounding dominant societies have integrated elements they became acquainted with into their lives, elements with corresponding impacts in social, economic, and cultural as well as spiritual and religious terms and fields. Many of them are struggling with internal contradictions produced by the experiences with new, different influences and impositions from dominant, colonizing societies. In any of those cases, present-day indigenous societies do maintain links to their own cultural backgrounds and heritage. New forms and contents of interpretation of adopted elements might rise and come to build part of present indigenous cultures. In the context of anthropological research and postcolonial theory, they are considered as "hybrid societies," as a result of colonial confrontation with its corresponding influence upon indigenous identity and culture.[7]

Considering perspectives to make a living under the conditions of globalization, a vast number of projects promoted by state and governmental programs in the economic realm are focused on the production of market products or products that might substitute the need to buy those products on the market. Other forms and modes of making a living developed by indigenous peoples and adapted to ecological and cultural conditions and knowledge of the communities very often are neglected, unknown, or ignored in the context of cooperation for development. However, new forms and practices of land use need to be perceived not just from an economic, productive perspective, but also in the context of history as well as social and cultural experiences and practice, in order to be able to estimate properly their mid- and long-term potential.[8]

Concepts of cooperation based on alternative indigenous proposals strengthening their present potentials to live with their territories still are not very much developed. Even though there are examples of initiatives that try to promote consciousness on the importance of indigenous values, knowledge, and practical experiences,[9] a great number of communities still negotiate proposals in adaptation to offers made by external actors in the context of the promotion of innovative land use neglecting their own approaches and practices within projects of cooperation.

[6] The term "colonialism," in this context, is related not only to the historical period of Spanish and Portuguese colonialism but also, in general and throughout history, to the experiences of indigenous societies in their contact with external agents, including times when the national states already existed in Latin America.

[7] See, among others, Babha (1994) and Kraidy (2005).

[8] For further details, see Bremen (2018: 17).

[9] See Bremen (2019).

Aspects of Ayoreo cosmovision

There is a broad variety of indigenous concepts and principles that need to be considered in every specific case that cooperation in territorial issues is proposed and implemented. As an example of indigenous cosmological principles that might help in approaching a basic understanding of how to be in the world and relate to the environment and territory, some remarks shall be offered on the Ayoréode of Bolivia and Paraguay.

According to the traditional cosmovision of the Ayoreo people, the living world is basically divided into two different spheres: the present world and the world of the ancestors—those who were first, before (see Figure 9 "Time-space in Ayoreo cosmovision"). This world of ancestors embraces everything that exists in the present world—humans, non-humans, and all phenomena one can think of—related to the physical, material world as well as to the world of feeling/sensing, the spiritual world, and the world of abstraction. This world is connected with the time when the phenomena of today's world "still were Ayoréode." In that sphere/stage/era, everything belonged to the same social community of humans sharing community life like everybody else. Like the members of present-day Ayoreo communities, the ancestors, in general, were also integrated into the clan system as it continues to exist in present times, defining social relationships between its corresponding members.

In certain moments within this original time of ancestors, the community or certain members of it became aware of the specific characteristics and qualities of an individual ancestral being and demanded the sharing of its benefits. Here is an example:[10]

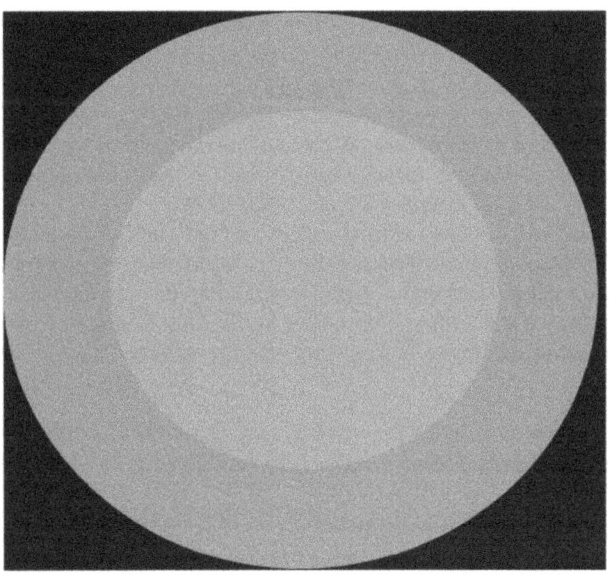

Figure 9 Time-space in Ayoreo cosmovision.

[10] For further details, see Bremen (1997) and Fischermann (1986: 448–9).

Beehive was a woman in times when she was still Ayore/human. She gave her sweet saliva prepared with water to her son-in-law when he came home exhausted. As it tasted good and strengthened the man, he shared it with his family. Although Beehive wanted to keep the secret, everybody in the community came to know about it and curiously asked for it. Beehive, therefore, decided to leave the community, turning into what is known as a beehive today. When she withdrew, she left the secret to the community on how to get in touch and find her.

The example may illustrate the basic principle of how the present world came into being. The original being lived as a member of the community of origin everybody belonged to. When the community discovers its specific qualities, the demand for sharing these qualities rises, pressuring the ancestral personality that disposes of these qualities. In order to fulfill this requirement of participation and distribution, the ancestor decides to leave the original social community, changing by metamorphosis into the being as it is known in present days. But that person does not just leave the community; he/she rather hands over the key principle to the people, to the community, so they can maintain contact and use the qualities and specific characteristics the ancestor is identified with.

These contacts with ancestors that in the present time constitute the existing world in any one of its manifestations are needed in order to make a living, find honey, game, places to cultivate crops, help in critical moments, advice in decision-making processes, etc. Contacts with the ancestor and his/her potential/power are established by the knowledge about his/her specific qualities as well as by songs and words he/she handed over to the community when he/she left the original community/sphere, turning into the being that is part of the present world.

That same principle is also valid for the surrounding world identified as *eami*:

EAMI

An indigenous word of the Ayoreo people to name the forest as well as the world in its life on earth. EAMI lives through the diversity of life of humanity and natural environment. The bigger diversity, the more lives EAMI.

EAMI is divided: into the world of scarcity and the world of abundance; into the closed world of taboos and threats and the open world of liberty and encounter.

EAMI are not just "ecosystems" in their organic life and functioning. The voices of bubbling springs, the groaning of branches in the wind belong to it as well as the hunter's fortune or the impact of a meeting with the nightjar.

Regarding the relationship to the environment, this basic cosmological principle of Ayoreo cosmovision gives evidence of the following aspects:

- The environment and all it consisted of is perceived as part of the original social community shared by the human beings as well as nature and all existing phenomena. They all joined the community as original humans.
- There is no qualitative distinction between culture and nature, as all natural phenomena belong to the human-like community of origin.

- Going beyond its mere physical composition, the environment is animated like all social beings and composed of a multitude of different beings, each one of them disposing of its own quality, will, and decision-making ability.
- As part of a common original social community, humans and non-humans share life as part of the environment, everyone with his/her specific role and quality ("Mitwelt").
- Economic action is considered as social interaction between social beings. There is no "management of natural resources."
- In their relationship with the environment, humans try to create favorable conditions with those beings they depend on for making their living in order to be able to participate in and have a share of their qualities.

Just as one example among a great variety and diversity of indigenous cosmologies, the mentioned extract of traditional cosmological concepts of the Ayoreo people gives evidence of the need for research in every setting and context of cooperation with indigenous peoples. For there will not exist either indigenous "participation" or "protagonism," as it is described in many projects, or the implementation of the "right to free, prior, informed consent" guaranteed by the UNDRIP, if cooperating actors are not aware of and do not know the specific principles of life-shaping that determine motivation and dynamics, capacity and potential of indigenous ways to be part of the world and make their living in and with it.[11]

Indigenous territoriality

The consideration of the described concept of "environment" might permit an approximation to indigenous concepts of territoriality and its practice under today's conditions.

In general, indigenous territoriality is not limited to the physical, geographical space defined by state authorities in the process of recognition of indigenous land rights. It rather expresses the special historical and cultural relationship of the communities to a geographically, socially, and spiritually defined region they live(d) in and belong(ed) to. In their relationship to this territory, many indigenous peoples—although not all—would not claim an exclusive ownership, as interpreted in the modern capitalist context of private property. They rather perceive themselves as belonging to their territory.

In most cases of state recognition of indigenous land rights, there is a conceptual and practiced difference between the legally defined and recognized territory and the spaces used and frequented by the corresponding indigenous people. However, in demands within legal frameworks and negotiations with the state related to achieving support and cooperation in land use and management, spatial reference is given almost exclusively to those fractions of land that are legally defined and recognized by national laws and their

[11] See further below.

regulations. When elaborated and approved development plans are implemented, contradictions often arise as indigenous communities might apply activities and resources beyond those spaces legally defined and recognized, but building part of the territory as it is perceived and experienced in practical life by the community.[12]

With the recognition of land rights and the transfer of land titles new dynamics might arise among indigenous communities, leading to conflicts between traditional concepts of territoriality and innovative modes and forms of production and land use by becoming state-recognized "land-owners."

Here are some examples of practical experiences related to this aspect:

- The space of land recognized by the state as a common land of indigenous groups and communities might be divided and delimited internally into fractions of land used and inhabited by individual families/groups, which claim exclusive rights upon those fractions. This might create contradictions with the traditional concept and mode of use of "natural resources"—or, in terms of the mentioned case of the Ayoréode, "relationship to social beings"—practiced by members of the communities that live on the same legally recognized land (hunting, gathering, traditional agriculture, etc.).
- Individual communities and their authorities might negotiate with non-indigenous third parties the use of space and the application of projects (for example, the installation of sawmills, the sale of natural resources, and the permission of their exploitation by third parties, etc.). This might raise conflicts with the practice of intercommunal sustainable use and management of legally recognized territories.
- New areas of production (like cattle-raising, commercializing wood, etc.) might be introduced, especially in cooperation with leaders of communities based on their new role as land-owners according to models practiced in the dominant surrounding society and its corresponding offers of promotion (investment, production, commercialization). This might be in contradiction to the land use and reciprocity rules applied by the inhabitants and families, members of the communities that hold the land title or, in the case of Brazil, have the permission and right of permanent occupation and use.

The Guarani struggle for their *tekohá*

Under today's specific living conditions of indigenous groups and communities, the struggle for the recognition of determined land rights has become crucial in order to secure possibilities to survive. In general, this struggle occurs within the legal norms and principles determined by dominant and ruling national societies and states, and the criteria corresponding to those principles. Indigenous peoples are obliged to adapt to those norms in order to achieve the recognition of their land rights. There is the

12 See Bremen (2017a).

need and the challenge to find some kind of compromise between these norms and the corresponding indigenous perception of territory and territoriality, since usually there are quite some differences of concepts and perceptions between both parties. Among the Guarani groups of Eastern Paraguay and the adjacent regions of Brazil and Argentina, the principal unit in socio-political as well as geographical terms is the *tekohá* or *tapyi guasu*. It is to be understood as the "micro-territory" of a group of relatives that shares life in it, considering its exclusive right of usufruct.[13] The term *tekoha* includes a socio-political connotation of cohabitation and organization according to Guarani cultural standards, called the *teko*. Consequently, the *tekohá* is "the place of being," the place to live the *teko porã*—the appropriate (divine) way of living on Earth.

Guarani people feel and know that they belong to a certain place and area where their way of being can be fulfilled, as an individual and as a group. The *tekohá* combines the belonging to a social group as well as to a place/area. In this combination, the principles of "being in the world" given by spiritual, divine powers can be accomplished and implemented.

In their demand for recognition of their land rights or, rather, their need to live in their *tekohá*, Guarani community leaders not only and not always refer to their legal rights recognized by national jurisdiction. Their arguments, rather, are based on principles of their cosmovision that justify and motivate their struggle. The right to live in their territory is derived from the need to live in their *tekohá* according to their destiny as it was determined and given by *Ñande Ru* ("Our Father").[14]

As they do not understand themselves as land-owners, but rather as fulfilling the divine destiny of belonging to a certain place/space on earth, their focus on struggling for their land rights is, in the first place, not so much oriented and motivated by demanding the fulfillment of constitutional rights, international conventions, laws, and commitments of the nation-state and its governing bodies. It is more derived from the destiny determined and given by their divine authorities of Creation. They consider having been put into their social group of relatives, families, and communities, which also are related to a specific geographical space where the Creator has put them in order to make their living, following the path toward the divine home where "Our Father" lives. It is in this place where they can fulfill and accomplish the destiny according to the divine plan of life.

Adapt to survive

In the context of cooperation and support of communities that are threatened in their way of life according to their proper principles and cosmovisions, the main focus is oriented toward the claims of fulfillment of the indigenous rights recognized by the legal framework of the states in which they are living. Besides the indigenous social

[13] See Lehner (2008: 5).
[14] For further details, see Bremen (2017b), Grünberg (2024), and Grünberg and Grünberg (2014).

organizations that exist according to their own, internal structures (leadership, authority, family, clans, moiety, community, local group, intercommunity networks, etc.),[15] there are indigenous organizations recognized by law that use the rationale of hegemonic, established legal and political reasoning as a tool against the aggression to which their communities are exposed. In fact, in national as well as international contexts of cooperation and negotiation with public and private entities, it is this type of indigenous organization that is mainly recognized as a counterpart and interlocutor representing the demands and interests of its members. In order to become a legal entity according to national law, indigenous groups and communities have to meet legal and administrative requirements defined by the state and its corresponding bodies, and its rules and legislation. There is the need to adapt to those requirements and the political practices implemented by the actors of dominant regional and national society in order to get a chance of corresponding respect and recognition.

However, there is a lot of disappointment and frustration, as indigenous communities cannot count easily on the practical application of the formally established legal system that constitutes the fundament of the state. Many indigenous peoples live in areas where political and economic practices are applied that are in contradiction to the National Constitution and indigenous rights guaranteed by it. In many cases, practical experiences with those dominant powers present in indigenous everyday life lead them to the need to find strategies of adaptation to this reality. As locally dominant powers impose their interests even with extra-legal tools threatening the existence of indigenous groups, it is very difficult for them to file charges against those dominant groups. And since they have to experience that official institutions, constitutionally legitimized to defend indigenous rights, often do not fulfill their function and duties, the communities constantly look for allies that might have the power and potential to support them in their interest of making their living according to their principles and cosmovision.

The need to deal with the real, dominant power forces active in their surroundings that have determining influences on the possibilities to make their living conditions indigenous ways to put into practice their own cultural principles. This is a big challenge for the communities and their members, as they are confronted with many threats and contradictions, even within the rationale of the nation-state. And it is quite frequent to observe that indigenous people seem to act against their own interests, giving their formal, superficial consent to a project that might harm the future of life in their territories. In their struggle for survival, leaders representing their communities give their consent to a project as they perceive advantage by the offers of money, work, or some other short-term benefits given by the stakeholders interested in the project and its implementation that affect local indigenous communities and their territories.

Considering these real living conditions, the indigenous right already mentioned of "Free, Prior, Informed Consent," recognized by the United Nations, with difficulty might be achieved.

[15] See Speiser (2013).

> Free, Prior and Informed Consent (FPIC) is a specific right that pertains to indigenous peoples and is recognised in the United Nations Declaration on the Rights of Indigenous Peoples (UNDRIP). It allows them to give or withhold consent to a project that may affect them or their territories. Once they have given their consent, they can withdraw it at any stage. Furthermore, FPIC enables them to negotiate the conditions under which the project will be designed, implemented, monitored and evaluated. This is also embedded within the universal right to self-determination.
>
> Food and Agriculture Organization of the United Nations/FAO

In general, this right is very complex and requires a whole lot of aspects to be accomplished. In the context of this article, only one of them might be mentioned. Besides the issues related to the right of information and the negotiating process with project stakeholders, there are all the topics concerning the diversity and differences caused by the cultural peculiarity lived by the indigenous groups affected. In general, projects affecting indigenous territories are based on a rationale not necessarily shared by the indigenous groups affected. Criteria to analyze an ecological region to define principles of "development" and change to be achieved by the implementation of projects often do not match with those of indigenous peoples. The extraction of natural resources, the definition of regional development plans, and the implementation of infrastructure projects and their possible impacts may not be shared by people that consider themselves as part of a social network with beings that compose nature and environment following other concepts of time and planning.[16] While they take this principle for granted, scientists, engineers, governmental agencies, and business companies ignore those principles in their "development planning," arguing on the basis of "modern" Western scientific concepts and knowledge and interests derived from economic principles that legitimize extractivism and the constant ongoing growth and expansion of the capitalistic mode of production. Although indigenous rights to live according to their own cultural principles are guaranteed by international conventions and declarations,[17] they are ignored in many practical circumstances indigenous peoples have to live with.

[16] See the Ayoreo case mentioned in this article.

[17] For example, the ILO-Convention No. 169 already mentioned, as well as the right to freedom of religion or belief guaranteed by article 18 of the Universal Declaration of Human Rights, article 18 of the International Covenant on Civil and Political Rights and the Declaration on the Elimination of All Forms of Intolerance and of Discrimination Based on Religion or Belief. For further details on this topic, see Bielefeldt and Bremen (2024).

Conclusion

> Wake up and look what I show you!
> Get up and pay attention
> to what is happening in the world!
> Ashes and dust are the future
> of the new generations.
> A cloud of dust becomes visible.
> I want to stop it,
> thwart it,
> protect our children against it.
> But the Spirit of the Dust
> pushes me aside.
> Like a suction, he pulls the children away
> towards the City of the Whites.
> They don't hear me anymore.
> They paint their faces,
> and their true face
> becomes darker and darker.
> It was one morning
> when this happened.
>
> Vision of Ibeua Chiqueñoro (Ayore), 2011

Indigenous life-shaping in the twenty-first century is still confronted with threats, discrimination, and marginalization. Although progress has been made in formal recognition of indigenous rights, their implementation on local and national levels is restrained. Communities and their members are expected to integrate into national societies and their hegemonic cultural, social, and economic settings, leaving behind their own knowledge and practices. Even within contexts of negotiation where indigenous rights are supposed to be respected, the gap between indigenous and dominant patterns and values still is big, as presented by the examples of Ayoreo and Guaraní peoples. However, indigenous movements in different countries are growing, not only with the purpose of fighting for their rights guaranteed by law, but also trying to find ways of encounter between the very different worlds they belong to in our days. They are looking for support and alliances with those sharing common interests in searching for meeting points and moments that might give the chance to cross bridges in order to find ways of life-shaping on the basis of respect and recognition, not only of the established rights of the others, but also of their knowledge and wisdom about life on this planet, that might offer contributions to the needs of building a common future.

References

Babha, H. K. (1994), *The Location of Culture*, London: Routledge.

Bielefeldt, H. and V.v.Bremen (2024), *Indigene Spiritualität als Menschenrechtsthema – Ethnologische, juristische und politische Aspekte,* Berlin: Bundesministerium für wirtschaftliche Zusammenarbeit und Entwicklung (BMZ)

Bremen, V.v. (1997), *Acerca de la utilización del saber indígena en la cooperación de desarrollo de orientación ecológica*, in: Desarrollo Agroforestal y Comunidad Campesina, Año 6, No.27: 2-7, Salta, Argentina: Boletín del proyecto GTZ -Desarrollo agroforestal en Comunidades rurales del NOA.

Bremen, V.v. (2017a), *Indigenous Deals – Cosmologies Negotiated in Environmental and Development Projects,* in: Sociologus 67 (1): 43-58, Berlin: Duncker & Humblot.

Bremen, V.v. (2017b), *Die Rückkehr der Guarani an den Ort des Seins*, Brasilicum 246-7: 20-23, Freiburg: KoBra.

Bremen, V.v. (2018), *Orientierungsrahmen zur Kooperation mit indigenen Völkern und Gemeinschaften Lateinamerikas*, Aachen/Berlin: Misereor/Brot für die Welt.

Bremen, V.v. (2019) (compilation), Tagungsbericht *Zukunft mit indigenen Völkern Lateinamerikas - Herausforderungen für die Zusammenarbeit*, Weingarten, 28.-30. Juni 2018

Fischermann, B. (1986), *Zur Weltsicht der Ayoréode Ostboliviens*, PhD dissertation, Rheinische Friedrich-Wilhelms-Universität, Bonn

Food and Agriculture Organization of the United Nations/FAO, *Indigenous Peoples – Free, Prior and Informed Consent.* Available online: https://www.fao.org/indigenous-peoples/pillars-of-work/free--prior-and-informed-consent/en (accessed 06.11.2024).

Grünberg, F.P. (2024), *Die inspirierten Worte – Die Welt der Guarani Südamerikas*, Wien/Berlin: mandelbaum verlag

Grünberg, F.P. and G. Grünberg, (eds.) (2014), *Los Guaraní – persecución y resistencia,* Quito: Ediciones Abya-Yala.

International Labour Organization/ILO (1989), *Indigenous and Tribal Peoples Convention (No. 169).* Available online: https://www.ilo.org/dyn/normlex/en/f?p–ORMLEXPUB:12100:0::NO::P12100_ILO_CODE:C169 (accessed 06.11.2024).

IWGIA (2024), *The Indigenous World 2024*, Copenhagen: International Work Group for Indigenous Affairs.

Kraidy, M. (2005), *Hybridity, or the Cultural Logic of Globalization*, Philadelphia: Temple University Press.

Lehner, B. (2008), *¿Qué es un Territorio Tradicional Indígena?*, Asunción: Servicios Profesionales Socio-Antropológicos y Jurídicos.

Speiser, S., ed (2013), *¿Quién habla por quién? - Representatividad y Legitimidad de Organizaciones y Representantes Indígenas. Un Debate Abierto*, Quito: Ediciones Abya-Yala.

United Nations (UN) (2007), *United Nations Declaration on the Rights of Indigenous Peoples (UNDRIP).* Available online: https://www.un.org/development/desa/indigenouspeoples/wp-content/uploads/sites/19/2018/11/UNDRIP_E_web.pdf (accessed 06.11.2024).

10

Spirited Five Fingers Model: A Way of Reconnecting to the Spirit of Reciprocity in Social and Community Work

Ksenija Napan

Wonderings

I always wondered how life on Earth would be different if humans, when they started exploring their environment, did so with genuine curiosity and a willingness to learn instead of conquering and overpowering. Then I check my memory, and if I go far and deep enough, I realize that most humans, when they travel and are hosted in a new place, respect their hosts, bring gifts, and remain open-minded and accepting. It seems that we all have it in us, as hospitality is a genuine trait indigenous to all humanity. I wondered what happened to that genuine trait that somehow got buried so deep that humans started striving to climb to the top of some imaginary hierarchy, which resulted in discrimination and exploitation. Would it be possible to start re-growing a culture of appreciation and respect, or is it too late?

When I share this with my friends, they roll their eyes and start playing the air violins and singing John Lennon's song Imagine calling me a hopeless optimist, but really, why not? Do we need a massive worldwide catastrophe to wake up? Aren't senseless wars, climate change and a pandemic enough? Is it possible that massive catastrophe accompanied by awakening is already happening whilst multiple grassroots movements of unity and respect are sprouting at various places in the world? How can indigenous wisdom that managed to survive tsunamis of colonization and has been buried deep in all of us help us?

In, on, and also with

When I was about 4 years old, I asked my mother if we lived "on" the planet or "in" the planet. She replied that we lived on the planet, but that did not sound right to me. Later, I learned about the Earth's atmosphere, realizing that we are in our planetary home, and as much as we are in her, she is in us, above us and beneath us. She gives us life, and we are supposed to reciprocate, not by being patronizing custodians or exploiters of her bounty, but by simply being one with her. We are Earth, and Earth is us. I am a human indigenous to the planet Earth, and after centuries of colonization, I may have forgotten that connection, but in my DNA, in my earthly elements-filled body, the living water circulates through my veins, living air oxygenizes my organs, living fire transforms food into energy, the living ground keeps me grounded and the living ether or life force moves me through the space and time. Māori saying *"Ko au te awa, ko te awa ko au"* meaning "I am the river, the river is me" (Te Tāhuhu o te Mātauranga—Ministry of Education, 2015) reflects this way of being, which granted personhood to the Whanganui River (National Library of New Zealand, 2017), meaning that she is perceived as an

> . . . indivisible and living whole comprising the Whanganui River from the mountains to the sea, incorporating its tributaries and its physical and metaphysical elements . . .
>
> Te Awa Tupua will also be recognised as a legal person. Reflecting the view of the river as a living and integrated whole, Te Awa Tupua will have its own legal personality with all the corresponding rights, duties and liabilities of a legal person.
>
> Parliamentary Counsel Office Te Tari Tohutohu Pāramata, 2014

Before I start exploring a vision of how social and community work practice can contribute to the development of a socially just, ecologically appreciative, culturally, and spiritually respectful world, I would like to acknowledge all the lives that got lost in wars, colonization, and violence around the world and all the pain and suffering many our brothers and sisters are experiencing now. Wars and violence have never brought peace to humanity, nor solved any problems. After all the slaughter, murders and genocide, humans need to sit at the table and negotiate. I always wondered why they had not sat around the table before the slaughter. I also would like to acknowledge my privilege that I can sit in front of my computer in peace on stolen land knowing that it is not likely that the bomb will drop on my home or that a random soldier will rape me in front of my children. I do not want to pussyfoot around these topics. I am making it graphic intentionally. I need to acknowledge and appreciate that the ground on which we stand (colonized or not) is not shaking and that we have a roof over our heads and clean water to drink (otherwise you would not be reading this chapter). War and violence make people crazy and crazily resilient. It is up to us who are not actively engaged in wars to speak love, peace, reconciliation and promote non-violence around the globe. When people are in fear, they are silenced; therefore, people living in peace should promote peace, people who understand the damages of colonization and who managed to develop the ethic of restoration (Jackson, 2021) can provide guiding lights from the mess humanity got our planet into.

I have experienced war in what was then called Yugoslavia in the nineties, and I know that war is the worst thing any community can experience. And with war comes colonization. This can be either overt, covert, or both. It sneaks under the skin and makes people forget who they are and what they stand for. Out of fear or convenience, humans reject their heritage for short-term gains, try to fit in, protect children, and walk with a bowed head, look up to the rich and powerful and slowly become like them, although not quite. However, deep down, the ancestral wisdom keeps pulsing. It visits in dreams and visions, resonates in art, songs and poems, and somehow, we get attracted to ideas related to our cultural, often forgotten heritage. We get goosebumps, and we feel that there are some things "we cannot not do", and then comes the time when we cannot perform any more, and our bodies tell us that there are some incongruences in our lives and we come out, warts and all; awakened to our true calling and our ancestors smile.

Culture

The process of becoming a competent social worker can facilitate that process and enable social workers to break the cycle of colonization by acknowledging their own Indigenous and respecting the Indigeneity of the people they work with. Although the so-called Western culture, on the surface, does not have much time for it, deep down it is longing for it. While on the surface, it is focused on economic prosperity, the illusion of white supremacy and its fragility, male dominance and emasculation that follows it, outcomes, progress, justification of oppression, ticking boxes and law and order; deep down the levels of depression, alienation, recurrent conflicts, overindulgence, addictions and deadly illnesses are pointing to failings of the so-called Global North (which is an economic, not geographic term).

The term Indigenous is understood in this chapter along the lines provided by Te Ahukaramū Charles Royal, Māori researcher, musician and composer.

> To be indigenous is to be born from the land where you live, and continually born and reborn through an intimate relationship with earth, sea and sky.
>
> Royal, 2007

This definition implies the intricate connection of personhood, connection with the land and the context where one's indigeneity can flourish.

When we are immersed in a culture that is not our own, our cultural heritage comes to the fore. In my native language (Croatian), culture meant being polite, aware of the context and being well-educated. Culture was not discussed much, as it was the water in which we swam. I started learning about my cultural background only when immersed in a context different from my native one. I was delighted to realize that my mixed cultural background provided a range of experiences and learnings from my family, books, history and even more from what got silenced, not talked about, and left to be forgotten. In my cross-cultural encounters, I was able to pull from a range of themes and experiences that enabled me to engage in respectful conversations of mutual learning and discovery. Digging deeper into my family heritage (but also personality, spirituality, a sense of life purpose and calling) enabled me to broaden my horizons and put my life into perspective. Exploring my calling, sense of being and presence while dialoguing with Māori about commonalities, differences and "in-between stuff" prevented me from being swallowed by the superficiality of capitalism that surrounded me when I moved from Croatia to colonized Aotearoa, New Zealand.

The idea of a false binary of individualistic and communal cultures never sat well with me. I grew up in socialism, and the benefit of the whole was always prioritized over the benefit of the one, however, individuality, personal values, attitudes, and achievements were equally acknowledged. I have not grown up in a context of binaries, as things were relative to the context. The culture I grew up in is what is by some called a high-context culture (Hall, 1976). However, communication was often clear, direct, and explicit, and individuality was highly valued, which, according to Hall (1976), characterizes low-context cultures. It dawned on me that Yugoslavia, where I grew up, was not only non-aligned politically but also socially and psychologically. Very soon after arriving in Aotearoa, New Zealand, I realized that due to the idea of biculturalism (Māori and settlers living in relative peace and as equal partners) Hall's binary of high and low context cultures did not make much sense anymore. We as mammals have a strong need to belong, as our survival depends on it, and at the same time, we as humans need to express our individuality as it expresses our unique contribution to the world. It is not "either–or" it is more "and & also."

The Spirited Five-Finger model

In response to all these conundrums, I have developed a Spirited Five-Finger model over the years of working as a social worker and educator to create something simple and embodied that social workers can take with them wherever they go. As social workers, our human instrument is our main tool. We use our brains to think, hearts to feel, eyes to see what's going on, ears to listen to our people and communities, noses to smell and our sense of taste to try the food we get offered, as it is a privilege and a sign of hospitality. We use our legs to walk to our clients' homes or walk alongside them when needed, and our bottoms to sit and reflect. We also need to be aware of our gut feelings and know how to tell the difference between our intuitions and our projections. We need to breathe freely and fully and allow the fire in our bellies to ignite our sense of social justice. Our hands enable us to give and collaborate, to touch, network, and manifest, to shake when we meet and to shake off when we accumulate too much grief or feel helpless.

The Spirited Five-Finger model attempts to go beyond binaries and enable social practitioners to figure out who they are, what they do in their practice, and how to use their unique strengths and abilities to improve the quality of life within the communities where they work and live. It is a reciprocal model where the social practitioner does not ask any questions that they are not prepared to answer themselves. It can be used for personal, professional, political, cultural, and spiritual growth and work with individuals, families, groups, and communities. It is also an evolving model that goes beyond the artificial binaries of global south and north, west and east, communal and individual, young and old, social worker and client, rich and poor, oppressor and oppressed. Having said that, it does not negate injustices and differences exemplified in these binaries. It presupposes that everybody has strengths (Saleeby, 1996) but also that we are one with our planet, including all beings, that intelligence is universal and that its manifestation depends on biology. In other words, every living being is doing their best considering their biological and contextual circumstances. This innate intelligence (revealed with the help of a life force that flows through us all) is manifested through our willingness to be good hosts to life.

The model is simplified as five fingers of the human hand, with its core in the middle of the palm.

The Thumb stands for ancestral wisdom, heritage, what has been handed down to us through our DNA, genetics, nature, but also nurture we experienced before coming to this world, while growing up and through life. The thumb is deeply rooted in the hand, and its opposing nature allows it to grasp, grab, hold on to, bend backward, and be used as our unique signature. Lines on our palms change as we age, but our thumbprint stays the same. The thumb symbolizes biological ancestry but also all elders who made an impact on us. Foster and non-biological parents, teachers, wise members of the community, role models, imaginary friends, and fictional characters have an equal role as our biological ancestors. A social practitioner needs to know where they are coming from, what is their place to stand (Grace and Burger, 2013), what they identify with and what basic values and principles guide their life. Without meaning and purpose, social practice becomes stilted, performative, and bureaucratic.

Social work is a reciprocal profession and knowing who you are and where you are coming from is equally important for social workers as it is for the people they work with. Questions that may be useful for this exploration are:

1. Tell me a little bit about yourself, and I am happy to answer what you would like to know about me. Who are you? Or would you like me to go first?
2. Where do you (I) come from (but only as a genuinely curious question, not as a judgment sometimes passed through this question on people who may look or speak differently from dominant inhabitants)?
3. What is important to you (me)? Why?
4. What would your (my) grandmother say about you (me) if she were to introduce you (me) to a stranger?
5. What are some of the values that have been passed onto you (me) from your ancestors and the people who brought you up?
6. Who are the fictional characters that shaped your (my) life?
7. Who or what helped you (me) to become the person you are (I am) today? How?
8. What are the stories you (I) would like to pass on to children?

The thumb also stands for the element of earth; it is grounding, so it relates to exploration of roots. Roots are important for all humans, and although some languages got forgotten, the stories, songs, proverbs, incantations, and lullabies that were passed on to the younger generations lived on regardless of how dangerous this may have been.

It answers the question of **WHO**. Who am I, and who influenced me?

The Choice Point is a sensitive point located between the thumb and the index finger. Humans can be victimized by their heritage, colonization, abuse, and trauma. They may ignore it, bury it, deal with it, or utilize it as an example of how to live or how not to live their lives. It is common that if trauma does not get addressed, it may become projected onto others through biases, prejudices, unawareness of privilege or lack of empathy. Many social work students chose social work because of unresolved trauma that may have been inflicted by colonization, violence, pain or trying to make up for what their ancestors have done. The abusers and the abused are part of the same dynamic that keeps humanity stuck in a never-ending cycle of violence. We often use the term "breaking the cycle" in social work, which relates to ending a pattern of behavior that has been harming generations. Internalized trauma lodges itself in the body (van der Kolk, 2015) and if we disregard it, the body starts sending us signals that become too strong to be ignored (Mate, 2019). The choice point is incorporated in this model as it empowers and enables social workers and their clients to become agents of change and transformation.

Questions that may inspire a conversation about choices:

1. When was the last time I did something for the first time?
2. What kind of person would I like to be in this challenging situation?
3. Can I imagine it? Describe it in detail.
4. Would I like to spend the rest of my life being depressed (disempowered, angry, ignored . . . .)?
5. What is the first small step?
6. When, if not now?
7. Is this something I can change or would complaining about it for 10 minutes help?
8. What is in your (my) power? What is not?

The Index Finger represents freedom from (abuse, discrimination, neglect . . .) and freedom to (make choices and live without political, personal, professional, cultural, and spiritual oppression). It also stands for sovereignty, having a right to nationhood (although nationhood has caused a lot of trouble on the planet, it would be too far-fetched at this point to ignore it), expression of sexual or political orientation, acceptance of human rights, freedom to travel and freedom of expression. Social workers need to be aware and cognizant of what freedoms they hold in their professional lives and the ability to explore freedoms or longings for freedom with individuals, families, groups, and communities they encounter. Conversations about the expression of freedom and longings for freedom can be used as a useful pointer toward directions when working with people. Colonization processes around the world have silenced this basic human need, imposing the hegemony of a singular view. Capitalism, under the guise of democracy, enslaved the precariat constructing the term "working poor" which manifests in my country by families living in their cars with children while having full-time jobs (Moir, 2023).

In the Spirited Five-Finger model, the index finger represents the element of air, as air flows freely and is essential for a fulfilling life. Exploration of how freely air flows through our lungs and the communities we live in is a useful inquiry that can enlighten social workers about how the sense of freedom and sovereignty is embodied and manifested in their families. Whose voices are heard and whose voices are silenced? Are children playing freely, or are they constrained? How does the family treat their pets and neighbors?

It relates to the notion of having a purpose, choosing where to go and what to do.

For social practitioners, questions can become introspective, whereas for clients they can, just like air, fire them up!

1. Do I feel free to express my views within my agency?
2. Am I an agent of social change or social control?
3. Do I breathe with full lungs at work?
4. Am I a bully to my coworkers, employees, or clients?
5. Do I feel bullied or micromanaged by my boss?
6. Do I feel free to express my spiritual beliefs and political views?
7. Do I impose my spiritual or political views on others?

The index finger answers the question of **WHAT** and helps to negotiate the general direction of our work.

The Middle Finger represents *mana*. According to Te Aka, Māori dictionary (Moorfield, 2023), mana is more than power or control. It signifies spiritual power and charisma manifested through a supernatural force in a person, place or object. It is an enduring, eternal strength that gives humans the ability to be an agent of divine force, but never a source of it.

Mana is about strengths and special abilities; it is power in its most empowering sense. It relates to the ability to effect change, one's calling, unique gifts and potencies. When working with individuals, groups, families, and communities, a social practitioner would answer (and ask) the following questions:

1. What am I skilled at?
2. What do I love doing?
3. Who do I respect or look up to?
4. Who respects me?
5. Who am I without colonization? Who am I without patriarchy?
6. What characterizes me when I am at my best?
7. When do I feel empowered? How come?
8. What do I do with ease?
9. Who moves my feet? How come?
10. What talents would my grandmother notice in me?
11. How does the power within me manifest without getting angry?
12. What is the worst that can happen?

By exploring a person's fears, a social practitioner can find out how they can use their positional power to create space for being empowered and minimize the negative effects of being a mere agent of the state.

Oranga Tamariki—Ministry for Children, a statutory child protection agency in Aotearoa, New Zealand has engaged Māori scholars and practitioners to develop the mana-enhancing paradigm for practice (Wi-Kaitaia et al., 2021). This model is grounded in indigenous wisdom and has been influenced by writings and dialogues between indigenous scholars and advocates for over fifty years. It is based on principles and values that are relevant to all. As a side note, I have always wondered why Western models of social work practice have done so much damage to indigenous communities, whereas indigenous models of practice are relevant and mana-enhancing to all. Western social workers can learn from Indigenous wisdom in the same way that Indigenous cultures around the world have learnt from the scientific findings of Western science.

For practitioners, exploring their source of power, developing the ability to discern various kinds of power they and their clients hold (legitimate, positional, charismatic, coercive, referent, expert, informational, reward, imaginal, moral, or connective) and how it is shared makes that power explicit, transparent, and able to be utilized for reciprocal and transformative social work practice. Understanding this inherent power as a force that can instigate and support change is essential for engaging in social work.

The middle finger in the Spirited Five-Finger model stands for the element of water that changes and transforms with persistence and gentleness. Strong waves of change and awareness of its power need to be in balance with the life-giving and soothing nature of water that exists in four aggregate states, permeating all life on Earth.

The middle finger represents the question of **HOW**; based on the strengths, competencies and abilities a social practitioner and a client possess.

The Ring Finger stands for love in the widest possible sense. Māori call it "aroha" and ancient Greeks called it "agape"—selfless, unconditional love for ideas, others, the universe, deities, professions, communities, and the work we do to bring forth the world. In professional circles, the term "unconditional positive regard" refers to that concept and enables social workers to empathically connect to their clients and be of service regardless of which circumstances brought them together.

It also represents fire as a transformative element that can ignite and empower, but also completely transform or destroy. Humans seek connection and need to belong at least in some aspects of their lives. As mentioned earlier, the dichotomous idea of individualistic and communal cultures promoted by Western scholars in the seventies is not helpful when we realize that it speaks more about the uneven distribution of material wealth where "individualistic" cultures can afford to pay for goods and services whereas "communal" cultures rely on their communities for survival. Humans are both individual and communal. The level of pain, alienation and suffering in highly industrialized (or digitalized) countries confirms that no person is an island, and everybody possesses individual uniqueness and idiosyncratic expression that contributes to their community.

Questions that relate to our sense of love and belonging, and can be used for reflection and personal and professional growth of social practitioners as well as in work for clients, are:

1. What is the fire in my belly?
2. What motivates me?
3. Who and what do I care about?
4. What kind of person would I like to be in this relationship?
5. Who is there for me?
6. Where do I belong?

7. What is one group I belong to but would not share at a job interview?
8. For whom or for what would I walk 10 miles in the snow?
9. What makes my heart sing?

In the Spirited Five-Finger model, the ring finger corresponds to clear boundaries, reciprocity, and balance in giving and taking. It is the heart of collaboration and the development of trust and respect in a professional relationship. It also stands for the questions **FOR WHOM** and for doing social work with love and appreciation.

The Little Finger is about joy and fun. Most Western models ignore fun as a basic human need, while most indigenous cultures have ceremonies and rituals as foundations of their uniqueness essential for rites of passage and acknowledgment of significant memories or hopes for the future. These ceremonies serve many more purposes but joy with fellow human beings is one of them. William Glasser (1999) is (to my knowledge) the only Western scholar who not only incorporated fun as a basic human need but has not presented basic needs hierarchically and claimed that they are all equally important. He challenged Maslow's hierarchy of needs by saying that if the need for survival was paramount and more important than the need for love and belonging, freedom or power, there would be no suicide in the world. According to Glasser, the need that is least satisfied is the one that appears to be most important for the person in their specific situation and needs to be addressed as for optimal functioning and a fulfilling life; all needs need to be in balance.

The ability to connect is essential for joy. Social workers need to be widely educated and knowledgeable about a range of fields: arts, sports, music, fashion trends, movies, theatre, alternative ways of being, slang, and cultures, as this will facilitate connecting with clients and their communities in a less formal and more light-hearted way.

The little finger is about humor, laughter, energy, and light-heartedness. It is also about sharing and playfulness.

Questions that can facilitate light-heartedness in social work are:

1. What do I do for fun?
2. If money were not an issue, what would I do?
3. What makes my heart sing?
4. Can I recall a situation when I was joyful without an apparent reason?
5. How do I play? Who are my playmates?
6. When was the last time I played?
7. Can I remember a situation when I could not stop laughing?
8. When was the last time I forgot about time?
9. What makes my community thrive?
10. Can I recall a situation when I, my family or community were in a state of flow?

The little finger relates to the importance of the joy generated within a community. Having fun alone is not as much fun as having it with others. It answers the question of **WITH WHOM** and for being light-hearted but not superficial.

The middle of the palm, or more precisely, a ticklish spot right in the centre of the palm, represents the essence of our being. It also answers the question **WHY**. It is the seat of the soul, tender and well-protected and all other aspects emanate from it. The hand can be open, and this is when it is giving or receiving, or it can close in a fist, when it is tensed, angry, scared, or protective. All basic needs and elements shared above are generated from the middle of the palm and are interrelated. None of the fingers stand alone nor can instigate change, we need them balanced and in tune with one another.

When working with clients, the metaphor of the spirited five fingers can be used to explore the basic needs of individuals, groups, families, and communities. When appropriate, a hand can be gently massaged, embodying needs and wants and enabling messages from the body to come through. When working with couples or groups, they can massage each other's hands and express their appreciation and challenges with each other related to the concepts presented. When working with communities, various perspectives can be explored and ideas for moving forward negotiated.

If we explore Rose Pere's Te Wheke model (Pere, 1991) as enhanced by Catherine Love (Love, 2004), the notion of individual selfhood has been expanded with the notion of wairua (loosely translated as soul, spiritual aspect of being, aspect of a person that survives physical death). This is particularly relevant for mental health, healing of physical ailments and relationships as the individual cannot be separated from their context and environment. The impact of colonization, oppression and discrimination on the human spirit was historically interpreted as mental illness or antisocial behavior leading to segregation and medicalization of socially constructed pathologies. Furthermore, indigenous people around the world were forced to verbalize their thoughts and feelings, neglecting the body and rituals necessary to process grief and pain caused by the appropriation of land and cultural appropriation is still happening around the world. The dominance of the rational mind in Western culture is only now giving way to more holistic approaches that often draw on indigenous wisdom and holistic practices. Western obsession with outcomes, efficiency, self-esteem and white supremacy, domineering the twentieth century is now giving way to an appreciation of ancient wisdom that managed to survive despite the brutal annihilation of anything not conceived in a prefrontal cortex or proven with positivist science. This is not to say that Western models have not improved the quality of life on the planet, it is just that in their arrogance and righteousness a large part of human nature got neglected and the planet herself became a resource to be exploited forgetting that we are living in and with her and when she is unwell, we are too. Dispirited social work is not worth the effort as its bureaucratic nature and obsession with filling forms and following prescribed formulae makes it incapable of dealing with the complexities of human contexts, cultures, and situations.

The Te Wheke model (Pere, 1991) explores the importance of belonging to a group without neglecting being an individual, the importance of past, present, and future in every here-and-now moment and the importance of spirituality and beliefs related to any social problem. Neglecting the spiritual nature and the meaning and purpose of every presenting issue will augment a disconnection between social practitioners and the people they are attempting to work with. As opposed to missionary work perpetuated through charitable social work throughout history, addressing spirituality in social work is not a religious endeavor. It is about going beyond and above the obvious and, at times using metaphors, storytelling, music, art, or sport to transcend the mundane aspect of the problem and resolve it through transcendence that corresponds to the way clients understand and perceive reality.

Wrapping it up

The Spirited Five-Finger model tends to integrate the personal indigenous of every human being by exploring the roots, purpose and meaning of every social work intervention, the unique strengths and abilities of a social worker and the people or communities they work with. It emphasizes that social work is a labor of love that needs to be done lightheartedly, but not superficially. It is a reciprocal model that integrates non-dogmatic and non-authoritarian spirituality, where social workers discover who they are, what they need to do, and how to gain the competence necessary to be useful, to do it with love, reverence, and joy.

References

Glasser, William (1999), *Choice Theory: A New Psychology of Personal Freedom.* HarperPerennial.

Hall, Edward T. (1976), *Beyond Culture.* New York: Anchor Books.

Jackson, Moana (2021), "Decolonisation and the stories in the land," *e-tangata.* May 9. https://e-tangata.co.nz/comment-and-analysis/moana-jackson-decolonisation-andthe-stories-in-the-land/.

Love, Catherine (2004), "Extensions on Te Wheke," *Working papers*, October 6: 1–85.

Mate, Gabor (2019), *When the Body Says No: The Cost of Hidden Stress*, Sydney, Australia: Scribe Publications.

Moir, Jo (2023), *Newsroom.* 09 14. Accessed January 3, 2024. https://newsroom.co.nz/2023/09/14/workers-living-in-cars-the-new-normal-in-queenstown,.

Moorfield, John C. (2023), "Te Aka Māori Dictionary," *Te Kōhure Textbook (Ed. 2): 238–240; Te Kōhure Video Tapes (Ed. 1): 6;*. Auckland, 12 27.

National Library of New Zealand (2017), *https://natlib.govt.nz/he-tohu.* 3 3. Accessed November 11, 2023. https://natlib.govt.nz/he-tohu/learning/social-inquiry-resources/cultural-interaction/cultural-interaction-supporting-activities-and-resources/change-maker-whanganui-river#:~:text=When%20the%20Bill%20was%20passed,and%20its%20iwi%20and%20hap%C5%AB.

Parliamentary Counsel Office Te Tari Tohutohu Pāramata (2014), *Te Awa Tupua (Whanganui River Claims Settlement Bill.* 85. Accessed December 27, 2023. www.legislation.govt.nz/ bill/government/2016/0129/6.0/whole.html#:~:text=Ruruku%20Whakatupua%20 provides%20for%20the,omissions%20before%2021%20September%201992.

Pere, Rose Rangimarie (1991), *Te wheke: A celebration of infinite wisdom.* Gisborne: Ao Ako Global Learning.

Royal, Te Ahukaramū Charles (2007), "*Papatūānuku—the land,*" *Te Ara—the Encyclopedia of New Zealand.* September 24. Accessed December 23, 2023. www.TeAra.govt.nz/en/papatuanuku-the-land/.

Saleeby, Dennis (1996), "The strengths perspective in social work practice: extensions and cautions," *Social Work*, 41(3): 296–305.

Te Tāhuhu o te Mātauranga—Ministry of Education (2015), *Māori History in the New Zealand Curriculum.* 2 12. Accessed December 27, 2023. https://maorihistory.tki.org.nz/en/videos/ko-au-te-awa-ko-te-awa-ko-au/.

van der Kolk, Bessel (2015), *The Body Keeps the Score: Brain, Mind, and Body in the Healing of Trauma*, New York: Penguin Books.

Wi-Kaitaia, M., L. Ruwhiu, and M. Eruera (2021), "Development-of-the-Mana-enhancingparadigm-for-practice," *Oranga Tamariki Ministry for Children.* Accessed November 11, 2023. www.orangatamariki.govt.nz.

11

Mobilities in the Global South and New Circuits of Affects

Maria do Carmo dos Santos Gonçalves

Introduction

The migrant is this Other who resists the classifications based on nationalist illusions (Benhabib, 2004). Categorizations of migration movements and subjects that move internally or internationally between countries are formulations widely applied by governments, international organizations, organized civil society and academia, operating processes of reification of the subject in mobility (Robertson, 2019). Would it be possible to understand the phenomenon of modern-day migrations from novel approaches capable of escaping from the fragmentation of the subjects in mobility, given by the search for their framing in predicates, such as migrants, immigrants, emigrants, refugees, displaced people, etc.? The several ways of defining subjects in mobility integrate systems that aim to control, through the classification of movement, people who cross geographical borders. They operate the control of their bodies, as well as the ways in which these subjects in mobility describe themselves and narrate their migratory experiences. Current migration policies, mostly centered on border control, reproduce the perspectives of methodological nationalism (Beck, 2018) applied to the processes of international human mobility.[1] Such policies tend to depersonalize and deprive the subjects in mobility of their attributes, classifying them into categories that may be as artificial and insufficient as the ways of naming such subjects from their presumed national identities. We continually come across various migratory typologies, such as multiple ways of adjectivizing migrations as economic, forced, illegal, irregular, returning, international, transnational, qualified, commuters, feminine, south-south, diasporic, etc. This range of ways of naming human mobilities expands when we include forced mobilities, classified as refuge, environmental refuge, internal displacement, and others.

1 Although the notion of methodological nationalism is quite useful to us to point out the participation of this perspective in the construction of classificatory categories of mobility of people, it is valid to note that the very notion of nationalism is also subject to a critical review. For a critical approach to nationalism and ethnicity, see Avtar Brah (2011).

To a considerable extent, these categorizations bear a deep relationship with colonial history itself, especially when one looks at the rapidity with which migrants are differentiated by racialized criteria (Gutiérrez Rodríguez, 2018). Race continues to be a colonial construct that is constantly mobilized to categorize mobile subjects, to describe their national identities, and to circumscribe these subjects within the established power relations in which they are placed in a position of subalternity. The ways in which people on the move are categorized as migrants, immigrants, refugees, displaced persons, and environmental refugees, among other ways of naming them, are social constructions permeated by discourses of subjection and with practical implications for their lives and destiny. Above all, they are elaborations strongly crossed by colonialism (Quijano, 2005, 2012), which defined cartographies, nationalisms, and identities based on the belief in race as a marker of biological differences (Mbembe, 2013) that tried to naturalize racial superiority to maintain relations of domination, exploitation, and division of labor (Mudimbe, 2019).

If the history of colonialism is the history of the construction of nations and the bases on which power relations would develop, then it is possible to say that the categorizations of mobility and, above all, their power to control mobility narratives operate a system with universalizing pretensions for classifying mobility phenomena and the subjects who move, strengthening colonial power and keeping the rules of domination unchanged.

From a discursive point of view, the categories manage processes of deconstruction of the subjects in mobility, fragmenting their narratives and appropriating those aspects of the migratory experience that can be instrumentalized to reify the condition of the migrant as a foreigner and safeguard the unity of the nation-state. Experiences such as the pain, mourning, suffering, fear, and anguish of people forced to migrate for economic, political, violence, human rights violations, natural disasters, wars, etc., are experiences that governments, international organizations, civil society organizations, and, to a large extent, academia itself discursively appropriate. This appropriation makes plausible the mobilizations and efforts of reception, humanitarian aid, and regularization of mobility. But, at the same time, it produces silenced subjects. Very rarely is the system of nationalism itself, which creates mechanisms for the exclusion and expulsion of people considered to be undesirable non-nationals,[2] and generates the so-called migratory crises,[3] put under suspicion.

It can be said that the definitions of mobilities and the subjects involved in them are still based on the same foundations that described the distinction between colonizers and colonized. Quijano (2005: 117) identified two historical processes that, according to him, converged and established the fundamental axes of the colonial power pattern.

[2] Nor is it common to reflect on how such crises are related to the colonial power that carried out, after the independence of colonized countries, a certain kind of erasure of the bonds that were established during centuries of relations with countries that today are considered expellers of people and that, in their great majority, are formed by former European colonies.

[3] Moreira et al. (2021) present an interesting critique of the notion of migration crises in which they try to shift the cause of the crises from the subjects in mobility to the contexts of social, economic, etc. crises that generate forced mobility.

On the one hand, there is the codification of the differences between conquerors and conquered in the idea of race, that is, a supposedly distinct biological structure that placed some in a natural situation of inferiority in relation to others. This idea was taken on by the conquerors as the main constitutive, foundational element of the relations of domination that conquest demanded. Consequently, the population of America, and later the world, was categorized in this new pattern of power. On the other hand, the articulation of all the historical forms of control of labor, its resources, and its products, around capital and the world market.

If race and the control of labor worked as articulating axes of colonial power, in the context of migration management today it is the control over mobility and national borders, to a large extent inherited from that system, that places the subjects in mobility under the same conditions of domination in which the "natural" subjects of the colonies found themselves in the colonial system. This management is only possible if, as in the past, the colonizers can themselves define and describe these subjects, and classify them. Moreover, if these categorizations become a kind of universal classification and are applicable to any subject who comes to move between one country and another, this process of appropriation and possession of fragments of the migration experiences of individuals and groups makes the bases that underpin local and national policies of reception, support, and socio-economic integration hegemonic. However, it also legitimizes expulsion actions in the destination countries of millions of people, especially in the global north. If the act of naming is fundamental to maintaining the power of domination and hegemony, then it is possible to understand the need of those who manage categories to control the narratives that people on the move develop about their own motivations and experiences of migrating.

The present work seeks to contribute to the dialogue around new narratives about current forced migrations, assuming the subaltern perspective of migrant subjects and the circuits of political affects (Safatle, 2016) that they can generate from their mobilities. To do this, we used qualitative research carried out with migrant people and groups in different periods (2016 and 2021) and contexts in Brazil (Rio Grande do Sul and Roraima). North-centered perspectives on migration tend to translate the mobilities of vulnerable populations in the global south (due to social, economic, and cultural exclusion, situations of violence, or climate change) and who have experienced helplessness in their errancy processes as a problem to be managed based on fear. This results in exclusionary and security-ist state approaches based on alleged national identities that have largely been forged by colonialism. This perspective also imprisons and instrumentalizes the knowledge produced about the mobility's issue so that the epistemologies that surround it end up reifying exclusionary practices. In this sense, we propose to explore the potentialities of marginal experiences of subjects on the move in the global south. Based on the migratory narratives of people on the move in territories considered peripheral to human experience, we mapped the political circuits of affections (Safatle, 2016). Such affections, such as those linked to solidarity, help us to address critically the categorical constructions universalized by international law or humanitarian agencies that, by categorizing and quantifying people on the move, disconnect them from their main attribute, that is, from their *humanitas* and their recognition as such.

Political affections and the constitution of subjects in mobility

Considering the power that categorizations exert over the processes of human mobility, it can be observed that for people on the move to be able to migrate in a regular way to the countries they have chosen to live in (or to which they have been compulsorily directed), they must be able to present material proof of their condition of helplessness. At the same time, they must be able to modulate the narratives of their expectations and migratory experiences to the discursive constructions that sustain migratory policies of reception in these countries, thus enabling their entry and permanence in territories of which they are considered unnatural.[4] It is also mandatory that the evidence be aligned with a life story in which there is an explicit testimony of this violence. This logic of legalizing mobility will also impose rules on labor migrants. In addition to proof of their labor qualifications, they will be required to provide certificates from their nation-state proving their suitability. From a subjective point of view, they must demonstrate that they are willing to contribute to the economic and social development of the country that has welcomed them as a workforce. Often, even if they can prove their suitability and ability to work, they will fill low-skilled jobs and suffer the same stigmatization as other migrants. If we examined migration legislation in different countries, we would see how extremely difficult it can be to find frameworks for regular migration. In this context, it is not difficult to understand why a large part of the mobility will take place on the margins of the legal migration routes that states have constructed. There is currently a great deal of mobility of people that takes place through routes labelled irregular. For these individuals, there are no records of the materialities dealt with bureaucratically by the state. Nor are their narratives considered legitimate. They don't exist for many nation-states. We can say that these subjects move beyond the prescriptions described by national legislation and policies.

Contemporary mobilities are not exclusively limited to laws. Whether regular or irregular, people on the move are also immersed in circuits of political affections. The circuit of affections concerns the way in which social emotions, desires, and feelings of belonging and identity (political-national, ethnic-racial, gender, and sexuality) are articulated in the management and co-constitution of the body politic, not only in the exercise of power and social control of the State, but also in its related circulation and distribution of goods and wealth (Oliveira: 74).

Regarding the relationship between political affections and migratory dynamics, it can be said that fear is constituted as the affection recognized and legitimized in the production of acts of migration. For Safatle (2016), fear as a political affect tends to construct the image of society as a body that tends to be paranoid, trapped in the securitarian logic that must immunize itself against all violence that endangers the unitary principle of social life. Immunity that requires the functional perpetuation of a

[4] It is worth remembering that the definition of a person as a native of a given national territory is strongly marked by the prerogatives of classification, Jus *Sanguinis and Jus Solis*, that are applied to define who can be considered a member of a given national state.

potential state of absolute insecurity comes not only from external risk but also from the immanent violence of the relationship between individuals (2016: 20).

That is, within a securitarian logic of the state, fear not only mobilizes the subject who migrates because of his risk condition and the precarious situation of his existence (Bahl, 2022), but it is also presented as a sine qua non condition for the subject who wants or needs to migrate to legitimize his/her existence in a territory in which he/she is described as unnatural. Fear makes international mobility acceptable, defining who can migrate, that is, creating the subject who is in a situation of helplessness, who fears violence, and who seeks support to continue existing. The fear not only of violence but also of the impossibility of immediate or long-term survival for social, economic, and environmental reasons. However, paradoxically, fear continues to act as the predominant political affection when this same subject is in his destiny. Fear circulates in the host society and is expressed through structural racism, the criminalization of mobility, detention and expulsion of people and groups considered "undesirable," and xenophobia and violent discrimination. Fear also mobilizes local political bodies that are created out of opposition to everything that seems to threaten a state of nature based on the illusion of a pure national identity. They are against migrants, their culture, language, and presence. They defend an excluding nationalism based on a colonial notion of national and racial superiority.

Thus, fear, as a political affection, is highlighted, and it imposes the horizon from which people in mobility can narrate their motivations and experiences of migrating. In this context, the narratives of people in mobility to be accepted must be plausible and coherent with the discursive elaborations that define each of the migratory categories in which they must frame their experiences. In this sense, for example, it is expected that a national of Afghanistan in Brazil or anywhere in the world can resort to requesting asylum if their individual narrative is consistent with the escape from persecution of the Taliban state and the well-founded fear/terror of death.

Such dispossession of the subjects from their own narratives and their framing in broad and generalized categories, in which authoritative speeches about mobility experiences prevail, prioritize fear as an essential political affection while limiting and mischaracterizing what is most fundamental in the subject—that is, his human condition that precedes his nationality. But, as Freire (1997: 77) affirms, "existence, because it is human, cannot be mute, silent, nor can it be nourished by false words, (. . .) with which men transform the world." There is a power of latent transformation in the act of migrating, of moving in the world, and above all, of speaking. Speaking is an act of existence that unveils the political affections that generate political bodies capable of transformation (Safatle, 2016). To decolonize narratives is to allow subjects in mobility to be included in the problematization of the world, to be able to "tell the world" from their experiences. Freire (1997) states that to exist humanly is to pronounce the world, to modify it. The pronounced world, in turn, becomes problematized to the pronouncing subjects, demanding from them a new pronouncement. It is not in silence that men make themselves, but in word, in work, and in action-reflection. But if to say the true word, which is work, which is praxis, is to transform the world, to say the word is not the privilege of some men, but the right of all men. Precisely for this reason, no one can say the true word alone or say it to others, in an act of prescription, with which he steals the word from others (Freire, 1996: 77, translated by me).

To highlight this perspective, we present two biographical narratives[5] that can contribute to the exercise of decolonization of narratives, giving space for a better understanding of how subjects and groups of migrants establish circuits of affection during their experiences of mobility and that are expressions of transformation.

Moises: a *vocero* (spokesperson) of the Venezuelan community in Brazil

Moises,[6] 55, is an asylum seeker who in December 2020 fled Venezuela for Brazil after his home was set on fire as a political reprisal and he suffered direct death threats.[7] Born in Sucre state, he is the youngest of nine siblings. His family had an excellent social status until his father separated and economically and emotionally abandoned his family. At this time, Moises was 7 years old and experienced his mother's struggle to ensure the survival of the family. In this context, they moved to Guiria, capital of the Sucre state. The mother went on to work two jobs, and they faced many difficulties. As their older siblings grew up and gained autonomy, constituting their families, they began to contribute with their mother in the sustainability of the family nucleus. As a teenager, Moises found it exceedingly difficult to attend school regularly. He reports that he was "rebellious" and liked, for example, to venture to swim in the sea. His mother then sent him to another city in the care of an older brother. At the age of 17, he became economically emancipated from his family, completed his studies, and took a job in a large steel mill. During this time, he took a technical course and began working as an occupational safety technician, a position in which he would work for many years. Before long, he became involved in labor union activity. Among the many activities as a trade unionist, he received the invitation of his community to exercise the function of *vocero* (spokesperson)[8] of this community, also assuming a function in the local council. It was in this function of community representation that Moises expanded his field of political activity and began a close relationship with the Catholic

[5] The biographies presented here are inserted in different research contexts. The story of Moises was collected in exploratory ethnographic research conducted by the author in 2021, on the border of Brazil and Venezuela. The story of Samba is part of the author's doctoral research in the period 2016–2020.

[6] The names of the interviewees are masked. We have altered all information that would allow their identification.

[7] According to data from the Federal Subcommittee on Reception, Identification and Screening of Immigrants, more than 717,000 Venezuelans entered Brazilian territory between 2017 and March 2022, with more than 330,000 people having remained in the country. As of March 2022, about 200,000 temporary or indefinite residency records of this population were active.

[8] *Vocero*, according to the *Ley Orgánica de los Consejos Comunales, 26 November 2009 of Venezuela* is the person elected through the process of popular election for the purpose of coordinating the functioning of the communal council and the instrumentation of the decisions of the *Asamblea de Ciudadanos y Ciudadanas*.

Church through the Gumilla Leadership Training Center,[9] linked to the Jesuit priests. Assuming this perspective of community action, Moises fostered in his community the opening of a unit of the Faith and Joy School.[10] At the age of 35, Moises became the father of a girl, the fruit of a relationship that did not last. According to him, it was difficult to reconcile family life with his social activism. After the death of President Hugo Chavez (2013) and the rise of a new government under the aegis of Nicolás Maduro, Moises began to have political conflicts related to his union and community performance. In December 2020, his home was set on fire by strangers, and two weeks after the incident, he was accosted on the street, assaulted, and threatened with death. The warning was clear: "Go away and don't come back." Moises then decided to leave for Brazil for fear of being murdered or that his family would suffer some kind of violence.

At the root of Moises' flight from Venezuela is the "well-founded fear of death" that justified his request for asylum and guaranteed him to remain in Brazil. Fear was mobilized as political affection by actors linked to the state who, by threatening the life of Moises in Venezuela, aimed to annul his action of contestation and the possible transformations of reality that he could do, putting at risk the maintenance of the power of the state. As in the case of Moises, it is the circulation of fear, used as a political affection, that currently causes the flight of millions of people inside or outside their national borders and directs them in the search for protection in other territories. In Moises' narrative, it is clear that it was fear that was the main affection mobilized in his flight from Venezuela. But it is also evident that throughout his life course, acting as a trade unionist and community leader, as well as later assuming a 'vocero' position for the Venezuelan community in Brazil, Moises set in motion other types of political affections that favored processes of change in his condition of personal and Venezuelan community vulnerabilities. In Venezuela, their commitment to expanding spaces for social participation, the political training of leaders and young people through the perspective of the Faith and Joy Schools, and the trade union struggle were activities in which political affections were mobilized, based on traditions of community organization that formed circuits of affections, such as the desire for justice, the search for equality and historical reparation, empathy, generosity, and solidarity. The affections that shaped democratic spaces of social and political participation, such as community

[9] Centro Gumilla is a Centro de Investigación y Acción Social (CIAS, or Center for Research and Social Action) of the Society of Jesus in Venezuela. It is part of the Catholic Church's effort to make Christian life a way of being in solidarity with the anguish and hopes of humanity, of living the cause of the poor of the earth, and of the longing for justice and peace founded on love. It is fundamentally dedicated to contributing to the realization of the justice that flows from faith. About the Center, see https://gumilla.org.

[10] *Fe y Alegría* is a movement of popular education and social promotion, born on March 5, 1955, in 23 de Enero de Caracas, Venezuela. Today it is present in 22 countries in Latin America, Africa, and Europe: Argentina, Bolivia, Brazil, Colombia, Costa Rica, Chad, Chile, Ecuador, El Salvador, Spain, Italy, Guatemala, Haiti, Honduras, Madagascar, Nicaragua, Panama, Paraguay, Peru, the Dominican Republic, Uruguay, and Venezuela. There, it serves more than 1,500,000 children, young people, and adults from urban, rural, and indigenous sectors with a wide variety of educational, community, human, and work training programs. See https://www.feyalegria.org/venezuela/nuestra-historia.

councils, also formed political bodies with the capacity to resist the authoritarian onslaughts of the government itself.

Among the Venezuelan migrant groups and communities in Brazil, we observed the circulation of affections with strong potential for transformation, such as solidarity, empathy, joy, and resilience. These affections are, above all, mobilized by groups in mobility, constituting agendas of struggle for improvements in living conditions in Brazil and for social transformations in their countries of origin. Moises' narrative of his experience in Brazil did not focus on his status as a refugee, but highlighted his career as a *vocero de la comunidad* (community spokesperson), a community leader who took on the task of mobilizing and representing vulnerable people and groups. In Brazil, he became the *vocero* of a community that, like him, experienced migration in the context of flight from precarious living conditions or political persecution. In this sense, Moises commits much of his time to the task of mobilizing, representing, and giving voice to people and their experiences of reinvention in Brazil. In addition to staying connected with the developments of Venezuelan politics, which, from his point of view, he defines as a brutal dictatorship that is at the origin of the country's mass migrations, Moises is involved with his community, in which he has become popular and known. People call him to ask for advice, guidance on aspects of practical life in Brazil, but also to hear about how they can be organized and have better guarantees of socioeconomic integration in the country. He works by publicizing courses and job vacancies and, at the same time, creating agendas with local politicians to present the demands of the migrant community in which he is inserted. He has developed a good relationship with Brazilian Catholic leaders and actively participates in the activities of the ecclesial community. Although categorized as a typical refugee case in Brazil, his narrative, activities, performance, and insertion in the context in which he resides go far beyond the condition of fragility, vulnerability, and impotence that discursively circumscribe the condition of a refugee. The classification as a refugee, understood as a condition caused by flight movement by fear of death, does not define the totality of Moises' mobility narrative. Nor does it explain his stay in Brazil. From his point of view, he continues to act as *a vocero and* mobilize together with his community of people in mobility circuits of affections that generate processes of change in the situation of social, political, and economic vulnerability in which they find themselves in Brazil.

Solidarity as a political affection (Bizon, 2020) is observed in the narrative of Moises when he describes the experiences lived in his community of migrants and with Brazilian people, where spontaneous acts of sharing food occur, the provisional welcome between strangers in a shared rented space, the indications of job vacancies, the accompaniment to the sick of the community, the empathetic listening of people who suffer loneliness and live with depression, the lending of money, and the sending of remittances to relatives of unknown people. Moises also describes the circulating affection of the joy that the community mobilizes, experienced in informal meetings, in listening to traditional songs of the country, in the celebration of weddings, births, etc., in the unexpected invitation to dine at the house of an acquaintance and share a good plate of soup, in having one's own house or obtaining a school certificate. These narratives of the affections of everyday life, in which varied affections are in circulation, contrast with the universalizing discourses centered on fear and move circuits of affections that produce new political bodies. The empathy that emerges among migrants through spaces of listening to their life stories, struggles, difficulties and dreams also creates the possibility of seeking social transformations within a shared horizon of the future. It makes them potent subjects, strengthening affections perhaps much more powerful than fear.

Samba: migrations and religious affections

Samba is Senegalese. He came to Brazil in 2012 with the support of the Sufi Islamic brotherhood, of which he is a member. His desire was the search for better living conditions. Samba lost his mother when he was just 3 years old. The orphanage put him in a very vulnerable position, with the result that much of his childhood was spent in the care of family members. As a teenager, he became involved in the family of a religious leader who provided him with work and, later, suggested that he could migrate to Brazil. After going through many difficulties on arrival in the country, including learning Portuguese, immigration regularization, and the search for work and housing, Samba established a small business of typical Senegalese meals whose main consumer audience was the Senegalese residents in the city where he lived in the interior of the state of Rio Grande do Sul. Despite his migratory status that framed him, first as an asylum seeker and later as a labor migrant, Samba presented himself as someone who liked to help, his main motivation being religion. His religious belonging was a constant factor in his life course. Although labor insertion was an important objective within his migratory project to Brazil, he often prioritized the religious mission, to the point of adopting an informal work regime to have more time to devote to his missionary activity. The informal work allowed Samba greater autonomy to be able to carry out the "aids" that he developed within his Muride religious community. For some "aids" or "easy-to-do," as he described it, Samba charged tiny amounts to cover his transport and meal expenses. However, the aids he provided could also be "of the word," as he described them. These included giving advice, saying prayers, or making "briefs" of protection. Samba also helped in the conduct of the collective activities of the religious group, such as assisting in the processes of reproduction of the rituals of his group in the festival of the *Gran Magal de Touba*, the main event of the Muride religious group. Muridia is also a religion of witness, a missionary religion through the propagation of an ethic that encompasses essential themes such as work, solidarity, equality, valuing the family, and discipline. The person would need to keep the "heart working well," according to Samba, so as not to do wrong things. He needed to preserve the bond with the sacred through prayer, participation in the community, work, and responsibility to the family.

Cheikh Bamba (1853–1927, Senegal), founder of the religious confraternity of which Samba is a devotee, also represented the center of his affection and religious convictions. Samba's narratives about his own personal trajectory and of the collective of Senegalese Muride migrants in Brazil were resignified through the relationship they established between their experiences as migrants and the life trajectory of the founder of the confraternity. The life story of the founder, the religious love circulating in the community, provided meaning and perspective to the migratory movement experienced by Samba and his brothers of Confraternity. The history of Cheikh Bamba's religious militancy was resignified by Samba in his narrative from the perspective of the struggle against colonialism. This struggle, according to him, was exercised peacefully within the social space of religious belief. According to Samba, the Europeans enslaved and demanded that Cheikh Bamba renounce his faith. He then exercised a form of peaceful resistance, grounded in faith and the belief that God had elected him to intervene directly in the history of Senegal and Africa more broadly. According to Samba's narrative, it was due to the religious resistance of Cheikh Bamba that Senegal came to its independence from France. The message and legacy of Bamba, according to his description, were carriers of universality and were concretized in the performance of the Muride brotherhood currently dispersed throughout the world. For Samba, narrating Bamba's life was an act of resistance through speech.

The way Samba tells his story is not limited to mobility in search of work or better living conditions but brings with it a circuit of affections linked to religious belonging that generates processes of social and cultural transformation from which he and his community of believers establish links with the local community and claim rights and spaces for social participation. The circuit of religious affection moved solidarity. Internally in the Muride community, it was expressed in the multiple forms of mutual support given by the reception and housing to migrants recently arrived in the city, loans, apportionment of passage values in the case of Senegalese who had to return to the country for health reasons, and campaigns to raise money to pay the costs of expatriation of the bodies of Senegalese who had died in Brazil. Religious affection also mobilized the Senegalese community in the fight against racism and for justice in cases of violence suffered by community members in which xenophobia and non-acceptance of the group's cultural and religious expressions were evident.

Final considerations

It is common to take the dimension of affections as something that concerns only individuals in isolation. However, we are social beings, and in life in society we are always affected. Affections circulate and are always politically appropriate to generate social effects, forming political bodies with potential for transformation.

Fear, as a political affection on which the legitimate power of the state to govern with sovereignty is based, by opening legal possibilities of reception of "others," foreigners, and migrants, does so by its own means and rules. To obtain regularization, the legal stay in a country of which one is not a natural, the foreigner must be dispossessed of his attributes and, above all, of his narrative, of the way he understands himself as a subject constituted in the world. This process of depersonalization understands that the subject fits into some of the pre-established categories of mobility and incorporates, through language, ways of defining and describing his condition of otherness. In this context, categorizations are presented as power relations permeated by colonial prerogatives. Within this dynamic, the subject in mobility must assume narratives that place limits on their experience and impose not only the reasons for migration but also where and when subjects can narrate their life stories. Thus, their experiences must necessarily be connected to the established migratory categories.

Considering the relationship between colonial power and its intersection with the circuits of political affections involved in the processes of human mobility, the task of decolonizing our way of narrating migrant lives is urgent. The universally categorized ways in which one seeks to enclose a plurality of human experiences of displacement serve much more to maintain the prerogatives of sovereignty of the nation-state than to effectively welcome people who leave their countries in search of a better life. And perhaps the most striking proof of this can be found in the visible exhaustion of models of global migratory governance based on fear of the mobility of the Other. The subalternity of the existence (Spivak, 2023) of subjects in mobility and their potential for transformation are presented as narratives, as those stories that are not told, or that are simply neglected.

Through their own narratives, people in mobility bring to light circuits of political affections that overcome the fear of the other, overcome protectionist barriers, and put into circulation other affections, such as solidarity, empathy, resilience, joy, etc. By presenting biographies of subjects in mobility through the narratives of these subjects, we seek to bring to light people who cannot be categorized in a universalized way. In this sense, understanding the circuits of affections that are mobilized by migrant people through their narratives is a way of also decolonizing political affections and seeking new horizons for an authentic hospitality, overcoming fear as the founding affection of mobility policies.

As it is possible to ascertain, in the narratives presented, the categories in which it is sought to classify and situate the experiences of subjects in mobility may be insufficient and function as a straitjacket that limits their possibilities of realization. There is a great paradox about how countries considered to be welcoming to people fleeing continue to mobilize fear as a political effect, to contain and repress mobility. The precariousness of the reception expressed by restrictive policies of entry and regular stay, by xenophobia, and by the meager initiatives of labor integration are also subtle and indirect forms of exercise of power through fear. In this sense, the challenge of decolonizing political affections that shape and design migration policies based on the fear of the other, of the foreigner, represents the challenge of giving greater power to the affections mobilized by migrants, bringing them into the projects of migratory governance that can no longer be exclusively thought of from the logic of national sovereignties.

Finally, the intention here was not to overestimate the power of migratory categorizations in terms of the possibilities they present as a response to migratory realities. The questions presented are part of a work in progress. By calling into question migratory categorizations and the hegemonic narratives that sustain forms of control over people on the move, we emphasize how migrants can construct political bodies from circuits of affect not limited to fear.

References

Asamblea Nacional de Venezuela (2009), *Ley Orgánica de los Consejos Comunales. 26 de noviembre de 2009. Años 199º de la Independencia y 150º de la Federación*. Caracas.

Bahl, E. and A. Worm (2022), "Precarious transnationality," in G. Rosenthal (ed.), *Transnational Biographies: Changing We-Images, Collective Belongings and Power Chances of Migrants and Refugees*, Göttingen: Göttingen University Press, pp. 173–219. https://doi.org/10.17875/gup2022-2208

Beck, U. (2018), *A metamorfose do mundo: novos conceitos para uma nova realidade*. Editora Schwarcz-Companhia das Letras.

Benhabib, S. (2004), *The Rights of Others: Aliens, Residents, and Citizens* (No. 5). Cambridge: Cambridge University Press.

Bizon, A. C. C. (2020), Acolhimento e solidariedade em contexto de pandemia: a experiência do banco de tradutores e intérpretes da Unicamp. *Migrações Internacionais e a Pandemia da Covid-19*, 584.

Brah, A. (2011), *Cartografías de la diáspora: identidades en cuestión*. Traficantes de sueños.

Oliveira Junior, N. H. F. (2016), *Tractatus político-theologicus: teoria crítica, libertação e justiça*. Editora Fi.

Freire, P. (1996), *Pedagogia do oprimido*. Editora Paz e Terra.

Gutiérrez Rodríguez, E. (2018), "The coloniality of migration and the 'refugee crisis': on the asylum-migration nexus, the transatlantic white European settler colonialism-migration and racial capitalism," *Refuge*, 34(1). https://doi.org/10.7202/1050851ar

Mbembe, A. (2013), *Critique de la raison nègre*, Éditions La Découverte, Paris.

Moreira, J. B. and J. H. O. M. de Borba (2021), Invertendo o enfoque das "crises migratórias" para as "migrações de crise": uma revisão conceitual no campo das migrações. *Revista Brasileira De Estudos De População*, 38, e0137. https://doi.org/10.20947/S0102-3098a013

Mudimbe, V. Y. (2019), *A invenção da África*: gnose, filosofia e a ordem do conhecimento. Editora Vozes.

Quijano, A. (2005), "Colonialidade do poder, eurocentrismo e América Latina," *A Colonialidade do Saber: etnocentrismo e ciências sociais–Perspectivas Latinoamericanas. Buenos Aires: Clacso*, 107–26.

Quijano, A. (2012), '"Bien vivir': entre el 'desarrollo' y la des/colonialidad del poder," *Viento sur: Por una izquierda alternativa*, (122), 46–56.

Robertson, S. (2019), "Status-making: rethinking migrant categorization," *Journal of Sociology*, 55(2), 219–33. https://doi.org/10.1177/1440783318791761

Safatle, V. (2016), *O circuito dos afetos: corpos políticos, desamparo e o fim do indivíduo*. Autêntica.

Spivak, G. C. (2023), "Can the subaltern speak?," in P. H. Cain and M. Harrison (eds.), *Imperialism*, London: Routledge, pp. 171–219.

12

Indigenous Social Security Systems for Endogenous Social Development in Southern Africa

Ndangwa Noyoo

Introduction

Southern Africa occupies the southernmost part of the African continent, as the region's name suggests. It is located in Africa—a continent with a plethora of development hurdles. Arguably, Africa has not reached a threshold where it can define its destiny, unimpeded by mainly exogenous forces, while on the other hand, its endogenous imperatives continue to be beholden to or defined by external influences. As a continent that has a painful and tortuous past, one would have thought that there would be agency and urgency, on the part of politicians, government bureaucrats, academics, and civil society actors, to steer Africa in the direction of indigeneity, self-reliance, and autonomous development (see Noyoo, 2019, 2022). However, this does not seem to be the case, as African countries continue to be defined by, first, their former Western colonial masters and other parts of the West, and second, by new global powers such as China and Russia (although its forerunner, the Union of Soviet Socialist Republics—USSR—was a global power during the Cold War era). What is clearly a misnomer in this scenario is that Africa has vast minerals and other natural resources as well as human capital, which are plundered and exploited by other continents, but it continues to remain the poorest continent in the world. In this technological age, Africa is still the main primary resource exporter and does not have substantial capacity for beneficiation, that is, to be able to manufacture products from its raw materials. One of the biggest challenges Africa faces, to date, is its inability to break out of the mold it was cast in by its colonizers. This chapter pays particular attention to southern Africa and argues that it has great potential to use indigenous social security systems as springboards for endogenous social development processes for the region's overall development and other forms of human progress.

It is argued in this chapter that due to colonialism and especially settler colonialism, in this part of Africa, southern Africa presents a unique case of continued bondage and subordination of indigenous populations by descendants of Europeans in their own lands, several years after independence or liberation. This situation militates against a holistic and endogenous form of social and economic development, which makes indigenous peoples mere spectators in the development processes in their respective countries, which they should actually own and define. Conversely, this region's pre-colonial past potentially offers an important platform to actualize an endogenous development process based on indigenous social security systems. Arguably, this approach is rarely factored into present-day regional integration efforts aimed at the social and economic development of southern Africa.

Methodology and conceptual underpinnings

In writing this chapter, the author was guided by an exploratory research approach in order to gain more insights into the issues raised in this discussion. Exploratory research in the social sciences can be understood and defined in different ways. However, at its core, it is research that attempts to discover something new and interesting by working through a research topic (Swedberg, 2020). Also, the author undertook a desktop analysis to augment the exploratory research approach. Desktop analysis is a method that explores data from existing documents and previous research—secondary data—to gather information on a particular topic. It can provide solid arguments and help a researcher elaborate a line of thought. It relies on data already collected from other people (Editorial Aela, 2022). The chapter incorporated a case study on Zambia to strengthen its arguments as well as to add more empirical evidence pertaining to the utilization, or not, of indigenous social security systems (ISSS) in southern African countries by bringing to the fore the Zambian example. There are several concepts that anchor this chapter. For starters, its discussions are guided by the concept endogenous. According to the Cambridge online Dictionary (2023), endogenous refers to something that is found or comes from within something, for example, a system or a person's body or mind. Therefore, this chapter first recognizes the importance of endogenous systems and other aspects of Africa that can genuinely propel an autonomous development process on the continent that could be poised to raise the living standards of all its marginalized and vulnerable people. Social development has been an issue of key concern since the creation of the United Nations, and, as such, permeates the latter's founding instruments, including the Universal Declaration of Human Rights (1948) and the International Covenant on Economic, Social and Cultural Rights (1966). Its conceptual discussion has been linked constantly to the notions of well-being and improvement of people's quality of life, according to the Economic Commission for Latin America and the Caribbean (ECLAC) (2018). For this chapter, social development offers a comprehensive macro-perspective that focuses on communities and societies, emphasizes planned intervention, promotes a dynamic change-oriented approach that is inclusive and universalistic, and above all seeks to harmonize social interventions with economic development efforts (Midgley, 1995).

In addition, ISSS underpin this chapter, and they are taken as an important concept by the author. According to Noyoo (2022), cited in Noyoo and Kleibl (2023: 89), ISSS are:

> Mechanisms of care and organic safety-nets emanating from Africa's pre-colonial past which survived colonial onslaughts and post-colonial State neglect, which are employed by Africans to support each other in times of need and during social and economic crises, and which assist the most vulnerable, marginalized, and excluded members of African societies to sustain their livelihoods. They are naturally occurring and intrinsic to particular communities across Africa and are characterized by strong bonds of solidarity and buttressed by value-systems, philosophies, cultures and traditions, which enjoin all members to relate to each other in a mutually obligatory and reciprocal manner.

It is worth noting that colonialism, especially settler colonialism and apartheid, served to erode the ISSS of Africa. Europe's brutal occupation of Africa and other parts of the Global South, accelerated the dismantling of indigenous structures that held African communities together before colonization. In addition, urbanization eroded indigenous systems and patterns of social interaction. Nonetheless, southern Africa became more embedded in the web of colonialism than other parts of Africa because it was dominated via a more insidious and localized type of colonialism known as settler colonialism. According to Vincente (2017: 4) settler colonialism is a relationship. It is related to colonialism but also inherently distinct from it. As a system defined by unequal relationships (like colonialism) where an exogenous collective aims to locally and permanently replace indigenous ones (unlike colonialism), settler colonialism has no geographical, cultural, or chronological bounds. It is culturally nonspecific. It can happen at any time, and everyone is a settler if they are part of a collective and sovereign displacement that moves to stay and that moves to establish a permanent homeland by way of displacement. Due to settler colonialism, some countries in the region, such as Zimbabwe, Namibia, and South Africa, could not define an endogenous informed development process, which is informed by, among others, ISSS, for decades due to a high number of settlers who superimposed their values and systems on the local ones. To date, white privilege continues to define human relationships in South Africa.

The context and its history

For purposes of this chapter, southern Africa refers to countries in the SADC but excludes the islands of Comoros, Madagascar, Mauritius, and the Seychelles. This approach has been employed for only analytical purposes. These countries are namely Angola, Botswana, the Democratic Republic of the Congo (DRC), Lesotho, Malawi, Mozambique, Namibia, South Africa, eSwatini, Tanzania, Zambia, and Zimbabwe. Almost all these countries have a shared history as well as linguistic and cultural ties. Thus, it is not incidental that the present political economy of southern Africa is linked to the historical trajectory of the area (Noyoo, 2013). In 2019, SADC's aggregate population was 354 million people. The DRC had the single largest population among member states at 87 million people, while South Africa and Tanzania had approximately 58 million each (African Futures, 2023). Seven of the 16 member states had populations below three million, including Namibia, Botswana, Lesotho, Mauritius, eSwatini, Comoros, and the Seychelles. It is estimated that the region's population will nearly double to 618 million people by 2043. This growth is likely to be led by the DRC, which will more than double its population to 173 million. Tanzania will overtake South Africa as the second most populous country in SADC with 106 million people by 2043 (African Futures, 2023).

Economically, despite a wide array of minerals and other natural resources abundantly found in all SADC countries, the region's economic performance is below par, and the region is still very poor. SADC has an aggregate economy that stood at US$924.3 billion in 2019. In 2019, extreme poverty levels (people living on less than US$1.90 per day) were at 51 percent (African Futures, 2023). According to the African Development Bank Group (2023) in its Southern Africa Economic Outlook Report, in 2022, the region's Gross Domestic Product (GDP) growth barely reached 2.7 percent, a level much lower than global and African averages of 3.4 percent and 3.8 percent. The report observes that the slowdown in South Africa has been mirrored in other countries within the region, such as Zimbabwe, Zambia, Malawi, Madagascar, and São Tomé and Príncipe, which have also experienced intense adverse weather events. Weighing down the environment further is the external debt burden, which is forecast to remain high across the region. In 2022, it stood at 48 percent (African Development Bank Group, 2023). Despite the SADC's main objectives being to seek to achieve economic development, peace/security, and growth, alleviate poverty, enhance the standard and quality of life of the peoples of the region, and support the socially disadvantaged through regional integration (SADC, 2023), it can be argued that the region has failed dismally on this score. It was envisaged by the SADC that these objectives were to be achieved through increased regional integration, built on democratic principles, and equitable and sustainable development (SADC, 2023). However, the region is sweltering under bad governance, corruption, and malevolent leadership in some countries, especially Zimbabwe, while rebels and jihadist insurgents are wreaking havoc in the DRC and Mozambique.

In the 1960s, 1970s, and 1980s, southern African countries coalesced together to spearhead the decolonization of the entire region by forming a political grouping known as the *Frontline States*. Noyoo (2013) reports that the *Frontline States* had been instrumental in the dismantling of colonial rule by providing logistical, financial, and moral support to the liberation movements in the region. This organ also raised global awareness about the liberation struggle. The preferred channels for such an articulation were mainly diplomatic and international, for instance, at the United Nations General Assembly, the Non-Aligned Movement (NAM), and the Organisation of African Unity (OAU). The chief architects of the *Frontline States* were Julius Nyerere of Tanzania, Kenneth Kaunda of Zambia, and Seretse Khama of Botswana. After the independence of both Angola and Mozambique, this formation was expanded, and presidents Agostinho Neto and Samora Machel of the former countries, respectively, became part of the leadership of the *Frontline States* (Noyoo, 2013). In 1980, the Southern African Development Co-ordination Conference (SADCC) was established in Lusaka, Zambia, to reduce the region's dependence on apartheid South Africa. Thus, the overtly political formation of the *Frontline States* was reconstituted into the SADCC, which was an economic group. After Namibia gained its independence in 1990, the SADC was founded and then replaced the SADCC. At the time, South Africa was engaged in intense negotiations that eventually paved the way for the democratic elections of 1994, which resulted in the demise of colonialism and apartheid (Noyoo, 2013). However, the peoples of southern Africa have a longer history, which predates the colonial and post-colonial eras. Thus, it is important for us to examine the pre-colonial era.

Southern Africa's pre-colonial period

According to Wright (2017), archaeological evidence in this region indicates that *Homo sapiens*, descended from earlier forms of hominin, were present in southern Africa at least 200,000 years ago, but many members of the southern African public reject evolutionary notions of the past. From about 200 BCE onward, groups of hunter-gatherers, pastoralists, and farmers were in constant contact in southern Africa. Wright (2017) points out that a widespread European settlerist view, based on deep-seated stereotypes of warring races and "tribes," is that communities in this region were permanently in conflict. However, historical evidence shows that in fact they interacted and intermingled in a range of different ways. Interactions became yet more complex from the mid-seventeenth century as settlers from Europe gradually encroached from the southwest Cape Colony into most of southern Africa. In some areas, settler graziers sought to wipe out groups of hunter-gatherers and to break up pastoralist groups and enserf their members; in other areas, particularly in the shifting colonial frontier zone, mixed groups, including settlers, made a living from raiding and trading (Wright, 2017). The foregoing analysis resonates with this chapter's thrust as it acknowledges the biased ethnocentric and Eurocentric views that have distorted southern African ethnic groups' dispersal, known as *mfecane* and *defaqane*, and solely attributed this phenomenon to the "cruelty" of King Shaka of the Zulu Kingdom. Such colonial and settlerist perspectives have distorted the history of southern Africa.

It is due to the history of the *mfecane* and *defaqane* that we see today almost similar polities of indigenous peoples across southern Africa, which can trace their roots to ethnic communities in present-day South Africa. Sobantu and Noyoo (2022) assert that in southern Africa, there was the dispersal of peoples from present-day South Africa who moved to Malawi, Mozambique, Tanzania, Zambia, and Zimbabwe. These dispersals, which culminated in the mid-1800s, were attributable to the rise of the Zulu nation under King Shaka. These authors point out that from the Ndwandwe, under the leadership of Zwide, two groups migrated to other parts of the region. One group from this area was led by Zwangendaba and moved to the area presently known as the Eastern Province of Zambia. Part of this group, which had become known as the Ngoni people, continued to Malawi and Tanzania, where they settled. In addition, from the Ndwandwe kingdom, a group escaped Shaka's wrath and migrated to Mozambique. The Gaza (or Gasa) kingdom was established in the highlands of the middle Sabi River in Mozambique in the 1830s by Soshangane, the Ndwandwe general who fled from Zululand after his defeat at the hands of Shaka. Soshangane extended his control over the area between the Komati (Incomati) and the Zambezi rivers, incorporating the local Tsonga and Shona peoples into his Zulu-type state and reducing the Portuguese to tributary status (ONWAR.COM, 2019). Noyoo and Sobantu (2022) assert that from the present-day Free State province of South Africa, a Sotho group led by their leader Sebetwane moved northward and settled in the kingdom of Bulozi (Barotseland) after subduing the Aluyi people who ruled this territory. This area is now referred to as the Western Province of Zambia. Lastly, one of King Shaka's generals, Mzilikazi, who was cited for insubordination and feared reprisals from his leader, fled with his followers to present-day southern Zimbabwe and established the Ndebele kingdom at a place called *koBulawayo*, which is present-day Bulawayo.

Marks (2023) makes an interesting point, which notes that the history of southern Africa cannot be written as a single narrative. Shifting geographic and political boundaries and changing historiographical perspectives render this impossible. Research into local history in the late twentieth and early twenty-first centuries has presented fragmented historical knowledge, and older generalizations have given way to a complex polyphony of voices as new subfields of history—gender and sexuality, health, and the environment, to name but a few—have developed (Marks, 2023). Archaeological and historical inquiry has been extremely uneven in the countries of the southern African subcontinent, with Namibia the least and South Africa the most intensely studied. Divided societies produce divided histories, and there is hardly an episode in the region's history that is not now open to debate. This is as true of prehistory as of the more recent past (Marks, 2023).

Presenting the argument

Whenever Western systems collapse or what is termed modernity fails to function in Africa, most Africans instinctively revert to their indigenous systems. We can thus infer from this that even though Western systems and modernity were superimposed onto indigenous systems by the colonialists, they are still not as rooted as the indigenous ones, which occupy a subterranean space across the continent. Africans seem to have more confidence in their indigenous systems even though "officially" they purport to be living modern lifestyles. A good example of when African indigenous systems seemed to have appealed to Africans of almost every class was at the height of the Structural Adjustment Programme (SAP) of the World Bank and International Monetary Fund (IMF). In the 1980s and 1990s, SAPs had resulted in the implosion of African economies and social systems. Instead of generating the much-vaunted economic growth, SAPS resulted in rising food prices, unemployment, and deterioration in health and education services while threatening the political stability of African countries (Cleary, 1989).

In addition, SAPs have a notorious propensity for undermining and weakening vulnerable groups in Africa, which were already enfeebled by a host of forces before the SAPs were introduced. Furthermore, they have a detrimental impact on child and maternal health. In particular, these programs undermine access to quality and affordable healthcare and adversely impact social determinants of health, such as income and food availability (Thompson et al., 2017). When the economic austerity measures resulted in a social crisis, Africans easily resorted to their indigenous systems. Previously, indigenous systems were wrongly thought to be only best suited for rural populations and those considered to be of a "lower class." However, SAPs had decimated the middle class, leading to the emergence of a group of new poor—a *nouveau pauvre*, drawn from formerly relatively prosperous people. Such people had earlier worked either in industry or government service. Often, they were teachers or health service employees (Cleary, 1989). In this environment of desperation, Africans from different walks of life and classes gravitated toward indigenous systems for their health, psychological, emotional, and other needs.

Another key example is when Africans were failed by modern banking and financial systems. In most cases, individuals at the lower end of African economies are unable to access finances or secure loans because they do not have collateral. Anecdotal evidence and data from empirical research point to the fact that many individuals and households have been saved from financial ruin by indigenous banking systems. In South Africa, these informal banking systems are known as *stokvels*, and in Zambia, they are referred to as *chilimba*. According to Mwaba (2021), *chilimba* is a type of Rotating Savings and Credit Association (ROSCA). *Chilimba* is indigenous, and it is the most widely used method of informal group savings in Zambia. In a *chilimba*, members make regular fixed contributions to a common fund in each meeting in a predetermined order. *The total common fund is then given to each member in turn*. The group members agree upfront on a standard fixed amount contribution, each meeting, frequency of contribution, and who collects the common fund in each meeting. New members can only be added at the end of a rotation cycle. There is usually a minimum requirement for recordkeeping. At each meeting, the money is collected and given to one member. Once a member has received the money (or *lump sum*), she/he must continue to contribute until all the members have had a chance to receive the total common fund once from a meeting group (Mwaba, 2021).

In a research study that tried to ascertain the economic contribution of *stokvels* toward the broader economic transformation of South Africa, while paying particular attention to municipalities, Bophela and Khumalo (2019) note that *stokvels* are a type of "rotating savings and credit associations" found in South Africa with an estimated value of R49 billion (as of 26 November 2023, 1 Euro equals 20.55 Rand) and a membership of 11.5 million in a country with a populace of 57 million (which is now 62 million) and an informal economy with a total value of R160 billion—yet remain a hidden sector of the economy. Their research revealed that money saving and investment, business opportunities and job creation, and policy development were the main economic contributing factors of *stokvels* in South Africa, especially at the municipality level. They recommend that policy-makers should formulate an economic transformation policy framework inclusive of *stokvels* and adopt their economic contribution factors (Bophela and Khumalo, 2019). Furthermore, the researchers contend that developing the *stokvel* industry could be crucial for poverty alleviation, reduction of unemployment, and broadening equitable access to the ownership of the economy and capital accumulation, and thus improving people's livelihoods and raising their standard of living (Bophela and Khumalo, 2019).

Case study on Zambia and indigenous social security systems (ISSS)

In this case study on Zambia, the chapter intends to cast light on that country's public policy-makers' approach to indigenous social security systems and how they understood them. The case study is based on empirical research that was undertaken by Noyoo and Sakala (2018), which sought to find out if ISSS were referred to when public policies were developed in Zambia. Being a former British colony, the research had attempted to ascertain if Zambia had decolonized its public policy-making processes 54 years after its political independence. This research was part of a larger study that had focused on ISSS in the regions of southern and west Africa. It was initially conceptualized in 2014 and comprised country-specific studies that were led by researchers and academics from southern and west Africa. The research question of the study was the following: Does public policy-making in Zambia incorporate ISSS? The research was guided by the following objectives:

1. To determine the existence of ISSS in Zambia.
2. To find out how ISSS were being utilized by ordinary people in Zambia.
3. To investigate and determine whether the Zambian government incorporated ISSS into mainstream government public policies and public policy responses.
4. To make recommendations on how ISSS could be incorporated into Zambia's public policies (Noyoo and Sakala, 2018).

The methodology that was employed for this research study was a qualitative one that relied upon an exploratory research design. The study had focused on two different population groups at the provincial and district levels. The individuals who had participated in this research were sampled from Lusaka, the capital city of Zambia, and the rural town of Katete in the Eastern Province of the country. The research participants were sampled using the purposive sampling technique because the researchers were interested in gathering information from people who had a certain level of understanding or knowledge of ISSS (Noyoo and Sakala, 2018). The first category of participants comprised government officials, whilst the second group was made up of ordinary citizens who were mainly semi-professionals or worked in the non-governmental sector. This inclusion criterion was used to have richer information and to ensure that the objectives of the study were met.

The research respondents' responses were varied, but they were able to proffer some plausible explanations to the researchers. For example, some Zambians did have knowledge of ISSS. However, it was interesting to note that some respondents had knowledge of the usefulness of ISSS, especially in times of distress or economic hardship. In terms of the research's theme relating to the incorporation of ISSS into public policy-making, all the respondents had asserted that ISSS were not incorporated into this process. The ordinary citizens' responses were also analyzed by referring to another research theme, namely the *utilization of indigenous knowledge to deal with shocks and insecurities*. When it came to the former issue, the respondents were of the view that ISSS were vital and played a key role in the circumstances where people's needs were not met by formal social security systems. The implication of this response is that Zambians can deal with financial shocks and insecurities by making use of ISSS. The respondents also mentioned the fact that in the absence of ISSS, it was very difficult for people to cope, particularly those who were mainly working in the informal sector. A significant number of the respondents observed that even for people in the formal sector, when the formal security systems were not available, the ISSS were very helpful in providing support to poor and vulnerable Zambians. Based on the foregoing issues, it can be discerned and argued that after five decades of independence, Zambian policy-makers still seemed to give more prominence and preference to Western forms of social security than indigenous ones. Therefore, there is still a need to decolonize the public policy-making space. Arguably, these findings from Zambia can be extrapolated to the rest of southern Africa, which seems to be steeped in colonial administrative processes and public policy-making approaches.

Advancing Afrocentrism for endogenous public policy and ISSS for social development

In this section of the chapter, attempts are made to move the argument toward Afrocentrism, which, in this case, should serve as a springboard for an endogenous public policy and social development that are predicated on ISSS. It is contended in this chapter that Afrocentric theoretical lenses and conceptualization are required in southern Africa for countries to arrive at an endogenous public policy agenda-setting, which should also act as the main artery for endogenous social development processes in Africa. Afrocentrism argues that for centuries, Africans and other non-whites have been dominated, through slavery and colonization, by Europeans, and that European culture is at best irrelevant—and at worst diametrically opposed—to efforts by non-Europeans to achieve self-determination (Britannica, 2023). For this reason, according to Afrocentrism, people of African descent need to develop an appreciation of the achievements of traditional African civilizations; indeed, they need to articulate their own history and their own system of values. According to Afrocentrism, African history and culture began in ancient Egypt, which was the birthplace of world civilization (Britannica, 2023).

It is important to note that the indigenization of social security systems to advance endogenous social development in southern Africa can only transpire if there is the decolonization of the public service and government bureaucracy in countries across the whole region. Decolonization of the public policy arena could be practically attained if countries in this region begin to discard some of their colonial heritages and start using their own indigenous knowledge in governance and state matters. This can only be attained if there is first political will to set this agenda and then, secondly, if there are intellectual and administrative capabilities to execute this mission to decolonize the state, government, and public service. This means that there should be efforts made to go back to the basics for this new type of thinking to percolate across countries. These basics in most African countries are to be found in their pre-colonial past. Thus, pre-colonial African history should not be deliberately distorted and, at times, buried, as is the case in some southern African countries. As long as there is political will, there must be the following keystones that have to be put in place for the decolonization agenda to unfold in southern African countries:

1. Decolonization of the curriculum—this must transpire from early childhood to tertiary levels.
2. Decolonization of political and administrative systems.

Conclusion

This chapter called for post-colonial governments of southern Africa to harness and implement endogenous social development, which is underpinned by indigenous social security systems. It also argued that public policy-making should be decolonized in southern Africa for the former processes to take root in the region. A call for the appreciation of Africa's pre-colonial past in the public policy-making space was made. Lastly, it was argued that for social development to be relevant and adequately respond to the needs of ordinary people in the region, it must be anchored in indigenous systems and philosophies.

References

African Development Bank Group (2023), Southern Africa Economic Outlook: Mobilising Private Sector Financing for Climate Change and Green Growth. African Development Bank Group.

African Futures (2023), Southern African Development Community (SADC). Retrieved from https://futures.issafrica.org/geographic/recs/sadc/

Bophela, M. J. K. and N. Khumalo (2019), "The role of stokvels in South Africa: a case of economic transformation of a municipality," *Problems and Perspectives in Management*, 17(4): 26–37. doi: 10.21511/ppm.17(4).2019.03

Britannica (2023), "Afrocentrism," retrieved from www.britannica.com/event/Afrocentrism

Cambridge online Dictionary (2023), "Endogenous," retrieved from https://dictionary.cambridge.org/dictionary/english/endogenous

Cleary, S. (1989), "Structural adjustment in Africa," Trócaire Development Review, 1989, 41–59.

Economic Commission for Latin America and the Caribbean (ECLAC) (2018), Towards a regional agenda for inclusive social development: bases and initial proposal. Santiago.

Editorial Aela (2022), Desk Research: How To Conduct Secondary Research Efficiently. Retrieved from https://aelaschool.com/en/reseach-en/desk-research-conduct-secondary-research-efficiently

Marks, S. E. (2023), Southern Africa. Retrieved from www.britannica.com/place/Southern-Africa.

Midgley, J. (1995), *Social Development: The Developmental Perspective in Social Welfare*, London: SAGE Publications.

Mwaba, C. L. (2021), "Reflections on informal financial groups and financial scandals surrounding them—how to know what you can trust," retrieved from www.fsdzambia.org/reflections-on-informal-financial-groups-and-financial-scandals-surrounding-them-how-to-know-what-you-can-trust

Noyoo, N. (2013), "Bringing social cohesion into the equation of regional integration: the case of Southern Africa," *Regions and Cohesion Journal*, 3(1): 94–110.

Noyoo, N. (2019), "Pushing for autonomous African development," in K. Kleibl, R. Lutz, N. Noyoo, B. Bunk, A. Dittmann, and B. Seepamore (eds.), *The Routledge Handbook of Post-Colonial Social Work*, London: Routledge, pp. 244–57.

Noyoo, N. (2022), "Critiquing Western development paradigms in the age of the Coronavirus (COVID-19): an African perspective," in M. de. C. Gonçalves, R. Gutwald, T. Kleibl, R. Lutz, N. Noyoo and J. Twikirize (eds.), *The Coronavirus Crisis and Challenges to Social Development: Global Perspectives*, Princeton, NJ: Springer, pp. 399–408.

Noyoo, N. and T. Kleibl (2023), "Kinship care, responsibility, and self-determination: exploring African individual and community rights in decolonised social work," in M. Madew, M. Boryczko, and M. Lusk (eds.), *Decolonised Approaches to Human Rights and Social Work*, Princeton, NJ: Springer, pp. 77–93.

Noyoo, N., and B. M. Sakala (2018), "Indigenous social security systems in Zambia: advancing a public policy agenda?," in N. Noyoo and E. Boon (eds.), *Indigenous Social Security Systems in Southern and West Africa*, Stellenbosch: Sun Media Press, pp. 87–104.

Noyoo, N., and M. Sobantu (2022), "Interrogating migrations and social transformations in Africa: a social work perspective," in R. Munck, T. Kleibl, M. de Carmo and P. Dankova (eds.), *Migration and Social Transformation: Engaged Perspectives*, Dublin: Machdohnil Ltd, pp. 208–28.

ONWAR.COM (2019), Ndwandwe Migration 1823–1831. Retrieved from https://onwar.com/data/soshangane1823.html

Southern African Development Community (SADC) (2023), 43rd SADC Summit of Heads of State and Government. Retrieved from www.sadc.int/sites/default/files/2023-08/43rd%20Summit%20Brochure.pdf

Swedberg, R. (2020), "Exploratory research," in J. Elman, J. Gerring, and J. Mahoney (eds.), *The Production of Knowledge: Enhancing Progress in Social Science*, Cambridge: Cambridge University Press, pp. 17–41.

Thomson, M., A. Kentikelenis, and T. Stubbs (2017), "Structural adjustment programs adversely affect vulnerable populations: a systematic-narrative review of their effect on child and maternal health," *Public Health Reviews*, 38(13): 1–18.

Veracini, L. (2017), "Introduction: settler colonialism as a distinct mode of domination," in E. Cavanagh and L. Veracini (eds.), *The Routledge Handbook of the History of Settler Colonialism*, London: Routledge, pp. 9–10.

Wright, J. (2017), "Southern African before colonial times," retrieved from https://oxfordre.com/africanhistory/display/10.1093/acrefore/9780190277734.001.0001/acrefore-9780190277734-e-92

13

Indigenous in Europe: The Challenge in Bringing Back Livonian

Valts Ernštreits and Gunta Kļava

Introduction

When speaking about colonial heritage and Indigenous people, Europe does not usually come to mind. However, various colonization-like processes have also taken place here, e.g., in the Baltic countries (Samalavičius, 2023: 13), though usually these are too close to home, too distant in time, or too recent and contemporary to be included in discussions on these topics. The experience gained and heritage left by such processes and policies keep haunting many European nations, and especially endangered ethnic communities, still determining and affecting their perspectives on sustaining their identity, language, and culture.

One such endangered community is the Livonians. They are an officially recognized Indigenous people of Latvia (since 1991[1]) and their written history as a nation dates back almost a millennium. While only a handful of Livonians remain today, they are somewhat like a living fossil and have managed to keep their language, culture, and self-determination throughout history despite harsh historical circumstances and suppression policies consistently directed at them. However, at the same time, the Livonians are a true twenty-first-century nation deprived of its land and daily contact among community members for almost a century, with these contacts becoming virtualized even before the Internet age.

Nowadays, the terms "revitalization" and "sustainability" are frequently employed in discussions of endangered languages and cultures. However, questions like "What is the expected result?," "How to accomplish it?," and, most importantly, "Who needs to be involved?" are the ones that determine whether revitalization and sustainability are feasible (Hinton et al., 2018: xxiv; Lewis, 2023: 109). For the Livonian, the key questions are: how do you define who is Livonian, how is the Livonian community identified today, and, especially, what determines who is seen as Livonian from the perspective of the Livonians themselves? Answering these questions might seem easy,

[1] Law of the Republic of Latvia "On the Free Development and Right to Cultural Autonomy of Nationalities and Ethnic Groups in Latvia," adopted in 1991.

but in the Livonian situation—a situation in which many Indigenous nations may find themselves eventually—it is actually quite complicated. However, determining the effectiveness of any revitalization or sustainability efforts—or whether there exist any future prospects at all—remains a core question in this work.

Looking through the lens of the Livonians themselves, this chapter examines the historical developments that have impacted and determined what it means to be Livonian from the first mentions in written sources through the last centuries and up to the present day. This chapter also takes a look at the effects those developments and policies have had on narratives, Livonian self-awareness, and the transformation of Livonian identity as well as describes how membership in the Livonian community is understood from the perspective of present-day Livonians. Based on the observations and interviews, in this chapter we try to tackle the challenge of finding the preconditions for bringing back Livonian and the Livonians based on the true voice of the Livonian people hidden under narratives developed over many centuries by colonial-like powers controlling Livonian lands, Livonian neighbors, and allies.

The Livonians throughout history

The written history of the Livonians as an ethnic group dates back more than a millennium. While the first possible written mentions of the Livonians may even date back to the beginning of the first millennium (Ernits, 2014: 14), the first certain mentions of the Livonians as an ethnicity are from the beginning of the twelfth century, when the Livonians are noted as an ethnicity in the chronicle *The Tale of Bygone Years* (Повесть временных лет), written by Nestor, a monk at the Pechersk Monastery in Kyiv (Ernits, 2014: 18; Zemītis, 2011: 75).

The first major source on the Livonians is, however, the *Livonian Chronicle of Henry*, which is also considered to be the first major written source on the history of the Baltic region. This chronicle records the conquest and conversion to Christianity of the Finnic and Baltic people of the region, starting with the Livonians—this is also referred to as the Livonian Crusade. Launched in 1099 and sanctioned by Pope Innocent III, this crusade took place in the lands of the Livonians and their neighbors with the goals of converting “pagans” to Christianity. This resulted in the establishment of a medieval state called Livonia, which was ruled by Germans but took its name from the Livonians (Tumzers, 2019: 25).

In describing the conquest of Baltic lands, the chronicle provides a wide range of data on the Livonians, the areas they inhabited, their language, livelihood, power structures, and even intangible heritage and customs. Along with being a valuable source of history, the chronicle and, especially, its interpretation have “become anchored in the general history of the Baltic region as well as in national histories” (Kivimäe, 2011: 77), and it has played a major role in forming identity narratives in the region, including those of the Baltic Germans, Latvians, Estonians, and also the Livonians.

Around the time of the events described in the *Chronicle of Henry* (ca. the thirteenth century), the Livonians inhabited major areas of present-day Latvia, mostly situated along waterways, effectively controlling and profiting from the trade routes crossing

Livonian territories that connected Scandinavia with medieval Rus' and southeastern Europe. With the arrival of German traders and later crusaders, the Livonians lost their dominance, as did other local tribes, such as the Latgalians, Semigallians, Selonians, and Curonians.

Once the German-ruled Livonian state started to take shape, new power centers emerged, altering the previous ethnic situation in the area of present-day Latvia (Auns, 2018), which led to both the gradual assimilation of the Livonian elite by the ruling Germans and the merging and assimilation of ordinary Livonians by the Baltic tribes of the region, contributing to the formation of a new language among the locals of the area—Latvian. This process took place over many centuries and led to the major Livonian strata still found nowadays in Latvian language and culture and also formed areas with expanded Livonian influence where the Livonians were assimilated later, e.g., the Livonian-like dialect of Latvian (referred to formally as the Livonic dialect) spoken on both sides of the Gulf of Rīga (Endzelīns, 1980: 250, 425, 482).

The German domination of the state of Livonia and its successors, which lasted until the establishment of the Republic of Latvia in 1918, excluded the Livonians from the scope of written sources and referred to them (as well as other local populations of the area) as *Undeutsche*, "non-Germans" (Tumzers, 2019: 30) until the eighteenth and nineteenth centuries. As a result, limited (e.g., brief mentions of presence) to no data exist about the Livonians for several centuries.

Nineteenth-century developments and World War I

By the mid-nineteenth century, when the first broader Livonian documentation efforts had begun, Livonian was still spoken by ethnic Livonian communities in two areas: east of the Gulf of Rīga by as few as 22 elderly people (Wiedemann, 1861: LII; Cimermanis, 1998: 26) and on the northern tip of the Courland Peninsula by some 2500 people (Wiedemann, 1861: LIII) situated in an isolated island-like stretch of 14 fishing villages known as *Rānda*, "the Coast," by Livonians and today as the Livonian Coast.

This area is quite special from a geological perspective, as it was formed on a former seabed and consists of forest-covered dunes separated by swamps, creating a unique landscape (Caune, 2011: 7–9), while also limiting possibilities for livelihood and access, which helped the Livonians maintain their isolation. As the Livonian Coast was only suitable for fishing and had extremely limited opportunities for agriculture or other sources of income, this led to the development of a coastal lifestyle different from that of inland Latvians. At the same time, this landscape isolated the Livonian area from outsiders, as there were only a few minor roads leading into the area. While some settlers came to the Livonian Coast from inland or across the sea, Livonian served as the common language for communication in daily life, and it was also the language used among fishermen while fishing.

The mid-nineteenth century also marked the first significant turning point for the Livonian community in recent Livonian history for two main reasons. First, the last Livonian enclave on the eastern coast of the Gulf of Rīga ceased to exist as a result of assimilation. Second, in 1859–60, an organized Livonian uprising took place on the

Livonian Coast against local German landlords, demanding a decrease in land rent fees, which resulted in the oppression of the uprising but more significantly in the replacement of the Livonians, who actively participated in it, with Latvians from inland. In total, almost one-third of the Livonians on the Livonian Coast were replaced by Latvians and forced to leave the Livonian area (Cimermanis, 1998: 29); among them were also Livonian intellectuals and the first national activists. This significantly dissolved the last relatively homogenous Livonian-speaking area, the Livonian Coast, thus setting preconditions for further assimilation.

During World War I, the Livonian Coast, which had been a peripheral and isolated area, became a frontline. In 1915, the Courland Peninsula was occupied by the Germans anticipating an attack by the Russian Tsarist army located directly across from the Livonian Coast on the island of Saaremaa (in present-day Estonia). Prior to the occupation, the inhabitants of Courland were evacuated *en masse* to distant regions of the Russian Empire, but those who stayed were ordered to leave the coast when the German occupation took it over and began erecting fortifications, a process which also involved partially burning down housing and building railway connections to support the front line. For the Livonians, this time is known as the *ulzajtõbāiga*, "expulsion time" (Viitso and Ernštreits, 2012: 9). As the Livonian Coast was almost entirely emptied of Livonians, some researchers at the beginning of the 1920s even worried that the Livonians might have been fully lost to the Great War; however, the majority of them returned (Blumberga, 2006: 205).

The interwar period

With the collapse of the large European empires and the establishment of nation-states—including Latvia, Estonia, and Finland—significant changes took place in the Livonian situation. The Livonians became a focus of the Finnic "kindred nations" movement (Prozes, 1998), especially because, due to the revolution and establishment of the Soviet Union, many Finno-Ugric nations became inaccessible for Western scholars and cultural activists.

Due to the overall romanticization of the national awakenings spreading across Europe and with the support of scholars and cultural activists from Finnic states combined with limited support by Latvian state institutions, the Livonian community's (ca. 1500 members; Ernštreits, 2023) cultural life flourished during the interwar period, a Livonian NGO—called *Līvõd Īt*, "The Livonian Union"—was established, and Livonian national symbols such as a flag and anthem were introduced. There were a number of Livonian musical groups and writers developing Livonian modern culture; Livonian was for the first time taught in schools as a voluntary subject; publishing of secular and religious books began; and a monthly newspaper in Livonian—called *Līvli*, "The Livonian"—was introduced. Shortly before World War II, the Livonian Community House was built on the Livonian Coast as a result of a unique collaboration between the Livonian community and Estonia, Finland, Hungary, and Latvia.

While the establishment of the new nation-states was mostly beneficial for the Livonians, it also created tensions. At the beginning of the interwar period, there was strong support in the Livonian community for the establishment of a unified administrative territory encompassing the Livonian area (this idea had already been proposed in 1906; Blumberga, 2020: 59). Although the establishment of such a territory was initially supported by the Latvian government, all further steps were stopped as a result of growing tensions connected with the demarcation of the Estonian-Latvian border and groundless rumors that the Livonians might use such a territory for the expansion of their autonomy, later demanding unification with Estonia.

The involvement of Estonian and Finnish activists in Livonian issues, which included discussions of the Livonian situation and the community's needs on a state level, also caused negative effects and distrust from Latvia, especially after the establishment of an authoritarian regime in 1934. This could be seen, e.g., in the restrictions placed on using the Livonian schoolbook published in Finland for teaching Livonian in Latvia, censorship of the Livonian newspaper, a ban on using place names in Livonian, supervision of Livonian activists by the police, etc. However, no major repressions or conflicts took place, and Latvia kept supporting certain activities, such as teaching Livonian in schools and the construction of the Livonian Community House.

World War II and the Soviet occupation

With the start of World War II and the occupation of Latvia by the Soviet Union in 1940, all Livonian activities were fully ended. The Livonian NGO was closed down, the Community House that opened just a year earlier was nationalized, the erection of Soviet military bases began on the Livonian Coast, and the first deportations of people took place.

The subsequent German occupation, while for the most part a relatively calm period for the Livonians, brought with it plans for emptying the Livonian Coast for the purpose of establishing a large military training base there (expeditions of ethnographers and linguists were already sent out with the task of documenting the area prior to it being emptied; Jääts, 2023: 265; Stinkule, 2008: 239). These plans, however, did not come to fruition, and by the end of the war, the situation seen during World War I was being repeated—the Livonian Coast once again became a frontline between German forces in Courland and Soviet forces on the Estonian island of Saaremaa.

Fear of the approaching Soviet forces caused a large part of the population to flee to Western countries, also from the Livonian Coast. As a result, a small Livonian diaspora was established in Sweden, the USA, Canada, and Finland, while the subsequent reoccupation of Latvia and the Livonian Coast by the Soviet Union in 1945 brought deportations of many of those who stayed.

The last major blow to the Livonian community was the establishment of the border area of the Soviet Union after the war. Although it covered all of the Baltic Sea coasts of Latvia, Estonia, and Lithuania, for the Livonian Coast, due to its unique geographical setting, it was catastrophic. The purpose of the border area was to restrict the entrance of foreigners and also to prevent people from fleeing the Soviet Union. Major restrictions were therefore introduced on individual boat ownership and coastal fishing; even access to the beaches was strictly limited and closely watched. This made fishing on the Livonian Coast impossible, which led to the Livonian community abandoning the coast and leaving for other places in Latvia, primarily its capital—Rīga, the closest town to the Livonian Coast—Ventspils, and elsewhere. As a result, since the 1950s, the Livonians have been exiled and scattered, with their largest communities nowadays residing in Rīga (ca. 50 percent), Ventspils (ca. 30 percent), and other places (Census, 2011). This led to a halt in the Livonian language's transmission due to factors such as relocation, land loss, lifestyle changes, and the loss of daily ties within the community. Additionally, the lack of Livonian-speaking domains and the Soviet Union's overall policy, which was not supportive of national issues, further compounded the problem (Ernštreits and Kļava, 2023: 68).

During the Soviet occupation, efforts to maintain Livonianness in exile somehow continued (Balodis, 2023), especially among those actively involved in Livonian cultural processes during the interwar period. In Latvia, manuscripts of books were prepared on Soviet topics and on Esperanto as an emerging idea of a global language in the Soviet Union in the hope that having such topics as their focus would help them get published, but ultimately this was not successful. Gatherings of Livonians took place disguised as birthday celebrations, exhibition openings, or concerts.

Such activists and events, however, were closely watched by repressive Soviet state institutions. This led to a major development at the beginning of the 1970s, when several Livonian community members were successful in legalizing Livonian cultural processes by simultaneously establishing two singing groups: "Līvlist" in Rīga and "Kāndla" in Ventspils. These singing groups, which did not actually have singing ability among their entrance criteria, served as a place for the Livonian community to legally gather and practice their language and traditional culture and, as a side effect, led to the development of a wider folklore movement in Latvia (Vēri, 1994: 245).

In the 1970s, indicating Livonian nationality in passports was banned in the Soviet Union, which led to protests from the Livonian community against Soviet authorities. These included open letters and organizing public events in the restricted border area on the Livonian Coast, which were banned by Soviet authorities but nevertheless took place.

The collapse of the Soviet Union and the present-day situation

Livonian activities restarted nearing the end of the Soviet Union in 1988 by re-establishing the Livonian NGO and later celebrating 50 years of the Community House in the summer of 1989. This was a groundbreaking event, which brought the Livonian community together on the Livonian Coast for the first time since its exile and began the tradition of the Livonian Festival—a key event for gathering the community, which has taken place annually since then and continues up until today.

The restoration of the independence of Latvia in 1991 led to major changes in state policy and the resurrection of the Livonians. Since then, the Livonians have been officially recognized as an Indigenous people of Latvia in several laws (Kļava, Ernštreits, 2022: 212). They are also mentioned in the Constitution of Latvia, and several state programs have been initiated to help safeguard and sustain the Livonian community, including the introduction of official Livonian-language signage in 2023 and a program for Livonian-language acquisition in informal education in 2024.

Since then, the Livonian community itself has been highly active by organizing initiatives and events, sustaining traditional culture and developing modern culture, using digital sources and platforms, and participating in policy developments on a local, state, and international level. At the same time, the Livonian community remains vulnerable and, for the most part, hardly identifiable, scattered, diverse, and therefore blurred, making gathering the community together, setting common goals and visions complicated, and, on many occasions, making the true voice of the Livonian community overshadowed by outside perspectives and narratives.

The Livonians as seen by others

Livonian identity, self-perception, and self-esteem have developed in the context of majority narratives. Crucial for developing those narratives has been the *Livonian Chronicle of Henry*, which has a defining role for all nations of the region, including the Germans who dominated the region from the Crusades until the mid-nineteenth century, but much more for the Latvians and Estonians as well as the Livonians themselves.

For some German noblemen (e.g., the *Lieven* family; Līveni, 2002), references to a supposed connection with Livonian noblemen described in the chronicle, such as the Livonian king-like (*Quasi-Rex*) chieftain Kaupo, who was the first to come from the Baltic region to visit Rome and have an audience with the Pope, were important and benefited their ancestry. At the same time, Livonian people still present in more recent times were treated with suspicion, just as in the opening parts of the *Livonian Chronicle of Henry*, where the Livonians are characterized as idol worshippers, stubborn and treacherous, cunning and hypocritical (Indriķa hronika, 1993: 47–53). This general attitude is illustrated by a local German landlord at the end of the eighteenth century, who claimed that Livonians are "more cunning than other morons" and act arrogantly toward others; therefore, he was moving and dispersing them, but even so their youngsters "still come together and gobble like turkeys" (Noriņa and Ozoliņa, 2023: 19).

For Latvians (and also Estonians), the chronicle served as a source for building their national stories, as after the Crusades, living conditions and the formation of national consciousness were largely directed by German elites (Tumzers, 2019: 41), and the local population stayed generally out of scope, becoming simple peasantry and non-Germans.

In the Latvian case, this can be seen best in one of the cornerstones of Latvian identity—the national epic *Lāčplēsis*, which is a combination of folklore, fiction, and events described in the *Chronicle of Henry*.[2] This epic not only had an important role in consolidating and defining Latvian national narratives but also has a significant role in defining the blurred positioning of the Livonians within those narratives.

2 For more on the context of the creation of the epic, see Kalnačs (2023).

On the one hand, the epic, using many references to the chronicle, tells the story from the perspective of the "Latvians." However, it takes place in areas and key settlements inhabited at the time of the chronicle by Livonians (e.g., Lielvārde, Ķegums, Turaida, Aizkraukle). One of its main characters (who is Latvian according to the epic) is named after the Livonian chieftain Kaupo (*Kaupa*), but in the context of the epic is also portrayed as the main traitor. On the other hand, of all local tribes described in the chronicle that later consolidated into the Latvian nation—the Baltic Latgalians, Semigallians, Selonians, and Curonians as well as the Finnic Livonians—the epic only mentions the Livonians as a separate ethnicity (different from Latvians), describing their subversion by approaching Germans near Rīga. However, the epic does not say anything about who they are, what their connection is with the land they are living on, or their relationship with the Latvians. Thus, the epic actually describes a confusing situation, and some of its key and contradictory narratives about the Livonians are still circulating in Latvian society to this day.

The first narrative: The Livonians do not exist as a separate ethnicity; they are part of the Latvians in the same way as people from other regions of Latvia. This narrative has been widespread even until the current time, particularly when the need for any special efforts to ensure the protection of Livonian language and culture is discussed. This narrative is also supported by the extensive appropriation of heritage elements clearly attributable to the Livonians, which instead are referenced as the heritage of *senlatvieši*, "ancient Latvians."

The second narrative: Livonians are traitors who cooperated with the Germans. This is manifested through the character of Kaupo, who, for the general public in Latvia, has been widely used as a synonym for "traitor," along with the second character from the epic, Kangars, who gets his name from the Livonian word for an elongated hill—*kāngar*. While this narrative is much less notable in public in connection with the Livonians (Rozeniece, 2005), it still has its effect in minor situations when Livonianness or the Livonian heritage of Latvia are discussed.

The third narrative: Livonians are strangers, "other people," non-Latvians, and have nothing to do with contemporary Latvia or Latvian identity. This narrative, reflected in the epic in combination with references to the "traditional" religion of the Balts, reflects the general idea that Latvians, who speak Latvian, which is a Baltic language, are formed from the Baltic tribes inhabiting Latvia at the time of the chronicle, while the Livonians, whose language belongs to another group, Finnic, have nothing to do with Latvia, despite the fact that contemporary Latvian language and culture are heavily influenced by Livonian language and culture, due to the assimilation of the Livonians.

This narrative is still widely present, even in academic research, in Latvia; the most notorious recent example is a book on the cultural history of Latvia, published in 2021, which does not mention the Livonians at all. This also reflects the common misperception in Latvia that Livonian had already gone extinct in the thirteenth century (when they are explicitly mentioned in the *Chronicle of Henry*), and no Livonians existed afterward or in modern times.

All of these narratives also explain political attitudes toward Livonian issues, especially during the interwar period, when the existence of the Livonians was generally perceived as a disturbing obstacle to the development of a homogenous Latvian nation.

A letter written by Latvia's Ambassador to Estonia in 1922 best illustrates this: "[our task is] setting national education, i.e., Latvian schools, in this area [the Livonian Coast] in particularly good condition, [in order] to assimilate the last remnants of the Livonians in Latvia" (Ščerbinskis, 2000).

Building narratives for the Livonians

The narratives of others—and the way Livonians appear in those narratives—have influenced the self-perception and self-manifestation of Livonians in several ways. While a sense of desperation and oppression can be seen in the mid-nineteenth century (Karma, 2014: 16), concerted efforts toward building the self-confidence of the Livonians can be seen after World War I, with these efforts mirroring those of the Latvians—and also Estonians—and sometimes strongly guided by outsiders as well.

In 1921, the first secular book was published in Livonian. It was compiled by scholars from Estonia, Oskar Loorits and Lauri Kettunen (Ernštreits, 2011: 55), and contained an introductory article written by Loorits that once again referenced the *Chronicle of Henry* but from a very different perspective. Loorits wrote: "Now is a time when every nation, no matter how small it might be, is becoming free, where large and small can exist together in the world, and where all languages are equally good, and the best language for everyone is, in any case, their native language. The Livonian nation should not be ashamed that it is so small today. But once long ago, the Livonian nation was also large and powerful and well-known among all the neighboring lands, there were times when the Livonians were much more powerful and larger than the Latvian nation."

This introduced the narrative of the Livonians being a continuation of a once powerful and great nation, and of them being equal to their neighbors who had just gained independent statehood. This was combined with the narratives, which were used for building the identities of nation-states like Latvia, Estonia, and Finland, narratives that were transferred to the Livonians and set the path for the creation of state-like symbols such as a Livonian national flag and anthem. These narratives also led to the construction of the Community House; the perception of the Livonian Coast as a land of its own and attempts to establish territorial autonomy there; and the development of education, literature, and other elements of modern culture in Livonian.

This transfer of narratives, just like for Estonians and Latvians, also stressed and highlighted the central role of language as well as the importance of land and ancient heritage in the identity narratives of Livonians.

The Livonians today

Due to the overwhelming exposure of the Livonians to the Latvian majority over the past centuries, their identity has been affected by numerous factors. Attempts to identify their community size have therefore resulted in estimates being based on the perception of those doing the estimating. The closest attempt to identifying the Livonian community on an individual level—and the only attempt made by the Livonians themselves—was a list compiled by Livonian pastor Edgar Vaalgamaa in 1935–7 (published in Blumberga, 2006: 301–52) at a time when the community still lived on the Livonian Coast. This survey identifies the inhabitants of the Livonian Coast and gives—in addition to other data—their nationality and language proficiency. Thus, it shows, with quite a high degree of accuracy, the state of the Livonian community prior to World War II.

Since then, the Livonian community's structure has changed greatly. To obtain data and an insight into how belonging to the Livonian community is perceived and manifested by the Livonians themselves nowadays, the University of Latvia Livonian Institute collected a considerable amount of data on these questions in 2020–22 through interviews and observations.[3] These interviews indicate shared components that are noticeable when individuals from inside (but also outside) the Livonian community describe what defines being a Livonian in the present day.

Being Livonian

One of the first indications noticeable in the obtained data is that Livonian identity is heavily intertwined with general Latvian identity, and both share a complex history of interaction. Historically, the Livonians contributed to the development of the Latvian nation by merging with other ancient tribes living in present-day Latvia and leaving the footprints of Livonian heritage on the defining elements of Latvian identity (Ernštreits, 2023). Livonians have also been in prolonged contact with and exposed to the surrounding Latvian population, especially following their exile from the Livonian Coast. This has impacted and continues to impact the formation of Livonian identity and is also reflected in the self-perception of Livonians.

As one participant in the interview described it, "I've actually grown up with these two identities [. .] I've always had two identities, I am Latvian, as I grew up in Latvia, but I also have this Livonian identity, and thinking about it, about traditions, and something like that even . . . It can't really even be distinguished, what is Latvian and what—Livonian"[4] (LI_K_09).

Keeping in mind this blurred borderline between Latvian and Livonian elements forming identity, a question arises about the key identifier marking one's belonging to the Livonian community. The language is named first both by Livonians themselves (despite the fact that the absolute majority of the community has no proficiency in

[3] Data collection was conducted as part of various projects implemented by the UL Livonian Institute, e.g., the postdoctoral research project "Applying the Common European Framework of Reference for Languages to Livonian: A New Opportunity for Endangered Languages" and the JPI Cultural Heritage project "Re-voicing cultural landscapes: narratives, perspectives, and performances of marginalised intangible cultural heritage (RCL: ICH)." Citations of the data from interviews and observations included in the article are coded; codes are non-consecutive since only those interviews are included that are quoted in this article.

[4] Notable, however, is the fact that in the interviews, Livonian elements included in Latvian identity, such as linguistic heritage in Latvian, although used on an everyday basis, are not generally felt as being part of Livonian heritage.

Livonian) and also by the Latvian majority. Livonians perceive language as a factor making them different from Latvians: "It is precisely with the language that we show we are something else, that we are not Latvians" (LI_K_04). Here, a certain role may also be attributed to the fact that Livonian and Latvian are from unrelated language families and are not mutually intelligible.

Language, however, is also closely connected with Livonian narratives and self-perceptions of earlier times. As one older Livonian not proficient in the language puts it, "Language is first, this forming of self-understanding, that you are Livonian, that it matters to you what is happening to the Livonians in the future" (LI_K_42). While today it clearly represents overall Livonian narratives and the symbolic role of the language in the community (Kļava, 2022: 130), such statements also have their roots in the actual position of the language in self-identification prior to the Livonian community's exile from the Livonian Coast.

Until their expulsion, the Livonians possessed quite a few aspects distinguishing them from their neighbors (Latvians and Estonians), such as a different lifestyle, but especially a rich and unique intangible heritage—from clothing to folklore and traditions. In the early twentieth century, the language—and proficiency in the language in particular—had nevertheless already been a key marker of belonging to the Livonian community, especially in the context of new settlers on the Livonian Coast arriving after the mid-nineteenth-century Livonian uprising. Language was considered not only an identifier but also a central element of the Livonian spirit or identity. There have been several recorded cases where Livonian people who grew up on the Livonian Coast, when talking about other people, used language proficiency as a synonym for being Livonian (a principle also implemented in Vaalgamaa's list). For example, when an elderly informant born in 1918 was asked whether her husband (born in the same village) was Livonian, she answered that his mother and father were Livonians but not him, as he did not speak Livonian. Similarly, another Livonian informant—born in 1910—did not count a certain person as a Livonian, saying that his mother and father were Livonians and spoke Livonian, but he was not, as he did not speak the language (INT_07; Kļava, 2022).

Language still possesses that key distinguisher role today,[5] although the way it is referenced has changed following the loss of the Livonian-speaking area. Nowadays, unlike before the community was exiled, the connection to those proficient in Livonian in one's family is of central importance rather than Livonian proficiency itself. It is not a person's own proficiency in the language but the relation to the historical community (which was determined by language proficiency) of language speakers that is the primary identifier of connection to the community. The different determinations used in the past and present sometimes collide, however, as also came up during the interviews: "My mother is Livonian, my father is not. But my mother is not completely Livonian, but she speaks Livonian and her mother also is Livonian. I count her as one" (LI_K_24).

[5] The role of Livonian as a strong Livonian identity marker is also reflected in cemetery inscriptions, which often contain epitaphs in Livonian or a person's name in Livonian without an explicit reference to the person being Livonian.

The role of language as a key identifier is strongly supported by the identity narratives of the majority Latvian community, where language plays a significant role. This is even reflected in the Constitution of Latvia and a number of laws (Osipova, 2021). On the one hand, this creates patronizing attitudes, as illustrated by the speech of the President of Latvia at the Livonian Festival in 2023: "While Livonian is spoken and studied, the Livonians will exist" (Observation notes), which creates additional pressure and, in fact, could even be perceived as shaming by the non-proficient Livonian community. On the other hand, it creates situations that are observed even today, for example, in schools, where teachers, upon hearing that a child gives their nationality as Livonian, ask that they confirm their fluency in Livonian. When they give a negative answer, the teacher will then publicly doubt the ethnicity of the child (INT_40). Similar doubts about the ethnicity of Livonians who do not speak Livonian are also found in other contexts and were mentioned by respondents, e.g., "When people ask, "Where can we see a Livonian or meet one?" I say that I am also one. And the first thing they ask is, "Do you speak Livonian?"" (LI_K_19) or "I have friends who say, [if you are a Livonian,] why don't you speak Livonian then?" (LI_K_24).

While connection to Livonian remains central to Livonian identity, other factors are also important for defining identity, such as one's connection to the Livonian Coast, personal choices and characteristics, and elements of intangible heritage. In addition to language, other lesser indicators of Livonianness are also mentioned in interviews, including songs and the environment, a connection to the Livonian Coast, a culture different from that of the Latvians, and a roughness created by the sea-bound life.

Referencing and manifestation

After the loss of the Livonian area, connections and relationships between community members have changed, as they do not meet each other daily nor are they involved in any common daily business. Their lifestyles, living environments, and communication circles are, in fact, greatly different, while meeting opportunities are very rare or nonexistent. Therefore, community members, in the majority of cases, are not familiar with each other and cannot identify each other directly; thus, indirect referencing of ancestry is widely used to signal to others about being part of the same Livonian community.

When two community members unfamiliar with each other meet for the first time, it is fairly common that they start a conversation about Livonian topics with a reference to their personal Livonian heritage and also explain their interest and right to express it (Observation notes). This is usually done by mentioning the particular Livonian Coast village or villages from which that person's family originates and/or identifying a person in their family who spoke Livonian (e.g., mother, grandfather, great-grandparents, aunt, etc.) or, alternatively, indicating a connection to the historical Livonian-speaking community in some other way, e.g., "My parents did not talk in Livonian to me" (LI_K_01); "Grandmother was accommodating Estonian researchers and told them Livonian legends" (INT_07).

Although the members of the Livonian community have lived quite a long time outside the Livonian Coast among Latvians, indicating such connections is still surprisingly relevant for being accepted by the community. Even if a person declares that they are Livonian or are to some extent proficient in Livonian, if that person's background is unclear and questions about their connection to the pre-exile Livonian speakers remain unanswered, then that person's belonging to the Livonian community is doubted, e.g., "Yes, he speaks Livonian and has compiled a Livonian book, but nobody knows what kind of Livonian roots he has" (LI_K_24).

The number of people with clearly detectable recent Livonian ancestry and connections to the pre-exile Livonian speakers is, however, significantly larger than that of those who consider themselves Livonian. Acknowledging that the Livonian and Latvian identities are heavily intertwined and that since the Livonian community's exile, its members have been scattered and disconnected from each other—and have been exposed to the Latvian majority while also continuing to develop in these circumstances—not all people with clear Livonian ancestry consider themselves to be Livonian.

Thus, an additional factor when determining membership in the Livonian community is one's awareness of belonging to the community and the manifestations of that awareness. This can be expressed in many ways, from listing Livonian as one's nationality in passports or noting it in public appearances to participating or showing interest in community activities or events, explaining Livonian history and culture to friends and colleagues, inspiring children and relatives to get involved in Livonian activities, or performing individual research or activities serving strictly personal interests in connection with the Livonians.

Concluding notes

The Livonians have shifted from being a localized community once bound to an isolated area, with a lifestyle and language different from its neighbors, to a scattered community, now separated from its historical territory for many decades, with Livonians today being diverse but disconnected from each other. The detachment of the Livonian community from its historical area, its overwhelming exposure to the majority, and the community's small size have heavily disrupted not only the sustainability of its intangible heritage and the Livonian language but also Livonian identity and self-perception, blurring community borders and burying Livonian voices under layers of outsider narratives.

There are many narratives that have affected and continue to affect Livonianness, above all those rooted in Latvian and Estonian discourse. Transferred to the Livonian case, these tend to put the language at the core of Livonian identity and cultural processes. Current research confirms that language is indeed located at the center of the Livonian community's ecosystem and has played a key role in ensuring the continuity of the Livonian community throughout time, and that even if community members are no longer proficient in Livonian, the language still carries symbolic value.

As identity is not frozen but instead variable and dynamic, emerging from individual experiences and conditions within society (Blackledge and Creese, 2016: 273), changes in the positioning of language as a key identity marker can be seen developing in the Livonian case over time. The research data show that before the expulsion, Livonian language proficiency functioned as a key marker of Livonian identity, and today, despite a lack of Livonian language skills, it is still language—but even more so, the symbolic role of the Livonian language and one's direct connection to the Livonian spoken in one's family—that is one of the strongest identifiers of Livonian identity.

As the boundary between the identity of the minority (Livonians) and the majority (Latvians) becomes increasingly blurred, it is language in particular that provides the needed distinction and takes on the role of the main marker of identity. It also holds a symbolic role in the self-respect of the community, which becomes magnified in the process of reclaiming one's identity (Craig, 1992: 23).

In summary, current research indicates that being a Livonian and a member of the Livonian community from the perspective of Livonians today is determined by two key factors. The first factor is a direct family connection with the Livonian Coast and the interwar Livonian-speaking community (in other words—with Vaalgamaa's list), which also ensures the community's acceptance. The second factor is one's awareness and expression of interest in being a community member via involvement in Livonian activities, whether these are organized or individual, active or passive.

Despite a harsh history and the passage of time, the Livonians have nevertheless survived as a nation up until today. Either with the help of outside narratives or by the matching inner voice of the community, making language a guiding star and a binding element of identity has worked for the Livonians. The challenge of sustaining the Livonians remains, however—the path to reaching this goal lies in finding the community's undistorted original voice reflecting its true hopes and color. And it really does not matter whether this path leads to brilliance or breakdown.

References

Auns, M. (2018), "Livonija un latvieši," in *Akadēmiskie raksti 2 sējumos "Latvija un latvieši"*, II sējums, 214–46, Rīga: Latvijas Zinātņu akadēmija.

Balodis, U. (2023), "Līvõd ēļ kōgaŗš ja mujāl. Lībiešu balss tuksnesī un citur," *Līvlizt Āigastrōntõz=Lībiešu Gadagrāmata*, 142–50, Rīga: Līvõd Īt | Līvu savienība.

Blackledge, A., Creese, A. (2016), "A linguistic ethnography of identity: adopting a heteroglossic frame," in S. Preece (ed.), *The Routledge Handbook of Language and Identity*, London: Taylor & Francis, pp. 272–88. https://doi.org/10.4324/9781315669816.

Blumberga, R. (2006), *Lībieši dokumentos un vēstulēs*, Rīga: Latvijas Vēstures institūta apgāds.

Blumberga, R. (2020), "Kārlim Staltem—150. Autobiogrāfija," *Līvlizt Āigastrōntõz=Lībiešu Gadagrāmata*, 57–66, Rīga: Lībiešu kultūrtelpa / Līvõd Īlma.

Caune, K. (2011), *Kangaru un vigu kompleksa hidroģeoloģiskie apstākļi Pitraga un Saunaga apkārtnē*, Master's Thesis, Rīga: Latvijas Universitāte, Ģeogrāfijas un Zemes zinātņu fakultāte.

Census (2011), *Latvijas 2011. gada tautas skaitīšanas rezultāti*, Centrālā statistikas pārvalde. Available online: https://stat.gov.lv/lv/statistikas-temas/iedzivotaji/iedzivotaju-skaits/publikacijas-un-infografikas/1299-latvijas-2011 (accessed June 15, 2023).

Cimermanis, S. (1998), "Par lībiešu saplūšanu ar latviešiem: Jautājuma izzināšanas ievirzei," *Latvijas Zinātņu Akadēmijas Vēstis*: A daļa, 3, 21–38.

Craig, C. (1992), "A constitutional response to language endangerment: The case of Nicaragua," *Language*, 68(1) (March): 17–24.

Endzelīns, J. (1980), *Darbu izlase*, III/2, Rīga: Zinātne.

Ernits, E. (2014), "Personal names and denomination of Livonians in early written sources," *Eesti ja soome-ugri keeleteaduse ajakiri. Journal of Estonian and Finno-Ugric Linguistics*, 5(1): 13–26. http://dx.doi.org/10.12697/jeful.2014.5.1.01

Ernštreits, V. (2011), *Lībiešu rakstu valoda*, Rīga: Latviešu valodas aģentūra, Līvõ Kultūr Sidām.

Ernštreits, V. (2023), *Lībiešu valoda*. Available online: https://enciklopedija.lv/skirklis/5259-l%C4%ABbie%C5%A1u-valoda (accessed September 29, 2023).

Ernštreits, V., Kļava, G. (2023), "Chapter 4 Experiences in Teaching an Endangered Language: Finding the Motivation and Means to Ensure the Acquisition of Livonian," In *Teaching and Learning Resources for Endangered Languages*. Leiden, the Netherlands: Brill. https://doi.org/10.1163/9789004544185_006

Hinton, L., L. M. Huss, and G. Roche (2018), "Introduction," in L. Hinton, L. M. Huss, and G. Roche (eds.), *The Routledge Handbook of Language Revitalization*, xix–xxx, New York and London: Routledge.

Indriķa hronika (1993), *Indriķa hronika*, trans. Ā. Feldhūns, Rīga: Zinātne.

Jääts, I. (2023), "Ränga ja Mägiste Liivi Retk," in M. Metslaid and I. Jääts (eds), *Keeruliste aegade kiuste. Gustav Ränga päevikud 1939–1948*, Tartu: Eesti Rahva Muuseum, pp. 265–71.

Kalnačs, B. (2023), *"Lāčplēsis," eposs*. Available online: https://enciklopedija.lv/skirklis/37565-%E2%80%9CL%C4%81%C4%8Dpl%C4%93sis%E2%80%9D,-eposs (accessed December 15, 2023).

Karma, T. (2014), "Kas nāca pēc neredzīgā?," *Jūrnieku svētās dziesmas un lūgšanas, 1845, 1914, 2014*, ed. V. Ernštreits, 7–20, Rīga: Ventspils lībiešu apvienība "Rānda."

Kivimäe, J. (2011), "*Henricus* the ethnographer: reflections on ethnicity in the chronicle of Livonia," in M. Tamm, L. Kaljundi, and C. S. Jensen (eds.), *Crusading and Chronicle Writing on the Medieval Baltic Frontier: A Companion to the Chronicle of Henry of Livonia*, 77–106, London: Routledge. https://doi.org/10.4324/9781315575186

Kļava, G. (2022), "Valodas loma lībiešu pēcteču identitātē un valodas nodošana nākamajām paaudzēm," *Valoda: nozīme un forma / Language: Meaning and Form* 13, Latvijas Universitātes Humanitāro zinātņu fakultātes Latvistikas un baltistikas nodaļas rakstu krājums, 124–38.

Kļava, G. and V. Ernštreits (2022), "Livonian in the linguistic landscape," *Eesti Ja Soome-Ugri Keeleteaduse Ajakiri. Journal of Estonian and Finno-Ugric Linguistics*, 13(1): 207–32. https://doi.org/10.12697/jeful.2022.13.1.08

Lewis, M. P. (2023), "Sustaining language use. Bridging the gap between language communities and linguists," in E. Derhemi and Ch. Moseley (eds.), *Endangered Languages in the 21st Century*, London and New York: Routledge, pp. 105–18.

Līveni (2002), *Latvju enciklopēdija 1962–1982*, 2. sējums, 540–1. Available online: https://web.archive.org/web/20071208120631/http://www.historia.lv/alfabets/L/LI/liiveni/encik/liiven_le_1962.htm (accessed January 25, 2023).

Noriņa, R. and L. Ozoliņa (2023) *Vidzemes lībieši: Informatīva brošūra*, Rīga: LU Lībiešu institūta atbalsta biedrība. Available online: www.livonian.lv/lv/blogs/informativa-brosura-vidzemes-libiesi (accessed September 15, 2023).

Osipova, S. (2021), "Latviešu valodas nozīme konstitucionālās identitātes kontekstā," *Latvijas Zinātņu Akadēmijas Vēstis*, A daļa, Nr. 4, 6–12.

Prozes, J. (1998), *Hõimuliikumine kahe maailmasõja vahel*. Available online: www.suri.ee/hs/prozes.html (accessed November 30, 2023).

Rozeniece, A. (2005), "Ar jaunrades un ticības atslēgu," *Latvijas Vēstnesis*, Nr. 92, June 10, 2005.

Samalavičius, A. (2023), *Baltic Postcolonial Narratives: Literature and Power*, Cambridge Scholars Publishing.

Stinkule, S. (2008), "Latvijas Nacionālā vēstures muzeja rokrakstu, zīmējumu, fotoattēlu un dokumentu kolekcija par Dundagas pagastu," *Kultūrvēstures avoti un Latvijas piekraste*, 238–58, Rīga: Latvijas Zinātņu akadēmijas vēstis.

Ščerbinskis, V. (2000), "Lībiešu jautājums Latvijas attiecībās ar Somiju un Igauniju 1920–1940," *Latvijas Zinātņu Akadēmijas Vēstis*, A daļa, Nr. 3–4, 6–7.

Tumzers, M. (2019), "Livonija viduslaikos. Pārdomas par kādu Eiropas vēsturisku reģionu," in A. Levāns, I. Misāns and G. Strenga (eds.), *Viduslaiku Livonija un tās vēsturiskais mantojums*, 24–41, Rīga: Latvijas Nacionālā bibliotēka.

Vēri, E. (1994), "Par lībiešiem un lībiešu valodu," *Lībieši: Rakstu krājums*, K. Boiko (ed), 227–48, Rīga: Zinātne.

Viitso, T. R. and V. Ernštreits (2012), *Līvõkīel-ēstikīel-leţkīel sõnārõntõz = Liivi-eesti-läti sõnaraamat = Lībiešu-igauņu-latviešu vārdnīca*, Rīga, Tartu: Latviešu valodas aģentūra, Tartu Ülikool.

Wiedemann, F. J. (1861), "Joh. Andreas Sjögren's Livische Grammatik nebst Sprachproben," in *Joh. Andreas Sjögren's Gesammelte Schriften*, Band II, Teil I, St. Petersburg: Kaiserlischen Akademie der Wissenschaften.

Zemītis, G. (2011), "Liivlased: vanim ajalugu (10.–16. sajand)" [Oldest history of the Livs (from 10th to 16th century)], in R. Blumberga, T. Mäkeläinen and K. Pajusalu (eds.), *Liivlased. Ajalugu, keel ja kultuur*, 75–104, Tallinn: Eesti Keele Sihtasutus.

14

The Hijab and Faith: Fashion and Modesty

Samira Adel Osman

Introduction

On the eve of the start of the World Cup in Qatar in November 2022, sporting institutions like FIFA, sponsors like a famous beer brand, and fans like the Brazilians were worried about something in common, and it was not the favorite country to win the world title. The concern was about some issues on the periphery of the sporting event: the ban on alcohol, issues related to the violation of human rights, the working conditions of immigrants, homophobia, and, of course, the oppression of Muslim women. Once the tournament got underway, tempers calmed and engagement with these issues waned as goals were scored.

Brazil was no different. In addition to sports coverage, the media also addressed these issues. In an article in one of the country's leading newspapers, an image of a woman in a black hijab, her eyes well lined with makeup, concentrating on the opening ceremony, took up more than half the front page of the special section on the World Cup. The title "Panos Quentes" (Hot Cloths), a play on words referring to clothing in the heat of the desert country, is also a Portuguese expression that could be translated as "pouring oil on troubled waters." Instead of calming tempers, the coverage was full of stereotypes, prejudices, and discrimination against Islam and women of the religion.

Muslims are a religious group that still has a modest presence in Brazil. We can assume that the presence of Muslims in the country is linked to the arrival of immigrants from Lebanon, Syria, and Palestine. This phenomenon began in the first decades of the twentieth century and intensified as a result of the civil conflicts in the region, such as the 1967 war in Palestine, the Lebanese civil war (1975–90), and the Syrian civil war (ongoing since 2011), creating an association between Muslim religious identity and Arab ethnic-national identity. Another form of this presence refers to the non-Arab Muslim immigrants who have arrived in Brazil from different regions of Africa and Asia, mainly concentrated in the South-Southeast and Center-West regions. They have migrated because of civil wars, and economic and environmental problems, and have found the opportunity to settle in the country as skilled workers for halal slaughter, an important agribusiness sector, as well as those who have recently arrived in Brazil, especially in São Paulo, from Afghanistan or other conflict regions. A third presence can be considered among converts of Brazilian origin who have joined the religion on an individual basis (Osman, 2019, 2018a).

From a numerical point of view, the data on the Muslim presence is contradictory. Officially, data from the 2010 census from the Brazilian Institute of Geography and Statistics (IBGE) is still in force, which puts the number of Muslims in Brazil at around 35,000 people (around 0.02 percent of the total population), while figures from PEW Research put the figure at almost 40,000 people. However, data from Brazilian Islamic organizations and in Latin America speak of millions of Muslims, which could also be an overestimate. There is no doubt about the Muslim presence in Brazil, which occupies an important position in Latin America, along with Argentina, Venezuela, and Colombia. This presence can be seen in the construction of mosques, charitable societies, centers for Muslim studies, schools, and cemeteries, which are more concentrated in the Southeast, South, and Center-West regions (Osman, 2019, 2018a).

Living as Muslims in Brazil: to use or not use the hijab

Regardless of the numbers, it is important to see how these people from a minority religion, in a peripheral Western country, predominantly Christian, have lived together and assimilated in Brazil, expressing their distinctive and visually marked religious practices in relation to women who cover their head and hair with scarves, veils, or hijabs. This chapter therefore looks at the experiences of young Brazilian-born Muslim women who have adopted the hijab and how this visual identity has affected or not their social, educational, professional, and cultural lives in the country. The first question to consider is why these young women, in their twenties and thirties, whose mothers, aunts, and grandmothers do not wear the hijab or started wearing it late, have begun to wear this religious marker as a way of reviving their Muslim faith and maintaining their identity.

First, I would like to consider aspects analyzed by Ahmed (1992, 2011), who has studied the issue of the resurgence of the veil among young Muslim women in the United States. In these works, Ahmed addresses the issue of the abolition of the veil from the historical perspective of women's emancipation movements as one of the consequences of the British occupation of Egypt (1882–1922). The veil became the ultimate symbol of female oppression in Islam, and its suppression became the path to modernization in these societies, and, as Hourani (1955) suggests, this movement that began in Egypt had repercussions in countries such as Syria, Jordan, Lebanon, and Iraq.

Indirectly, it is possible to infer that Muslim immigrant women in Brazil benefited from this movement and arrived in the country without wearing the veil, which I have verified myself in my research on the subject (Osman, 2007, 2011). Their explanations were accompanied by statements such as being young (at the time of their arrival) and the obligation to wear it, which was often linked to certain moments of religious practice (prayers, the pilgrimage to Mecca) or to a certain stage in life (marriage and the arrival of children and grandchildren). When they began to wear some form of head covering, it could not be considered the hijab, because of a lack of access to this garment either in remote decades or because of an adaptation expressed in the vocabulary. When asked about the garment, these women would say it was an *ishar* (a linguistic adaptation of the French *echarpe*), a kind of scarf that covered the hair in a less rigid way. It is the daughters and granddaughters, the second and third generations of those first immigrants, who have now taken to wearing the hijab and using this term.

In order to understand the issue of wearing the hijab among Muslim women in Brazil, I have been conducting oral history interviews with different generations of Muslim women in the city of São Paulo. Specifically for this chapter, I conducted six interviews, in 2023, with young women from Muslim families of immigrant origin, second and third generation, between the ages of 20 and 25 who have adopted the hijab. The objective is to understand their motivations for wearing it, the facilities or difficulties they face, and the religious issues that led to this decision.

These young women were brought up as Muslims, observing the practices considered the most common, such as the prohibition of alcoholic beverages and pork meat, the annual fasting during Ramadan, charity toward their neighbors and the observance of daily prayers. Some of them attended Islamic schools for part of their schooling, which was pointed out as making it easier for them to live together and maintain their practices by not feeling different or strange to others. Most of them are part of larger family and community circles, living with other families of the same origin and religion, and they attend an institution such as a mosque or Islamic association with the aim of studying, understanding, and practicing the Islamic faith more fully. They are engaged or married to young Muslims whose religion is one of the fundamental factors in their marriage. Some of them wore the hijab before their marriage and others after.

The hijab and faith

The main question asked of these girls was what motivated them to wear the hijab, and they all replied that it was a personal decision, not imposed by their fathers, mothers, husbands, brothers, or religious leaders. "I decided not to wear the hijab before, now I have decided to wear it" (Naima El Orra, 25); "I decided to wear the hijab" (Mariam Osman, 21). Religious obligation, a better way to practice the religion, to live and spread the faith, to feel fully Muslim, to connect spiritually with God, and to complete religious practices and pillars were the explanations that led to the personal decision. Most of them started wearing the hijab a few years ago, and from the beginning of their adulthood, although they expressed the wish to have started wearing it from the beginning of puberty, in accordance with some religious interpretations that identify the obligation to wear it at this stage.

They say that although they wanted to, it was their families, especially their mothers, who stopped them, because either they did not wear the hijab themselves or because they felt it was unnecessary at that stage of their lives. "I put on the hijab when I was about 13 years old and my mother did not want me to, even though she wore it herself. I always tell this story about when I started wearing the hijab to break with the idea that it is our parents who force us" (Fátima Fayad, 23). And also, "I thought that when I reached puberty my father would force me to wear the hijab, and I heard this from people outside my home, but I only started wearing it when I was 21 as my own decision" (Rena Jaber, 22).

The family and community to which these girls belong may or may not be a motivation for wearing the hijab. For some of them, the fact that their grandmothers, mothers, aunts, sisters, cousins, and friends wore the hijab was an incentive and a way of making their decision easier. Sometimes they take on the role of encouraging other girls and sharing the experience, understanding the difficulties of adapting to a new dress code and empowering themselves as a support in case of difficulties. "There are many different situations, sometimes you want to wear it but there is no support. I always wanted to. When you are part of a Muslim society where everyone wears a hijab, it is easier, there is more acceptance. You are not strange, you are not different" (Naima El Orra, 25). And:

> I think that nowadays people are looking for more information about religion, with the internet everything has become easier because you can follow and connect with other people from the same faith. Islam is also growing, there are more mosques, religious classes and lectures, all of which help people to seek the truth and find themselves. In addition, having a family and a group of friends who want the same things as you, who have the same goals, helps a lot in making this decision, it makes a difference.
>
> Soraia El Orra, 24

It can also be seen as female and even feminist solidarity in this process.

> My concern is that women wear the hijab by choice, because only those who wear it know the weight and the feeling of wearing it, what it is like, so if there's any possibility of someone imposing the hijab on a woman, I am against it, I think it is terrible, I will not tolerate this sexist attitude. A man will never know what it is like to wear the hijab, it is not his place to speak. No man has the right to impose something that he does not feel, because a girl who wears the hijab carries this mark by herself, especially in a country where we are a minority, so he cannot impose anything. He must respect the person's time and not interfere.
>
> Fatima Fayad, 23

Thus, while affirming their freedom of choice to wear the hijab, they refuse to impose it on any woman, as the men in the family are a support in the decision, not the reason for wearing it. "My fiancé was happy to see me wearing the hijab, he encouraged me in some way, but he was not the reason for my decision, which was personal and genuine" (Soraia El Orra, 24). The pressure is seen as misguided:

> My sister does not wear the hijab. When our brother tells her, even joking, that she should put on the veil and follow my example, I intervene and tell him to stop, because it should be her decision, she should not put it on because I did or because someone told her to. She cannot be pressured or forced. She has to study, understand, choose, decide; a decision that is individual and between her and God

only. It is a religious obligation, but no one can force you to wear it, not even your father or husband. It cannot be a sexist imposition.

Rena Jaber, 22

By choosing the hijab, these girls feel fully Muslim because "the hijab doesn't change who you are, it completes you" (Naima El Orra, 23), but they are no better than other Muslim women who don't wear it. For these girls, there is no contradiction, discomfort, or divergence in living with this diversity. They do not feel superior or more religiously correct than other women, and they see this diversity as part of the heterogeneity of the Muslim immigrant community in Brazil, because "everyone has their own time and their own way of wearing it or not wearing it. (Mariam Osman, 21) and "There are different ways of being Muslim: a Muslim woman with a hijab or a Muslim woman without a hijab will still be a Muslim" (Fátima Fayad, 23). The hijab is a way for these women to feel fully Muslim, to fulfill a duty, to be closer to God, and even to understand and defend other Muslim women who do not want to wear it. The choice to wear the hijab is linked to a vision of identity and a way of being recognized as a Muslim in non-Muslim societies, and is seen as a cultural marker in the integration of these women beyond their own social circles, such as the streets, schools, universities, and working spaces.

Adherence to the hijab also implies a change in the way they dress. As these women point out, it is necessary to adhere to a new dress code that goes beyond covering the head and hair. They have to give up transparencies, shorter or tighter clothes, or clothes that leave arms and legs exposed. The choice of wider, looser, longer clothes becomes a complement to the hijab, but no less elegant or detached from fashion and style. For some, it was a matter of giving up what they were wearing and adapting to a new style; for others, it was a matter of improving, as they were already dressing in a way that was closer to what is considered modest. There are also those who feel that the practicality of the clothing is more in line with religion. Opinions on the abaya vary: "I don't think it's necessary to wear it" (Mariam Osman, 21); "I know it's the right thing to wear, but I don't wear it" (Fátima Fayad, 23). On the other hand, "I wear the abaya, combining it with jackets, hijabs, because you can combine religion, style and fashion" (Soraia El Orra, 24); "I wear the abaya because I think it is more correct, but I did not wear it at first" (Rena Jaber, 22).

Although they all prefer to use the term hijab because it has a religious meaning and is more appropriate to the Arabic language (Mernissi, 1991), they also come across the words scarf, veil, and even cloth. The term headscarf or veil is considered appropriate when it comes to clarifying or facilitating understanding. However, there are terms that are considered offensive or inappropriate, including cloth. "I don't like it, but sometimes people say cloth because they don't know the right term" (Rena Jaber, 23); "It offends me when they say cloth, it sounds like I picked it up off the floor and put it on my head" (Naima El Orra, 25).

The hijab: prejudice and others

Burqa, another term used, in which case it is not a question of offense, but of total ignorance and the greater need to explain the differences and their uses. It may sound contradictory that the term burqa, which is more specific to the context of Afghanistan, is considered a common and popular way of referring to the hijab in Brazil. However, when one considers the media, the aftermath of 9/11, the Taliban and their recent return to power, which also led to the arrival of refugees in Brazil, it is understandable that the term burqa prevails alongside a negative and prejudiced view. A national media that echoes international coverage (Said, 1997; Osman, 2015, 2018b) fuels this use: "I was curious, but I had a lot of prejudices that I must have absorbed from the media, which construct a very negative picture of Muslims and Islam. On the subway, a man approached me and said I did not need the hijab and asked me if I knew who Osama Bin Laden was" (Rena Jaber, 23).

> Sometimes I hear statements that associate me with more violent countries such as Afghanistan, Pakistan, Saudi Arabia, Iraq, and even "from Arabia" or "from those people." The least bad is Dubai. They ask me where I am from and if I have to be born in my country to be a Muslim, because they confuse religion with nationality and have even said "go back to your country." If they hear from the media about Chechen, they will say that I am from there, because they only read or listen to the big headlines, nobody reads the smaller explanations.
>
> Naima El Orra, 25

She added:

> I think it's a lot of prejudice that comes from the image that the media has created, associating the idea of terrorism, female bombers and these things, so people end up having a much more stereotyped view of us. Every time I have felt attacked because I wear a hijab I felt it as an influence of that, but also a certain teasing; you hear "watch out, you are going to blow up the mall," for example.
>
> Fatima Fayad, 23

There is also a link with 9/11 and Osama Bin Laden. As Ahmed (2011) notes, the impact of 9/11 on Muslim immigrants needs to be considered. If, on the one hand, the events increased Islamophobia, the attacks on Muslims and especially on women wearing the hijab, on the other hand, they provoked a wave of curiosity to get to know this other: familiar and distant, equal and different, exciting and feared. In Brazil, this reaction was no different, leading the media in particular to search for the stranger living next door. To paraphrase an Islamic leader in the United States, 9/11 introduced Islam to Brazilians.

As noted by Abu-Lughod (2012) and Lalami (2015), salvationist and pitying discourses also fall on Muslim women in Brazil and, as observed by these young women, come from other women who feel uncomfortable with the hijab of others. The idea of female oppression in Islam is materialized in the hijab, which is seen as a symbol of male domination over women's bodies. Being in a non-Muslim country is seen as a safe haven for these women to liberate themselves, to feel saved, protected, and emancipated: "I feel that Brazilian women who are not Muslim give me permission to take off the headscarf. They tell me to take off the headscarf because we are in Brazil, because here I am free and not obliged. It is as if they were my saviours, you know?" (Naima El Orra, 25). In other words, "I always hear that in Brazil I am free, so I can take 'it' off my head" (Fátima Fayad, 23).

In the case of Brazil, there is another element to consider: the broadcast of a soap opera called "O Clone" less than a month after the attacks, which was rebroadcast 10 and 20 years later, after the events of 9/11 and their repercussions in the following decades. Human cloning was the central theme of the soap opera, but it became secondary to Islamic culture, based on issues such as forbidden love, marriage, polygamy, religion, tradition, customs, the veil, etc., which, oscillating between didacticism and Orientalism, offered viewers a humanization of people associated with terrorism and fanaticism as a result of the recent events. These girls also realized how the soap opera had an impact on the perception of the hijab in Brazil, making the garment visible to ordinary people. At the same time, it also reinforced misconceptions in which the veil, in the terminology used in the soap opera, was portrayed as a fashion and style accessory, and even as sensuality and fetishism along orientalist lines, as discussed by Said (1990) and Mernissi (2006, 2011).

Although they perceive these biased, discriminatory, and prejudiced views, the young women interviewed do not see these attitudes as Islamophobia, as has occurred in some European countries or in the United States. They see these attitudes as curiosity, ignorance, and even xenophobia, since the veil is seen as a national as well as a religious symbol. This is why they have heard direct expressions such as "go back to your country," denying their nationality and the Brazilian identity to which they belong, or even almost anecdotally:

> On the first day of classes at the Faculty of Dentistry, a teacher approached me, worried about how I could watch the classes without speaking Portuguese. I looked around and asked her: "Are you talking to me?" She was so happy and congratulated me for speaking Portuguese and I said, "Thank you, but it is my duty because I am Brazilian like you." She didn't think I was Brazilian because I was wearing a hijab.
>
> Fatima Fayad, 23

Islam and Brazil: a possible marriage

Although they identify and report these episodes, their assessment is that it is possible to be Muslim in Brazil, overcoming stares, curiosity, lack of information, common sense, and stereotypes, because these young women actually feel welcomed rather than discriminated against, especially because of the principle of secularism in force in the country. They say, "I do not understand the reason for the prejudices that exist outside Brazil. I have never experienced prejudice or discrimination, but I notice the looks, the curiosity, but it does not bother me because I do not have to take off my hijab" (Mariam Osman, 21). Again:

> There is curiosity, respect, prejudice, but compared to what happens in France, for example, I feel privileged. They may not like me because of my religion and the way

> I dress. However, here we have freedom of expression, we have rights, and I have never suffered anything other than bad looks, a bit of oddity, nothing more. Here I feel safe to wear the hijab and the abaya, even if some people get upset, I have the right to dress according to my religion.
>
> Rena Jaber, 22

Or: "I think Brazil is a good place to live, with or without the hijab" (Fátima Fayad, 23).

It is important to highlight the legislative issue: on the one hand, laws or legal rulings that allow Muslim women to issue official documents (identity card, driver's license, and passport) while wearing the hijab, and on the other hand, laws with commemorative purposes that highlight this presence. From the point of view of issuing documents, these women have the right to be identified and photographed while wearing the hijab according to the Brazilian Constitution, from 1988, which prohibits depriving Brazilian citizens of their rights on the basis of their religion. Legislation on the issuance of identity documents also does not prohibit the wearing of the veil.

Thus, on the basis of a spontaneous movement started by women who wanted to issue documents using the hijab, and with legislative support when this right was denied, the main cities with the largest Muslim populations, such as São Paulo, Rio de Janeiro, and Foz do Iguaçu (in the state of Paraná), have facilitated this right for women. Occasionally, these women are asked to provide a statement from a religious leader attesting to their faith, which can be obtained online.

Two types of situation are reported:

> My passport already has a photo of me wearing a hijab, as does my driving licence, but I have not yet had my ID renewed. I got a statement from the mosque, but they did not even ask me anything. I was very happy because I thought they would make it difficult and they did not even bother. I had no problems at the airport either. The first time I went to Lebanon, nobody called me. I have never had any problems in Brazil because of my religion, sometimes there are looks, curiosity, a certain strangeness, but there is no law that prohibits me, on the contrary. It is a question of accepting religion and differences, of coexistence, not a law that forbids us to be Muslims.
>
> Rena Jaber, 22

On the other hand, when the law was not in force, the vulnerability and violence of the obligation to remove the hijab as an imposition of authority were evident.

> When my mother came to Brazil from Lebanon, she had to take off her hijab to get her documents, and it was so invasive and violent for her that she never forgot the face of the man who took the photo! My sister did not have to take off her hijab, but she was asked to pull it back and show her ears, so it was a very strange photo. When I went to take the ENEM, the national university entrance exam, they took me to a separate room with only women. I was not prevented from taking the exam because of my hijab.
>
> Fátima Fayad, 23

On the other hand, in the same cities and throughout the country as federal legislation, several laws celebrate the presence of the Muslim community in Brazil, following the initiative of the existence of World Muslim Women's Day, celebrated on March 27, and World Hijab Day, celebrated on February 1. The oldest law dates back to 1992 and was enacted in the state of São Paulo, establishing September 14 as the "Day of the Muslim Community," and a municipal law in the city of São Paulo was enacted in 2007. May 12 has been celebrated as the "Day of Islam" in the state of São Paulo since 2009, as a municipal law in the city of Foz do Iguaçu since 2013, and as the "Day of the Muslim People" by state law in Paraná since 2017. At the federal level, the "National Day of the Muslim Community," celebrated on January 24, came into force in 2022. The content of these laws revolves around the recognition of the Muslim presence and Islamic cultural heritage in the formation of the country; the intention to reduce prejudice and religious intolerance against Muslims; and the promotion of respect and coexistence between people and their religions.

Conclusion

These young women are also connected to the new media and use social networks, either as followers or influencers, to break stereotypes and create a new vision of Muslim women and the Islamic religion in general (Karhawi, 2018). So-called Muslim influencers, some of whom already have millions of followers, are seen as playing an important role in explaining the religion, minimizing prejudices, and breaking stereotypes that limit these women to a religious social role. They clear up doubts about religion and the hijab in particular, answering questions such as "Do you take your hijab off to shower?"; "Can you take your hijab off after marriage?"; and "Did your husband force you to wear the hijab?" They also share their opinions on fashion and lifestyle, personal care and beauty, marriage and motherhood, and professional life in a way that reconciles their religious choices.

On social media, the hijab is seen from a modernizing and emancipatory point of view of Muslim women, giving a new meaning to clothing and the Islamic way of life in the contemporary world, as well as the dictates of fashion, beauty, and style, and so social media plays a role in demystifying the hijab.

> The interesting thing that influencers are doing is putting us on the map. We are a community, we exist, we are here, and our lives are not as different as you think, we go to the gym, we buy clothes, we study, we work, it is not as closed as you think. They are breaking the paradigms and the vision that people have of us. I myself am a 23-year-old Muslim girl who wears a hijab, who has Tik Tok and Instagram accounts and who interacts with young people my age who realize that I live a life like theirs.
>
> Fatima Fayad, 23

The issue of fashion and the aesthetics of the hijab deserves a brief analysis. While the political debate is increasingly about restrictions on the hijab, the issue is different in the fashion industry. Modest fashion, or dressing modestly, is associated with both

Islam and Pentecostal or neo-Pentecostal evangelical Christianity, and since the capitalist market has no prejudices or restrictions, they adapt to and profit from this niche. Western brands, stylists, commercial networks, individual entrepreneurs, and digital influencers are dedicated to producing and promoting this style of dress. In this case, the dresses transcend religious issues, as they can be worn by women who are concerned about the dress code of their religion, or by women who follow the style without belonging to any faith, thus normalizing its visualization in the West (Kiliçbay and Binark, 2003; Lewis, 2015).

For the German fashion designer Wolfgang Joop of Marks and Spencer (Knipp, 2016), who produces burkinis and plans to open a branch in Tehran, the justification for producing Islamic fashion is, paradoxically, to understand it as an "expression of female emancipation," while stating that "the whole attitude, the culture, the oriental fashion has something very, very erotic and very mysterious." Thus, the same orientalist stereotypes are used to defend Islamic culture and women's clothing, as Said (1990) and Mernissi (2006) have rightly pointed out.

In the contradiction between dressing modestly, adhering to the hijab as a form of modesty and protection, and guaranteeing their presence in the public sphere in accordance with religious precepts, women are also caught up in the world of business. This, in turn, disciplines their bodies and their faith, creating a tension between aesthetics and fashion, the market and consumption, and faith and religion, which together face challenges that call into question the revival of religion in the contemporary world. Challenges that women in hijab have to face, which can be summed up in the following expression:

"If my hijab offends you, I'm sorry, but it's none of your business!"

References

Abu-Lughod, L. (2012), "As mulheres muçulmanas precisam realmente de salvação? Reflexões antropológicas sobre o relativismo cultural e seus outros," *Revista Estudos Feministas*, 20(2), https://doi.org/10.1590/S0104-026X2012000200006

Ahmed, L. (2011a), *A Quiet Revolution: The Veil's Resurgence, from the Middle East to America*, New Haven and London: Yale University Press.

Ahmed. L. (2011b), "Veil of ignorance," https://foreignpolicy.com/2011/04/26/veil-of-ignorance

Ahmed, L. (1992), *Women and Gender in Islam: Historical Roots of a Modern Debate*, New Haven: Yale University Press.

Hourani, A. (1955), "Are women inferior beings? The changing outlook in politics, education, employment, society, the home," The Unesco Courier, 35–7. https://unesdoc.unesco.org/ark:/48223/pf0000069853

Karhawi, I. (2018), *De blogueira à influenciadora: motivações, ethos e etapas profissionais na blogosfera de moda brasileira*. (Tese de Doutorado). Universidade de São Paulo, São Paulo, Brasil.

Kiliçbay, B. and M. Binak (2003), "Cultura do consumo, islã e política do estilo de vida: o véu como moda na Turquia contemporânea," *Revista Eco-Pós*, 6(1), https://doi.org/10.29146/eco-pos.v6i1.1145

Knipp, K. (2016, June 8), "Estilistas descobrem a moda islâmica." www.dw.com/pt-br/estilistas-descobrem-a-moda-isl%C3%A2mica/a-19314595#
Lalami, L. (2006), "The missionary position," *The Nation*. www.thenation.com/article/missionary-position
Lewis, R. (2015), *Muslim Fashion: Contemporary Style Cultures*, London: Duke University Press.
Macedo, S. and V. Damasceno (2022, November 21), Panos Quentes. *Folha de S. Paulo*. pp. 1–2.
Mernissi, F. (2006), *El Harén en Occidente*, Madrid: Espasa.
Mernissi, F. (2011), *Beyond the Veil: Male-Female Dynamics in Modern Muslim Society*, London: Saqi Books.
Mernissi, F. (1991), *Women and Islam: An Historical and Theological Enquiry*, Oxford: Basil Blackwell.
Mule, P. and D. Barthel (1992), "The return to the veil: individual autonomy vs. social esteem," *Sociological Forum*, 7(2), https://doi.org/10.1007/BF01125046
Osman, S. A. (2007), *Entre o Líbano e o Brasil: dinâmica migratória e história oral de vida*. (Tese de Doutorado). Universidade de São Paulo, São Paulo, Brasil.
Osman, S. A. (2011), *Imigração Árabe no Brasil: histórias de vida de libaneses muçulmanos e cristãos*, São Paulo, Xamã.
Osman, S. A. (2015), "Mulheres Muçulmanas no Brasil: o fardo da pena," in S. D. Souza and N. P. Santos (eds.), *Estudos Feministas e Religião*. Tendências e Debates (237–59), Vol. 2. Curitiba: Prismas, 2015.
Osman, S. A. (2018a), Être musulman arabe entre émigration au Brésil et retour au pays: identité culturelle ou religieuse ? *Brésil(s)*. http://journals.openedition.org/bresils/3273
Osman, S. A. (2018b), O islamismo na mídia brasileira: entre humor, exotismo e terrorismo. Inside C. Victor, M. Goktepe, R. Chiachiri, Y. Elemen. *Posições diante do Terrorismo: religiões, intelectuais e mídias*. São Paulo: Labrador. p. 45–78.
Osman, S. A. (2019), Presença muçulmana no Brasil: breve síntese histórica. *Hamsa* [Online], 5 | 2019, http://journals.openedition.org/hamsa/343.
SAID, E. W. (1990), *Orientalismo. O Oriente como invenção do Ocidente*, São Paulo: Companhia das Letras.
SAID, E. W. (1997), *Covering Islam. How the Media and the Experts Determine How We See the Rest of the World*, New York: Vintage Books.
Tamayo-Acosta, J. J. (2014), *Islam: sociedad, política y feminismo*. Madrid: Dykinson.

Part Three

Perspectives

15

Thirty Years of Democracy in South Africa: Reflecting on Social Work and Climate Change Challenges through a Ubuntu Lense

Kefilwe Johanna Ditlhake

Introduction

After thirty years of democracy in South Africa, decolonization and indigenization continue to occupy an important place in social work education and practice by many policymakers, academics, and practitioners. The origin of social work and community development in South Africa is influenced by the social, political, and socioeconomic conditions underwritten by apartheid doctrine, racial segregation, and colonial legacies. Social welfare and social work mirrored the ideologies and systems of colonial power (McKendrick, 1987; Patel, 2015). Education systems were defined and based on racial grounds and social inequality. Due to the legacy of colonialism and apartheid, Western-Eurocentric social work theories or models have influenced education and practice in South Africa and other African countries in general, marginalizing indigenous African philosophies such as ubuntu as well as indigenous knowledge. Midgley (1981) denotes this as cultural imperialism. Additionally, Dei (2000: 113) points out that indigenous knowledge does not reside in a "pristine fashion" and highlights the need to challenge imperialist ideologies. Cajete et al. (2023: 92) suggest decentering cultural imperialism and that its limits "can only be dismantled through a fundamental change in awareness of diversity, in which universality is replaced by pluriversality." The challenge of the marginalization of the indigenous worldviews is the impetus that led to the renewed interest in the decolonization of social work curriculum, education, and practice in South Africa. Decolonization encompasses epistemic decentering and what some scholars refer to as "epistemic disobedience" (Mignolo, 2010: 160) or "epistemic reconstruction" (Quijano, 2007: 176). The decolonization agenda calls for the inclusion of indigenous African philosophy of ubuntu, values, and theories in the education and practice of social work and community development. Central to the decolonization debates is the need for the "recognition of indigenous epistemologies as legitimate and unique scientific paradigms" (Cajete et al., 2023: 92). In 2021, ubuntu was selected as a theme for the

2021 World Social Workday to acknowledge indigenous African social work, worldviews, and ways of life. "In South Africa before 1994, social work and social welfare were fragmented, exclusive, and organized on racial grounds that mirrored apartheid policies" (Ditlhake, 2021: 28). The social welfare system and social work profession have been closely linked to apartheid discriminatory policies, as their existence was associated with addressing the "poor white" problems (McKendrick, 1990). With transformation and the calls for indigenization after the end of apartheid in 1994, methodologies in African research contexts still tend to be based on Western epistemology (Keane, Khupe, and Seehawer, 2017). While considerable progress has been made for social inclusion and diversity, the challenge of embedding and incorporating ubuntu African philosophies in education and practice remains paramount. This reality check and challenge may be referred to as an identity crisis in African social work, which is persistent after the end of colonial rule and apartheid in South Africa. Social work was criticized for struggling to retain its relevance in developing countries or sub-Saharan Africa with its Western principles and approaches (Mupedziswa, 2001). It is against this background that the paper examines the perspective of ubuntu in social work and community development. Specifically focusing on the notion and the nature of ubuntu and fostering ubuntu as a decolonial practice in social work and community development related to disaster situations.

Legislation and policy context fostering diversity and inclusive ubuntu Indigenous worldviews

Decolonization and the end of the apartheid system in 1994 resulted in a transformation of the social, economic, and political structures of society in South Africa. Ubuntu was used in the 1990s during the transition period as the guiding ideal for transitioning from apartheid to a democratic government (Christian, 2011). Ubuntu appears in the Interim Constitution of South Africa (1993) and it is stated that "there is a need for understanding but not for vengeance, a need for reparation but not for retaliation, a need for *ubuntu* but not for victimization." After South Africa gained independence in 1994, the constitution of the Republic of South Africa, 1996, as the supreme law, recognized the importance of the indigenous worldviews, or ubuntu. The constitution as the legal framework provides opportunities for reconceptualizing indigenous worldviews for social work and community development in South Africa. The ubuntu indigenous worldviews and customary law are acknowledged in section 211(3) of the constitution and section 27 of chapter two of the Bill of Rights. Furthermore, the White Paper for Social Welfare (1997) policy framework recognizes that the developmental approach to social welfare in South Africa in social work and community development is premised on the principle and indigenous values of ubuntu and further emphasizes that in the area of child protection, the traditional and indigenous systems of foster care will be recognized provided that the best interest of the child is protected. However, the Code of Ethics of Social Workers in South Africa by the South African Council of Social Service Professions (SACSSP) has considerably not changed and remained engrained in Western worldviews and epistemological traditions. Despite the legislative mandates and policy frameworks, the African indigenous worldviews, knowledge systems, and epistemologies continue to carry the challenge of marginalization.

In the South African context, the end of colonial rule and attainment of independence in 1994 have seen social work academics, practitioners, policy reformers, and academics in higher education institutions challenging Western-Eurocentric-centered theories and education as inappropriate and alien to local indigenous values, culture, and needs. Decolonization and indigenization of the African worldviews, therefore, in social work and community development, call for an end to "the domination of Western epistemological traditions and histories" (Molefe: 32). However, despite the legislation and policy mandates after 1994, the Western worldviews and epistemological traditions dominate the theory of knowing, thus perpetuating marginalization of indigenous worldviews. In pursuit of the decolonization agenda in social work, perspectives provide an opportunity for a paradigm shift and rethinking theories and methodologies that are African and re-center indigenous worldviews and knowledge.

What is the African philosophy of ubuntu?

There are diverse meanings and definitions of ubuntu, African philosophy, and indigenous value systems. The lack of a single definition and explanation of ubuntu is connected to ubuntu as a living practice and being open to change over time. The literature suggests that ubuntu's ideal is often explained with the Bantu proverb *umuntu ngumuntu nga bantu* (*in Setswana motho ke motho ka batho ba bangwe*); this translates in English loosely as "a person depends on other persons to be a person" (Ditlhake, 2020a; Matolino and Kwindingwi, 2013; Mkhize, 2008; Karenga, 2004; Ramose, 1999; Samkange and Samkange, 1980). In a Western aphorism, Hailey (2008) notes that the individual is expected to say, "I think, therefore I am."

Despite the multiplicity of the various definitions, there is also a broad agreement in the explanation of the concept of ubuntu. The commonality in the definition of ubuntu is humaneness, group solidarity, communitarianism, indigenous values or ethics, and African philosophy. Recent literature offers the following definitions:

- For Dikoko and Mokhatla (2006), ubuntu is premised on" deep respect for the humanity of another." Ubuntu denotes "group solidarity and communitarianism" representing a shift from the Western values of individualism. The virtues of ubuntu include inclusiveness, community, family, belonging, human dignity, and respect.
- Mugumbate and Chereni (2019) define *ubuntu* as "a collection of values and practices that Black people of Africa or of African origin view as making people authentic human beings. While the nuances of these values and practices vary across different ethnic groups, they all point to one thing—an authentic individual human being is part of a larger and more significant relational, communal, societal, environmental, and spiritual world."

- Ubuntu is an "African humanistic and ethical worldview where communities as part of common humanity are necessarily part of what makes us human" (Ditlhake, 2020: 163).
- Mokgoro (1998: 1) described ubuntu as "the world view of African societies and a determining factor in the formation of perceptions which influences social conduct. It has also been described as a philosophy of life that in its most fundamental sense represents personhood, humanity, humaneness, and morality; a metaphor that describes group solidarity where such group solidarity is central to the survival of communities with a scarcity of resources, where the fundamental belief is that a person can only be a person through others."

Ubuntu as a social value and living practice

Ubuntu is an undefinable concept. There is no single definition of ubuntu. Portraying a single concept or definition of ubuntu will be a mere simplification or representation. Mokgoro (1998) reminds us that any attempt to define ubuntu will be a simplification of its complexity. Hence, it is in this context that the concepts associated with ubuntu are identified and set a circumstance that shows how ubuntu is a living practice or living ethical thought and will always change as it does not always remain constant and static. As Ditlhake (2020: 165) notes, ubuntu changes over time and recognizes differences of experience, diversity, and humaneness. It is in this context that we speak of the concepts that are closely associated with ubuntu. Mokgoro (1998: 1) argues that the "meaning of ubuntu becomes clearer when its social value is highlighted." Underlying the diversity of the definitions of ubuntu are the key concepts associated with ubuntu as a living practice. The following are the conceptions associated with the social value of ubuntu.

As a greeting, ubuntu in different African languages is revealed as a starting point in engagement with people. In Setswana, the usual greeting is translated as *dumelang,* and it is used to affirm. In isiZulu, it is *sawubonana*, meaning "I see you." This indicates that the starting point linked to ubuntu is the affirmation of our existence and acknowledgment of our humaneness. As Ditlhake (2020: 169) puts it, "humaneness is an important ubuntu ethical value in the sense of perceiving human needs, dignity, sensitivity towards others, desirable quality, humbleness, interest or aspirations and that it will therefore be important in social work education and community development." Humanness "regards being, or the universe, as a complex wholeness involving the multilayered and incessant interaction of all entities" (Ramose, 2009: 308) and is associated with ubuntu, and it is also referred to as *botho* in the Setswana language. Ubuntu as a humanitarian ethic and the worldview deep-rooted in communitarianism become crucial from this perspective as they disengage from Western individualistic ways of knowing and interacting with people, communities, and their social environment. It is in this context that the dictum "A person is a person because of other people" (Ramose, 1999; Teffo, 1994) means "*motho ke motho ka batho*" or "*umuntu ngumuntu ngabantu*". Meaning co-existence and human interdependence in social interaction. Nussbaum (2009b: 231) identifies the ontological and epistemological aspects of ubuntu in the explanation of "*motho ke motho ka batho ba bangwe*," and it captures both the ontological and the ethical dimensions: "to be a human be-ing is to affirm one's humanity by recognizing the humanity of others and, on that basis, establishes humane relations with them. Ubuntu is understood as being human (humaneness); a human, respectful, and polite attitude towards others constitutes the core meaning of this aphorism." Ubuntu recognizes individuals' human dignity and worth regardless of their social positions in society.

The interconnectedness and group solidarity are central features of ubuntu in the African conception, and the philosophy of human nature is crucial to survival issues in terms of people relating and finding each other. Several scholars (Ditlhake, 2020; Ramose, 2001; Mugumbate and Chereni, 2019) have pointed out that group solidarity and interconnectedness are the fundamental features of ubuntu and community to strengthen the solidarity to which a person belongs. The person's responsibilities to others are as important as my responsibility to the person's self-interest. A person cannot be human without the community to which one belongs. The community has a wide range of meanings and encompasses the collective group and social sticking together, or social cohesion. For instance, Nyerere's (1968) philosophy of Ujamaa reflected the notion of familyhood and communalism. This view is in line with the sentiments expressed by Ramose's (2001) assertion that a person derives their *botho* through relational connection with other people in their communities. Ditlhake (2020: 329) explains that the concept of community is "viewed as encompassing a complex system of interaction that includes historical and ecological components, as well as the political, economic, social and cultural perspectives." In the context of community development, it is about acting collectively to overcome challenges and differences and to meet basic needs that people require to pursue the comprehensive vision of the good for people and communities. Communality, collectivism, and group solidarity prioritize the collective group over the individual interest and the relational identities that shape ubuntu African living practice.

Ubuntu (*botho*) and the knowledge base of social work

Social work is a helping profession that utilizes professionally qualified personnel who use their knowledge, values, and skills to help people deal with social problems (McKendrick, 1987; Mupedziswa, 2005). The social context in which social work is practiced varies from one country to another. The bulk of the social challenges are developmental in nature and are related to poverty, unemployment, and other socioeconomic challenges. The major concern for modern social work in South African society and in Africa relates to the remedial model rather than the developmental approach. The origin of social work in South Africa is rooted in colonialism and was imported from Britain (McKendrick, 1987), and indigenous worldviews, knowledge, and methodologies were marginalized. The tentacles of colonialism and apartheid legacy influence the modern-day challenges of social work. Western Eurocentric worldviews and epistemology predominate the theory and individualistic ways of knowing. Decolonization and transformation of social structures led to the transformation of social welfare to a developmental social welfare as mandated in the White Paper for Social Welfare (1997). South Africa adopted ubuntu for its reform and as a guiding philosophy. Since the attainment of the democratically elected government after 1994, the knowledge landscape, despite the profession continuing to struggle for its identity, relevance, and appropriateness in terms of education, theory, and practice. As Mwansa (2011: 1) notes, social work lacks relevance in terms of its philosophical value and ideological base.

Ubuntu is also known in the Setswana language as *botho,* representing the personhood or mannerism and the positive character of the person and toward others. Community interest remains the foundation of the African philosophy in which community interest takes precedence over individual interest or self-interest. As Ramose (2001) and Mbigi (2000) state, ubuntu is deeply engrained in community or community-hood. Ubuntu, the African worldview, indigenous knowledge, and values have been widely investigated in education (Xulu, 2015), philosophy (Ramose, 2001), law (Mokgoro and Woolman, 2010), and psychology (Mkhize, 2004). However, in social work and community development, research studies focusing on ubuntu and community development practice are limited and evolving. The ubuntu as an expression of humanity resonates with the value of social work, the value of human rights, the value of social work (Ife, 2008; Patel, 2015), and the theory of human rights (Fox, 2001: 275; MacFarlene and Weiss, 2000: 112). Values are the knowledge base of the social work profession. Ubuntu as an indigenous value system should form part of the social work knowledge base, and "ubuntu is the foundation of the value dimension of the relationship building in communities" (Ditlhake and Maphosa, 2021: 146). Fostering ubuntu as a decolonial practice in social work and community development is fundamental to ensuring contextual relevance and effective execution of social work and community development practice.

Centering ubuntu in community development and climate change and natural disaster

Climate change and natural disasters are ravaging communities and present a modern challenge in modern society for social work. The United Nations (1992: 3) refers to climate change as a "change of climate which is attributed directly or indirectly to human activity that alters the composition of the global atmosphere, and which is in addition to natural climate variability observed over comparable time periods." In South Africa, communities have been hardest hit by natural disasters related to climate change. The centrality of social work's role in social work in assisting families and communities to recover resonates with ubuntu's collective value of humanity, human rights, and social justice. As Nyahunda et al. (2019) state, natural disasters and climate change fall within the psychosocial role of social work and mandate, as the floods and heavy rains result in the displacement of people and affect the well-being of marginalized people in communities.

Climate-related disasters are ravaging developing countries, leaving rural communities suffering from vulnerabilities and shocks. Social work, as a profession that aims to improve the quality of life and the welfare of people, has a key role to play in assisting communities. The ubuntu African indigenous worldview in community development provides the potential to facilitate social work involvement in community activities aimed at addressing social problems by collaborating and partnering with communities as key stakeholders and other relevant partners in the field of social welfare. Collaborative partnerships at the community level and ubuntu are an important form of community connectedness, mutual reciprocity, and collective decision-making (Ditlhake, 2022). Social work as a value-based profession that recognizes differences in experiences of diversity when working with groups, families, and communities recognizes experiences of diversity that resonate with ubuntu. Collective humanity, interdependence, group solidarity, and interconnectedness in community action play fundamental roles when communities are threatened and affected by natural disasters because of climate change, such as floods and earthquakes. The macro-level social work fits well with the philosophy of ubuntu as it is embedded in community and group connectedness. Community development plays an important role in the improvement of the quality of life and sustenance of humanity. It is for this reason that humanity, as the value of ubuntu, finds its motivation in social work and community development. Using ubuntu in community development to achieve socioeconomic needs presupposes a dual model of intervention that combines remedial and developmental approaches for the improvement of quality of life. Community development plays an important role in communities affected by natural disasters such as heavy rains and floods caused by climate change for humanitarian action and the benefit of the sustenance of humanity.

Using ubuntu in community development to achieve socioeconomic needs presupposes improvement of quality of life. It is for this reason that ubuntu provides an ethical framework and guide in a complex, challenging situation at the community level. Ubuntu's principle of humanity has a profound resonance with human rights and social justice in social work (Patel, 2015; Ife, 2013). The humanity principle also plays a crucial role in intervention with communities in a disaster situation, fostering relationships and interacting with stakeholders at the level of communities when there are vulnerabilities in families and communities. The centrality of ubuntu in a disaster situation places collective humanity at the community level in community development intervention. When a natural disaster affects people at the community level, challenges such as loss of human life, displacement of families, and economic and infrastructural disruptions become major challenges (Mhlanga et al., 2019).

Over the past few years, South Africa has experienced earthquakes, tornadoes, and floods, as well as wildfires in recent years. In 2017 and in September 2023, the weather storm affected seven of the nine provinces, with KwaZulu-Natal at the epicenter. The deadly storms and floods were reported to be one of the worst weather storms that happened in the history of South Africa. Natural disasters and climate change affected most of the poor people in communities. The deadly floods destroyed many houses and damaged infrastructure such as roads and buildings. Many families were displaced, which resulted in trauma among individuals and families. This imminent threat to life, resulting from either natural disasters or climate change, triggered deadly floods in South Africa. The core elements of ubuntu include the spirit of connectedness, solidarity, interdependence, compassion, caring, and humanity, which promote family and community integration (Ditlhake, 2020). These principles discourage interventions that materialize only in response to natural disasters or post-disaster. Instead, ubuntu is the social value and a way of living. More work could be done in the absence of disaster by embodying the practice that is firmly grounded in ubuntu and the needs of the communities and society that it serves, encouraging group solidarity, interdependence, mutual respect, and reciprocity.

Community development and social work's role in climate change-induced disaster

The debates about climate change and disaster are silent in scholarly work and community development in South Africa. The contribution of social work to climate change has been recognized at an international level (Dominelli, 2011) and social workers are trying to keep up with the changing contextual situation and to understand the science behind climate change. Dominelli (2011: 435) asserts that for social workers to remain relevant and appropriate in this role, social workers "can promote clean, renewable energy to enable people's living standards to rise, without increasing greenhouse emissions at the unsustainable rates set by carbon-based technologies." Climate change not only affects physical infrastructure but also has social and economic impacts on the daily lives of individuals, families, and communities. The profession has an important supportive role to play to deal with the repercussions of the disaster prompted by climate change. The professional responsibility of social work is to work with diverse people in their social environment from social justice and human rights perspectives. In community development, social workers are well placed to contribute to climate change by embedding the indigenous value of ubuntu, intervening by mobilizing people, and offering support to communities affected by disaster brought on by climate change. The increasing disasters induced by climate change in communities across various provinces highlight the pivotal role of social work to address socioeconomic situations, traumatic events, and marginalized and vulnerable communities. Community development is interventionist in nature and applicable to people experiencing social and economic challenges in the aftermath of climate change-related disasters. Different community development models and frameworks for partnership and collaboration exist for intervention in community development (Ditlhake, 2021; Weyers, 2013). Ubuntu philosophy and indigenous worldview involve a communal way of life in contrast to the Western ideology of individualism. Given the increase of the heavy rains, floods, and climate change that caused devastation in communities. Many people are displaced and traumatized. Collective action informed by spirit of ubuntu by giving, sharing and helping hands are the essence of ubuntu in the context of community development. Community development refers to interventions with individuals, groups, families, and communities at the community level that advance humaneness, caring, human relations, collective action, and respect for human dignity principles foregrounding ubuntu. For example, intervention and relief are offered to displaced individuals and families affected. Many shelters, sports halls, and community halls are used as temporary measures for providing relief to families who lost their houses, particularly immediately after a disaster. In a spirit of ubuntu and caring, some families and individuals are accommodated by friends, relatives, and neighbors in the community. Assessment and intervention guide social workers to provide psychosocial support to people affected and traumatized by disasters.

Conclusion and recommendations

Disaster and climate change are among emerging community developmental challenges in South Africa. In this regard, social work intervention has an important role to play, especially in the context of climate change preparedness, which is increasingly posing a major threat to the survival of individuals, families, and society communities. The centrality of ubuntu as an African indigenous knowledge system undergirds the reappraisal of social work as a significant influence toward contributing to the full comprehension of local needs in the context of community development and offers possibilities for contributing to the welfare of families, individuals, and communities. Social workers can directly work with communities, engaging the stakeholders and disaster management teams to address the challenges imposed by climate change. Ubuntu and community development offer opportunities for collective action to generate solutions to climate change challenges and address psychosocial issues as well as socioeconomic imperatives in a spirit of ubuntu interdependence.

References

Cajete, G., K. J. Ditlhake, K. Kiewitt, T. Keibl, R. Lutz, and N. Schirilla (2023), "Indigenous knowledge—Indigenous science," in M. Kottig, S. Kubisch, and C. Spatscheck (eds.), *Geteiltes Wissen-Wissensetwicklung in Disziplin und Profession Sozialer Arbeit, (GGS). Knowledge and social work: Volume VIII: Social work of the South*, Opladen: Paulo Freire Verlag, pp. 159–176.

Christian, N. B. (2011), "The historical development of the written discourse of Ubuntu," *South African Journal of Philosophy*, 30(3): 303–29.

Dei, G. (2000), "Rethinking the role of indigenous knowledges in the academy," *International Journal of Inclusive Education*, 4(2): 111–32.

Dei, G. J. S. and A. Asgharzadeh (2006), "Indigenous knowledge and globalization: an African perspective," in A. A. P. Abdi, P. Korbla, and G. J. S. Dei (eds.), *African Education and Globalization. Critical Perspectives*, Oxford: Lexington Books, pp. 53–78.

Dikoko, D. and P. v Mokhatla, (2006), (CCT62/05) [2006] ZACC 10; 2006 (6) SA 235 (CC); 2007 (1) BCLR 1 (CC) (August 3, 2006).

Ditlhake, K. J. (2020a), "Ubuntu/Botho culture: an indigenous African value system and community development," in U. Straub, G. Rott, and R. Lutz (eds.), *Knowledge and Social Work: Volume VIII: Social Work of the South*, Opladen: Paulo Freire Verlag, pp. 159–176.

Ditlhake, K. J. (2020b), "Centering decolonisation analysis for Indigenous social work education and practice in South Africa," in U. Straub, G. Rott, and R. Lutz (eds.),

Knowledge and Social Work: Volume VIII: Social Work of the South, Opladen: Paulo Freire Verlag, pp. 209–20.
Ditlhake, K. J. (2022), "Multidisciplinary community development: a substantive theory for community partnerships processes," *African Journal of Governance & Development*, 11(1.1): 164–89. Retrieved from https://journals.ukzn.ac.za/index.php/jgd/article/view/2562
Ditlhake, K. J. and N. Maphosa (2021), "Fostering healthy human relationships at community Level in post-apartheid South Africa," in T. Kleibl, R. Lutz, N. Noyoo, B. Bunk, A. Dittmann, and B. Seepamore (eds.), *The Routledge Handbook of Postcolonial Social Work*, New York: Routledge, pp. 143–56.
Dominelli, L. (2011), "Climate change: social workers' roles and contributions to policy debates and interventions," *International Journal of Social Welfare*, 20: 430–8, doi: 10.1111/j.1468-2397.2011.00795.x.
Fox, F. (2001), "New humanitarianism: does it provide a moral vanner for the 21st century?," *Disasters*, 25(4): 275–89.
Gray, M., J. Coates, and M. Y. Bird (eds.). (2008), *Indigenous Social Work around the World*, Cornwall: Ashgate.
Ife, J. (2008), *Human Rights and Social Work. Towards Rights Based Practice* (revised ed.), New York: Cambridge University Press.
Ife, J. (2013), *Community Development in an Uncertain World: Vision, Analysis and Practice*, New York: Cambridge University Press.
Keane, M., C. Khupe, and M. Seehawer (2017), "Decolonising methodology: who benefits from indigenous knowledge research?," *Educational Research for Social Change*, 6(1): 12–24. doi:10.17159/2221-4070/2017/v6i1a2
MacFarlene, S. N. and T. Weiss (2000), "Political interest and humanitarian action," *Security Studies*, 10(1): 112–42.
Maris, C. W. (2020), "Philosophical racism and ubuntu: in dialogue with Mogobe Ramose," *South African Journal of Philosophy*, 39(3): 308–26.
Matolino, B. (2015), "A response to Metz's reply on the end of Ubuntu," *South African Journal of Philosophy*, 34(2): 214–25.
Matolino, B. and W. Kwindingwi (2013), "The end of Ubuntu," *South African Journal of Philosophy*, 32: 197–205. doi: 10.1080/0 2580136.2013.817637.
Mbigi, L. (2000), *In Search of the African Business Renaissance*, Johannesburg: Knowledge Resources.
McKendrick, B. W. (1987), *Introduction to Social Work in South Africa*, Pinetown, South Africa: HAUM Tertiary,
Mckendrick, B. W. (2001), "The promise of social directions for the future," *Social Work/ Maatskaplike Werk*, 37(2): 105–11.
Metz, T. (2010), "Human dignity, capital punishment, and an African moral theory: toward a new philosophy of human rights," *Journal of Human Rights*, 9(1): 81–99.
Mhlanga, C., T. Muzingili, and M. Mpambela (2019), "Natural disasters in Zimbabwe: the primer for social work intervention," *African Journal of Social Work*, 9(1): 46–54.
Midgley, J. (1981), *Professional Imperialism*, London: Social Work in the Third World.
Mignolo, W. D. (2010), "Epistemic disobedience, independent thought and decolonial freedom," *Theory, Culture & Society*, 26(7–8): 159–81.
Mkhize, N. (2004), "Psychology: an African perspective," in D. Hook (ed.), *Critical Psychology*, Cape Town: University of Cape Town Press, pp. 24–52.
Mkhize, N. (2008), "Ubuntu and harmony: an approach to morality and ethics," in R. Nicolson (ed.), *Persons in Community: An African Ethics in Global Culture*, Scottsville, South Africa: University of KwaZulu-Natal Press, pp. 35–44.

Mokgoro, Y. and S. Woolman (2010), "Where dignity ends and Ubuntu begins: an amplification of, as well as an identification of a tension in Drucilla Cornell's thoughts," *South African Public Law*, 25(2): 400–7.

Molefe, T. O. (2016), "Oppression must fall: South Africa's revolution in theory," *World Policy*, 33(1): 30–7.

Mugumbate, J. R. and A. Chereni (2019), "Using African Ubuntu theory in social work with children in Zimbabwe," *African Journal of Social Work*, 9(1): 27–35. www.ajol.info/index.php/ ajsw/article/view/184222.

Mupedziswa, R. (2001), "The quest for relevance: towards a conceptual model of developmental social work education and training in Africa," *International Social Work*, 44 (3): 285–300.

Mwansa, L.-K. (2011), "Social work education in Africa: whence and whither?," *Social Work Education*, 30(1): 4–16.

Nussbaum, B. (2009b), "Ubuntu: reflections of South African on our common humanity," in M. F. Muurove (ed.), *African Ethics: An Anthology of Corporative and Applied Ethics*, Scottsville, South Africa: University of KwaZulu-Natal Press, pp.100–10.

Nyahunda, L., F. K. Matlakala, and J. C. Makhubele (2019), "Role of social workers in mitigating the effects of climate change in Makonde communal lands, Zimbabwe," *Journal of Social Sciences and Humanities*, 16(9): 172–83.

Nyerere, J. K. (1968), *Freedom and Socialism: A Selection from Writings and Speeches, 1965–1967*, Oxford: Oxford University Press.

Osei-Hwedie, K. and M. Rankopo (2008), "Developing culturally relevant social work education in Africa: the case of Botswana," in M. Gray, J. Coates, and M. Y. Bird (eds.), *Indigenous Social Work around the World*, Cornwall: Ashgate, pp. 203–19.

Oyowe, A. O. (2013), "Strange bedfellows: rethinking ubuntu and human rights in South Africa," *African Human Rights Law Journal*, 13: 104–24.

Patel, L. (2015), *Social Welfare and Social Development in South Africa*, Cape Town: Oxford University Press.

Quijano, A. (2007), "Coloniality and modernity/rationality," *Cultural Studies* 21(2–3): 168–78.

Ramose, M. (1999), *African Philosophy through Ubuntu*, Harare: Mond Books.

Ramose, M. B. (2001), *An African Perspective on Justice and Race*, Forum of Intercultural Philosophy.

Ramose, M. (2003), "I doubt, therefore African philosophy exists," *South African Journal of Philosophy*, 22(2): 113–27.

Ramose, M. (2004), "In search of an African philosophy of education," *South African Journal of Higher Education*, 18(3): 138–60.

Ramose, M. B. (2009), "Ecology through Ubuntu," in M. F. Murove (ed.), *African Ethics: An Anthology of Comparative and Applied Ethics*, Pietermaritzburg: University of Kwazulu-Natal Press, pp. 308–14.

Rankopo, M. J. and K. Osei-Hwedie (2011), "Globalization and culturally relevant social work: African perspectives on indigenization," *International Social Work*, 541: 137–57. https://doi.org/10.1177/0020872810372367.

Republic of South Africa. Ministry for Welfare and Population Development (1997), *White Paper for Social Welfare. Government Gazette*, No. 1108 (August 8, 1997), Pretoria: Government Printers.

Samkange, S. and T. M. Samkange (1980), *Hunhuism or Ubuntuism: A Zimbabwean Indigenous Political Philosophy*, Salisbury, Rhodesia: Graham Publishing.

South Africa. Constitution of the Republic of South Africa, Act No. 108 of 1996. Pretoria: Government Printers. Laws.

Teffo, L. J. (1994), *The Concept of Ubuntu as a Cohesive Moral Value*, Pretoria: School of Philosophy.

Xulu, V. S. (2015), "The place of African renaissance in South African education," *The Journal of Pan African Studies*, 8(2): 151–248.

16

Decoloniality as Total Disentanglement with Modern Capitalist and Neoliberal Complex: Social Work Perspective

Mbazima Simeon Mathebane

Introduction

Decoloniality as an epistemic and political discourse and movement stems from the general acceptance of the adverse reality of coloniality in ex-colonial contexts. Thus, decoloniality is a direct response aimed at uprooting coloniality (Ndlovu-Gatsheni, 2015; Sewpaul, 2016; Harm-Smith and Rasool, 2020). Coloniality refers to the persisting patterns of power relations emanating from colonialism that continue even after the fall of the colonial administration (Maldonado-Torres, 2007). This concept was coined by Anibal Quijano and later expanded by Walter Mignolo and Nelson Maldonado-Torres (Ndlovu-Gatsheni, 2015). The earlier scholars revealed a hidden connection between coloniality and modernity as two sides of the same coin (Maldonado-Torres, 2007; Mignolo, 2007; Quijano, 2007). As eloquently captured by Mignolo, coloniality represents the "darker side" of modernity (Mignolo, 2011). Mignolo further characterizes coloniality as an invisible power structure and epistemological design. He argues that coloniality lies at the center of the present Euro-North American-centric modern world, with race as its organizing principle. It is through racial organization that human beings are hierarchized, resulting in unequal global power relations. Most interestingly, Mignolo (2011) identifies a singular Euro-North American-centric epistemology as a key driver of modern coloniality. He lambasts such modern colonial epistemology for claiming universality, disembodiment, truthfulness, secularism, and scientific density.

Decoloniality as a direct response to coloniality

Decoloniality is born out of a realization that the modern world is an asymmetrical world order that is sustained not only by colonial matrices of power but also by pedagogies and epistemologies of equilibrium that continue to alienate people of the Global South. Several decolonial scholars presented varying arguments as proponents of decoloniality. On the one hand, Ndlovu-Gatsheni (2015) avers that coloniality cannot be viewed as an episode through an exclusive epistemic and political view but rather as an occurrence of epic proportions requiring consideration of the global coloniality of thinking, knowing, and doing. Consequently, even attempts made through the decolonization at the turn of the twentieth century failed to yield a "postcolonial world" but fueled global coloniality, as argued by Grosfoguel (2007).

On the other hand, Harm-Smith and Rasool (2020)'s call for deep transformation toward decoloniality posits that a one-sided focus on epistemic decoloniality alone that ignores the material realities of ongoing coloniality in all spheres would lead to the perpetuation of the very structures of modern colonial oppression and injustice. Thus, ex-colonial nations require a genuine "postcolonial" dispensation marked by the birth of new humanity as conceptualized by Frantz Fanon (Ndlovu-Gatsheni, 2015). The call for "de-westernization" advanced by Mignolo (2011) forms part of the plethora of decolonial options and interventions. Such a call is consistent with the African Union (AU)'s call through its Agenda 2063 (African Union (AU), 2013). The Agenda 2063 calls for a paradigm shift from coloniality to pan-Africanism and the African Renaissance (AU, 2013). While within the broader Global South context, we witnessed a decolonial call for South–South formations as exemplified by the Brazil, Russia, India, China, and South Africa (BRICS) constellation of economic power with the object of shifting economic power from the West (Ndlovu-Gatsheni, 2015). At the BRICS summit in August 2023 in South Africa, the bloc announced that Argentina, Saudi Arabia, Egypt, the United Arab Emirates, Ethiopia, and Iran had been invited to join the bloc from January 2024 (BRICS, 2023).

Ndlovu-Gatsheni emphasized the need to move the center, which addresses the problem of Euro-North American centrism.

The crisis of modernity

The difficult socioeconomic conditions imposed by modern capitalist and neoliberal political systems in ex-colonized nation-states herald a crisis of modernity. In his critical analysis of neoliberal policies and liberalization of economic relations, Lorenz (2016) argues against the adverse implications of the limited role of the state in protecting citizens against injustices. He argues that the harmful effects of modernity stem from a deliberate privatization of social responsibilities of the state and the strengthening of its role in maintaining public order and control. Similar sentiments are shared by Grosfoguel (2007) in his characterization of modernity as a racially hierarchized, patriarchal, sexist, Christian-centric, heteronormative, capitalist, military, colonial, imperial, and modern form of civilization and Ndlovu-Gatsheni (2015) in his enunciation of the "underside" of modernity, which he summarizes as a fundamental lie that posits modern colonization as tantamount to civilization. He postulates that decolonial thinkers aver that modernity unfolded as a phenomenon that colonized time, space, and being and was constituted by rhetoric of progress, civilization, emancipation, and development, on the one hand, and on the other, by the reality of coloniality (Ndlovu-Gatsheni, 2015). Consequently, slave trade, imperialism, colonialism, apartheid, neo-colonialism, and underdevelopment remain the hallmarks of modernity. Similar to Lorenz (2016)'s thesis, Ndlovu-Gatsheni advances that rationality and technology have not succeeded in managing and overcoming obstacles to human freedom in modern societies.

The notion of freedoms and human agency as propounded by modernists assumes that individuals are free to act independently and take responsibility. This is regardless of their state of brokenness perpetrated by past colonial and apartheid governments. The state is exonerated from its public/social responsibility for redress, and thus the need for social justice is rendered obsolete. Ironically, while promising liberty, the capitalist neoliberal philosophy and social thought breed what Giddens (1991) terms a fundamental process of "disembedding." The disembedding effect of the capitalist neoliberal philosophy is demonstrated by a sense of hardship and confusion experienced by ordinary people in modern times as they grapple with the tasks of establishing their own order, finding their own meanings in life, and identifying their own goals without a frame of reference to traditional norms. Consequently, the project of modernity is undergirded by a fundamental sense of ambiguity. The ambiguity is manifested in the discrepancies between modernity's key promises and the means for their realization. The concern for social unity and cohesion becomes a private matter, and the growth of new networks, symbolized by the spread of electronic means of communication, creates a marketplace of opportunities for those capable of adjusting to it at the expense of the former public sphere under the guidance of the state (Giddens, 1991). Or rather, the state interfaces with those new private networks more in the form of control and surveillance rather than as a provider of integration, leaving it to the private market to define boundaries of belonging economically. The above demonstrates a crisis of modernity as well as its polarizing effects.

The purpose of this chapter is to provide a critical analysis of the need to disentangle ex-colonial people from the harmful effects of colonial modernity and capitalism as a critical prerequisite for a true and meaningful decolonial turn capable of producing a new world order. The following sections will present an overview of decolonial theory as a theoretical frame for the unfolding discussions. This will be followed by contextualization of the limits of the modern capitalist and neoliberal state and the conceptualization of the call for social work as a profession and academic discipline in the Global South to disentangle itself from colonial modernity as a viable and sustainable decolonial option. A call is made for transformational social work. The chapter will demonstrate how the entanglement of ex-colonial nation-states with modern capitalism and neoliberalism stands as an impediment to their hopes for a decolonial turn. Finally, a critical discussion of all the above and conclusions are presented.

Theoretical framing

The decoloniality theory was used as a theoretical framework. Decoloniality pushes for transcendence, which refers to wholeness over narrow conceptions of being decolonized, and gestures toward liberation from coloniality as a complex matrix of knowledge, power, and being (Ndlovu-Gatsheni, 2015). Decoloniality introduces "the decolonial turn" as a long-existing "turn" standing in opposition to the "colonizing turn," which birthed and continues to undergird Western thought. Decoloniality provides a range of critiques of modernity, arguing that modernity and coloniality are inextricably intertwined. As purported by Ndlovu-Gatsheni (2015), decolonial theory is traceable to those thinkers from the zones that experienced the negative aspects of modernity, such as Aime Cesaire, Frantz Fanon, William E.B. Du Bois, Kwame Nkrumah, and Ngugi wa Thiong'o. Decolonial theorists begin their critique as far back as 500 years, covering Spanish and Portuguese colonialism. Decolonial theorists push forward an analysis predicated on questions of power, epistemology, and ontology as foundational questions in the quest to understand the unfolding operations of Euro-North American-centric modernity. Thus, decolonial theory gestures toward pluriversality and new humanism. As noted by Maldonado-Torres, decoloniality is not a singular theoretical school of thought but a family of diverse positions that share a view of coloniality as the fundamental problem in the modern age. It seeks to expose Euro-American deceit and hypocrisy and to stand up to subjecthood through judging Europe and exposing technologies of subjectivation (Maldonado-Torres, 2007).

At another level, decoloniality involves retelling the history of humanity and knowledge from the vantage point of those epistemic sites that received the "darker side" of modernity found in the Global South, including retelling the story of knowledge generation as involving borrowings, appropriations, epistemicides, and denials of the humanity of people of the Global South as part of the story of science (othering). It is also a call for democratization of knowledge, de-hegemonization of knowledge, de-westernization of knowledge, and de-Europeanization of knowledge (Ndlovu-Gatsheni, 2015). At the core of decoloniality is the agenda of shifting the geography and biography of knowledge, bringing identity into epistemology—who generates knowledge and from where and for whom? Decoloniality's point of departure is the existential realities of suffering, oppression, repression, domination, and exclusion. Decoloniality enables the unmasking of racism as a global problem as well as how knowledge, including science, was used to justify colonialism. Finally, decoloniality accepts the fact of ontological pluralism as a reality that needs ecologies of knowledge to understand (Ndlovu-Gatsheni, 2015).

The limits of modern capitalist and neoliberal states

A central argument advanced in this chapter is that the difficult socioeconomic conditions imposed by modern capitalist and neoliberal nation-states remain a stumbling block toward true emancipation. In his critical analysis of neoliberal policies and liberalization of economic relations, Lorenz (2016) argues against the adverse implications of the limited role of the state in protecting its citizens against injustices. He argues that the adversities result from a deliberate privatization of social responsibilities of the state and the strengthening of its role in maintaining public order and social control. Interestingly, Lorenz (2016: 4) further laments that "Neoliberal policies not only privatize formerly public services but also focus one-sidedly on discourses of individual autonomy and responsibility."

This assertion by Lorenz is an interesting one, and indeed helpful in illuminating challenges faced by the ex-colonized people of the Global South.

Critical interrogation of the role of social work in the Global South

There is consensus amongst decolonial scholars in social work that the profession and academic discipline have been historically entangled with modernity, coloniality, and the nation-state (Soydan, 1999; Payne, 2005; Dominelli, 2010; Smith, 2014; Lorenz, 2016; Mathebane, 2020). Some scholars took it further and asserted that the profession had effectively become "handmaidens of the nation-state" (Dominelli, 2010; Mathebane, 2020). Thus, social work is a product of modernity in a dual sense: on the one hand, it has come about because of the specific problems and insecurities produced by modernity, and its task became to remedy those and to contribute to the securing of a new social order. On the other hand, as a modern profession, it utilizes the central characteristic of a modern approach to problem-solving, which is the application of scientific principles that critically and reflexively analyze given scenarios to find rational explanations and solutions (Soydan, 1999).

Consequently, and consistent with modernity and neoliberalism, the profession tended to adopt a "blame the victim" mentality (Ferguson, 2008; Mathebane, 2020). Those experiencing social challenges were viewed as responsible for their misfortune (Ferguson, 2008). The negative label often given to those that sought assistance was "dependency" versus self-reliance and individual autonomy. However, Lorenz (2016) challenged such negative conceptualizations and argued that dependency needs to be seen in its wider significance, particularly for modernity. It is a necessary concomitant (rather than an obstacle) to the differentiation of lifestyles and the individualization of life projects, just as heightened interdependency is a concomitant of the division of labor on which the success of modern production processes relies (Lorenz, 2016).

Notwithstanding several transformative policies adopted by postcolonial nation-states, in the main such policies remain neoliberal in form and content. Case in point being the South African government that adopted a developmental approach to social welfare (Patel, 2015). The developmental approach is characterized by extensive state intervention in welfare services and social security, being inclusive and non-discriminatory, and targeting mass poverty and unemployment (Patel, 2015). This approach has been touted as progressive and transformative. However, its neoliberal nature allows for the continuation of the status quo with regard to economic structures and systems, which leaves much to be desired. Neoliberalism refers to new (neo) forms of opening up economic and financial systems so that business activities and the flow of financial resources move across national borders without regulations or restrictions from governments (Taylor and Triegaadt, 2020). It is based on the thinking that free market capitalist production systems create a competitive environment that reduces the costs of products and in turn reduces the prices at which these products are sold, thereby generating huge profits for shareholders and companies (Taylor and Triegaadt,

2020). Among the ways through which companies' costs are cut include paying very little for labor and not putting into place health and safety standards for workers or getting tax exemptions from governments (Taylor and Triegaadt, 2020).

A call for transformational social work

The current undesirable socioeconomic conditions, as discussed in this chapter, call for a different form of social work. We require a form of social work capable of confronting the structural and systemic barriers to genuine socioeconomic development. Such a form of social work would have to emerge from postcolonial contexts of the Global South. Social work as a profession and academic discipline would have to undergo a complete overhaul and reorientation that would allow for its ultimate disentanglement from the modern capitalist and neoliberal complex (Lorenz, 2016). This would naturally follow a process rather than an event. Such a process would include, but not be limited to, deliberate and forthright reflections on social work's historical position and roles within varying political contexts (Rose, 1996; Soydan, 1999). In light of the above, there is no question as to whether social work may assume a political role or not. By virtue of its founding commitment to social justice and social development, social work has an inherent political role.

The author's preliminary assessment and views on the social work transformation process are that it would ultimately ignite a renewed critical examination of the profession's political role in society, which is desperately needed in the Global South given the prevailing poor socioeconomic conditions. As propounded by Soydan (1999), the social work profession should effectively turn its interventions with various client systems into liberation and transformational platforms that affirm social citizenship, ensure rights, and promote social equality. As observed by Rose (1996), the nation-state has been rapidly diminishing its social role and significance in addressing socioeconomic issues such as poverty, unemployment, and inequality. I concur with Smith (2014) in arguing that social work has been intrinsically bound up with the nation-state as its handmaiden. Thus, there is a need for social work in the Global South to reconsider critically its position in this regard and assert its independence and neutrality as a profession. In this way, social work would be able to transform its professional methods effectively, so that they may confront head-on the fundamental political and ethical questions of social justice and equality, including structural reforms as intimated by Schmid (2015). Schmid decried social work methods and research that tend to focus more easily on individual dimensions of social problems instead of raising contextual structural issues such as poverty and lack of support systems (Schmid, 2015).

Conclusion

In conclusion, it is clear that coloniality cannot be separated from the broader wave of Euro-North American-centric modernity and its harmful impact on human histories in the Global South. As described by Cesaire (2000), colonialism is a disruptive, "decivilizing," dehumanizing, exploitative, racist, violent, brutal, covetous, and "thingifying" system. While coloniality itself cannot be viewed as an episode, but it is a major part of what Walter D. Mignolo termed "global designs" that became entangled with "local histories" (Mignolo, 2007). Globalization is today still driven by coloniality on a world scale (Ndlovu-Gatsheni, 2015). Additionally, the issues are fueled by the crisis of modernity, as argued by Giddens (1999) and amplified in social work by Lorenz (2013). Thus, a call for a total disentanglement with global colonial modernity stands as a necessary prerequisite for genuine, meaningful, and sustainable decoloniality. From such a move, pluriversality will emerge and propel the new humanity. For the social work profession and discipline, as argued by Lorenz (2013), modern societies require new forms of social contracts, which social workers can pioneer in their daily encounters with vulnerable individuals, groups, and communities. Taking their plight seriously and exploring the dilemmas they face with their active participation furnishes a vital contribution to the constructive advancement of "the social question." There is a need to work back to the social question, which is traditionally conceptualized as a "creative tension between autonomy and mutual dependency" (Lorenz, 2016: 15). This positionality develops an ethical sensibility that values entanglements of humans, non-human life, and the natural environment. It provides new insights into environmental destruction, human-animal relations, gender inequality and male dominance, as well as indigenous and settler/colonial issues and critical and green social work.

References

African Union (2013), *Agenda 2063: The Africa We Want*. https://au.int/en/agenda2063/overview. Accessed on June 15, 2023.

Brazil, Russia, India, China and South African (BRICS) (2023), https://brics2023.gov.za/#. Accessed on December 12, 2023.

Cesaire, A. (2000), *Discourse on Colonialism: Translated by Joan Pinkham*, New York: Monthly Review Press.

Dominelli, L. (2010), "Globalization, contemporary challenges and social work practice," *International Social Work*, 53(5): 599–612.

Ferguson, I. (2008), *Reclaiming Social Work*, London: Sage Publications.

Giddens, A. (1991), *Modernity and Self-Identity: Self and Society in the Late Modern Age*, Stanford, CA: Stanford University Press.

Grosfoguel, R. (2007), "The epistemic decolonial turn: beyond political-economy paradigms," *Cultural Studies*, 21(2/3): 203–46.

Harms-Smith, L. and S. Rasool (2020), "Deep transformation toward decoloniality in social work: themes for change in a social work higher education program," *Journal of Progressive Human Services*, 31(2): 144–64.

Lorenz, W. (2008), "Paradigms and politics: understanding methods paradigms in an historical context," *British Journal of Social Work*, 28: 625–44.

Lorenz, A. W. (2013), "Recognizing the face of the other: difference, identity and community," *International Journal of Social Welfare*. doi: 10.1111/ijsw12027, 22-279-286, ISSN: 1369-6866.

Lorenz, W. (2016), "Rediscovering the social question," *European Journal of Social Work*, 19(1): 4–17.
Maldonado-Torres, N. (2007), "On coloniality of being: contributions to the development of a concept," *Cultural Studies*, 21(2–3): 240–70.
Mathebane, M. S. (2020), "Quizzing the 'social' in social work: social work in Africa as a system of colonial social control," *Journal of Progressive Human Services*, 31(2): 77–92, doi: 10.1080/10428232.2020.1732273
Mignolo, W. D. (2007). "Introduction: coloniality of power and de-colonial thinking," *Cultural Studies*, 21(2–3): 159.
Mignolo, W. D. (2011), *The Darker Side of Western Modernity: Global Futures, Decolonial Options*, Durham, NC: Duke University Press.
Ndlovu-Gatsheni, S. J. (2015), "Decoloniality as the future of South Africa," *History Compass*, doi: 10.1111/hic3.12264.
Patel, L. (2015), *Social Welfare and Social Development*, South Africa: Oxford University Press.
Payne, M. (2005), *Modern Social Work Theory* (3rd ed.), London: Lyceum Books.
Quijano, A. (2007), "Coloniality and modernity/rationality," *Cultural Studies*, 21(2–3): 168–78.
Rose, N. (1996), "The death of the social? Re-figuring the territory of government," *Economy and Society*, 25, 327–56. doi: 10.1080/03085149600000018.
Schmid, J. (2015), "Overlooking the most vulnerable: the child welfare research agenda 2005–2010," *European Journal of Social Work, 18*, 6–16. doi: 10.1080/13691457.2014.903232
Sewpaul, V. (2016), "The West and the rest divide: human rights, culture and social work," *Journal of Human Rights and Social Work*, 1: 30–39. https://doi.org/10.1007/s41134-016-0003-2
Smith, L. (2014), "South African social work education: critical imperatives for social change in the post-apartheid and post-colonial context," *International Social Work*, 51(3): 371–83.
Soydan, H. (1999), *The History of Ideas in Social Work*, Birmingham: Venture Press.
Taylor, V. and J. D. Triegaadt (2020), *The Political Economy of Social Welfare Policy in Africa: Transforming Policy through Practice*, Oxford: Oxford University Press.

17

Indigenous Youth and the Recomposition of Community Life: Some Experiences in Oaxaca, Mexico

Eduardo Bautista Martínez and María Leticia Briseño Maas

Introduction

The present work stems from an alternative vision for defining a research agenda that allows us to comprehend the participation of indigenous youth as agents of transformation in complex realities characterized by cultural diversity and social inequalities, as is the case in Oaxaca, a state in southeastern Mexico that embodies the plurality of cultural expressions from 16 ethnic groups with their respective native languages. This situation is similar to what is observed in other regions of the country and Latin America. It is considered that Oaxaca's sociocultural composition cannot be encompassed by the definition of Mexican homogeneity conceived as a single mestizo culture. The lack of understanding of Oaxaca's complexity and plurality by elites and rulers leads to the multiplication of social and political conflicts in the territories. Therefore, it is necessary to comprehend different worldviews, diverse forms of social and political organization, and how singularities intertwine within the diversity of local, regional, and global contexts.

The analysis focuses on indigenous youth not as passive victims of a chain of multiple oppressions but as social actors transforming their own adverse circumstances. They are gaining increasing visibility through activism that promotes creative resistance beyond the dominant policies of centralized governments. This situation motivates us to understand how indigenous youth integrate into various community-based organizational processes aimed at addressing issues related to cultural revaluation, education problems, the promotion of new dissemination spaces, as well as their involvement in community assemblies as instances for decision-making, deliberation, participation, and problem-solving in their immediate environment.

In the national history, it's important to mention that the relationship between ruling elites and indigenous peoples has led to persistent political conflict, laden with tensions arising from colonial perspectives that have ranged from the extermination of peoples to their integration through a single language and culture: the Mexican mestizo. Dominant ideas of a homogeneous national culture have been sustained for decades in a linear narrative of "progress," assuming that for indigenous communities to achieve better living conditions and escape their perceived "backwardness," they must leave behind their forms of organization, languages, local knowledge, and ways of communal life, considered "archaic" in the framework of "civilized" lifestyles.

The objective of this work is to understand the work of indigenous youth in certain communities of Oaxaca, in order to strengthen their local knowledge, their forms of social and political organization, and their relationships with the environment and their territories, through alternatives of intercultural education and communication. These alternatives serve as strategies and practices for living better in conflicting and adverse conditions arising from a context of social inequality, discrimination, racism, and economic precarity. Specifically, experiences from two indigenous municipalities in the Sierra Norte of the state of Oaxaca are examined: Guelatao de Juárez and Santa María Tlahuitoltepec. Both municipalities base their social and political organization on internal normative systems, and through this, the participation of indigenous youth in these communities is analyzed. The experiences seek alternatives for the reconfiguration of communal life, offering perspectives that enable a better understanding of the complexities and specificities of the state of Oaxaca.

The multiple forms of inequality in Oaxaca

Oaxaca is a federal entity in the southeast of Mexico that boasts the greatest cultural, linguistic, and natural resource diversity in the country. Statistics record an indigenous population surpassing 32 percent of the total population, spread across 16 ethnolinguistic groups with their respective variations (DIGEPO, 2018). Paradoxically, this cultural diversity is permeated by deep social inequalities that place the indigenous population at the lowest levels of lag and lack of opportunities, particularly among indigenous women in the region (Briseño, 2023).

This panorama awakens an interest in analyzing the initiatives of indigenous youth to overcome the multiple conditions of oppression they endure. However, indigenous youth do not constitute a homogeneous and isolated social group. Instead, they present a diversity of situations, needs, and demands that we need to recognize in their specific relationships: indigenous youth in cities, in rural areas, in the various roles attributed within these same realms, and the conditions that, while common, are permeated by differences among them. As Judith Butler (2006) points out, in these relationships, we must identify the impossibility of thinking of oneself separately: "We, who are relational, do not exist apart from these relations, and we cannot think of ourselves separately from the decentering effects that relationality implies" (p. 215).

The conditions of inequality that indigenous youth experience cannot be understood from a homogenous viewpoint, but rather through their intersections or concurrences, from the multiple social dynamics that each context generates. In this regard, Pérez (2019) states, "(. . .) under certain conditions of time and space, the various forms of inequality (inequalities) take on specific contents, and respond to social and power relations that are also particular. This means that these can coincide in the same social subject, interacting with each other, and even supporting each other, but they can also not do so" (p. 32).

In Oaxaca, the conditions of inequality faced by indigenous youth in environments dominated by white or mestizo youth must be added to the discrimination they experience due to their youth and their poverty-stricken living conditions in rural or peripheral areas, distant from urban centers and cities. Additionally, it is necessary to distinguish indigenous young women who navigate socio-cultural realities where the prevailing gaze is masculine and patriarchal. The challenges faced by these young indigenous women are diverse, but with the idea that transformation is built every day, in the everyday fabric, they can access spaces traditionally occupied by men. The motivation of many of these women is not necessarily a defined feminist ideology, but is shaped by conditions, needs, and expectations in the face of the multiplicity of structural changes and their permeability in their regions and communities.

Therefore, in understanding the different ways of being indigenous youth in complex realities, it is important to revisit the debate on intersectionality. According to María Lugones (2008), one way to properly think about this connection between race, class, and gender is "intersectionality." Cubillos (2015) states that intersectionality is a counter-hegemonic political and epistemological tool that contributes to critically rethinking how we interpret reality, a product of a matrix of multiple, simultaneous, and inseparable oppressions that cross through the social conditions of youth sectors. Intersectionality allows us to move beyond the unitary and partial perspectives that abound in social analysis of isolated categories of inequality, to transition toward a comprehensive understanding of structural and simultaneous oppressions.

It is important to transcend the criteria of modern rationality that embodies a singular thought, which considers that social inequality results from internal dynamics that persist within indigenous communities and their limitations in achieving better living conditions on their own. The persistence of this interpretation continues to unfold within the workings of the state apparatus, through the implementation of various programs and policies that have operated vertically and centralized in the forms of community life in Oaxaca. The narrative that rural and indigenous communities can only achieve better living conditions by overcoming their perceived "backwardness" persists. This narrative overlooks their cultures, languages, local knowledge, and ways of communal life, which are considered "archaic."

In the case of Oaxaca, this perspective draws strongly from post-revolutionary governments since the 1920s and the post-war world of the mid-twentieth century, where it was important and strategic to consider the construction of national societies in peripheral countries according to the industrialized European countries. Authors like Hettne (1982) showed that the proposed strategies to achieve these "advanced" life conditions had to be designed based on a Western paradigm of "development."

With ideas from thinkers like Rostow (1960), it was conceived that the possibility of surpassing traditional societies could be achieved through "stages" to attain civilized life, where schooling was characterized by integrationist and monocultural logics. In this regard, Bermúdez (2020) points out, "(. . .) the integrationist model considered indigenous languages as an obstacle to progress. To solidify this idea within educational policy discourse, an ideology of indigenous social ascent through schooling was established" (p. 82).

Nevertheless, the local experience in Oaxaca creates room for maneuver and shows us that inequality cannot be understood or addressed from a homogenous perspective. Instead, it should be approached through its intersections or concurrences of multiple historical and structural oppressions that converge and interact on social actors, who also have the capacity to resist and construct alternatives. Pérez (2019) states that the proposal of intersectionality precisely allows us to denaturalize these categories, providing them with historical and contextual content in diversity. This means addressing the structural, political, and symbolic dimensions of inequality and the specific forms that power relations take at the local and regional levels. Hence, inequality cannot be seen as an inherent condition of indigenous communities, but rather as a historical construction generated by the imposition of centralized policies that have affected the internal dynamics of communities, where economic poverty contrasts with the rich culture and biodiversity of the territories.

From this perspective, the interest arises in understanding the ways indigenous youth seek to address common issues through their participation in community mechanisms of organization and deliberation, such as community assemblies, systems of responsibilities, religious festivals, tequio (community work), etc., which are elements of the political proposals of indigenous municipalities in Oaxaca. It's important to emphasize that indigenous youth are those who feel the effects of social marginalization and discrimination to a greater extent, as their ethnic identity is combined with economic status, and for indigenous women, gender comes into play, making them more exposed to specific instances of double or triple discrimination due to being women, indigenous, and lacking personal economic resources. This generates a multiplicity of oppressions marked by gender, class, ethnic origin, and sexual orientation (Cervantes and García, 2020).

Indigenous youth, far from viewing themselves as victims or merely recreating structural disadvantages, deploy various local strategies based on assuming their own autonomy and agency to overcome and prevail over adverse situations of multiple inequalities through the recognition of local knowledge, linguistic and cultural rights, aimed at generating changes in their communities and regions by identifying "what to do?," "how to live?," and "what and how to build?" To this end, self-managed experiences built over the course of the last four decades in indigenous communities in Oaxaca are recovered.

Pedagogical experiences in two municipalities

The discussion of concurrent or intersecting social inequalities leads us now to outline some paths in an attempt to understand the active role that young people play as agents of transformation in community dynamics and norms. This is particularly important in Oaxaca, where forms of political organization are based on what are known as internal normative systems, referring to specific forms of local governance. This includes mechanisms such as community assemblies for decision-making, "tequio" or community service without economic compensation, under a concept of power as a service to the people. Obligation, service, reciprocity, and prestige are principles of the political organization in these communities, which have shown resistance to the intrusion of political parties existing in the liberal democracy of representation.

In this context, indigenous youth are participating in community processes, while there has also been a progressive inclusion of women in the system of leadership positions. This inclusion has been prompted by circumstances such as the higher migration rate of men but also by the recognition of and right to female participation in the community. This recognition involves acknowledging the forms of social and political organization in experiences within the municipalities of Santa María Tlahuitoltepec and Guelatao de Juárez.

Firstly, the community of Santa María Tlahuitoltepec is located in the Mixe region, specifically in the highlands of Mixe, in the Northern Sierra of Oaxaca. The Mixe people call themselves "ayukjay," which means "people who speak eloquently like the jungle" or "people of the elegant language." There are more than 125 Mixe towns spread across seventeen municipalities, although many Mixe people also live in cities like Oaxaca and Mexico City, as well as outside the country, in the United States.

On the other hand, Guelatao de Juárez is a community where its inhabitants speak the Zapotec language, located approximately 60 kilometers from the state capital. What stands out in both municipalities is the importance of cultural reclamation through indigenous language and writing, as well as the diversity of worldviews that revolve around the collective work of each community and their numerous relationships with various contexts. The organizational systems vary from one community to another and are in constant transformation, making them not fixed situations.

Variations in political organization methods depend on responses to governments, political parties, and transnational corporations affecting the territories, as well as internal dynamics caused by the mobility and migration of the population. This is particularly true for young people, who move to the state capital, other cities within Mexico, and the United States. They may do this in pursuit of job opportunities, economic income, or higher education.

Migration disrupts the organization of indigenous peoples and communities, their organizational systems, and it is no coincidence that cultures are opening up to modify traditional gender roles that influence indigenous women being able to leave to continue university studies and, particularly, who enter disciplines considered for men. Cultural changes and the economic needs of families encourage the processes that lead women from indigenous peoples to continue their higher education.

Up to this point, history may seem favorable; however, it is essential to take into account the conditions in which they migrate or move to the cities where educational institutions are located, and the way in which they face precarious situations to remain there, to achieve the goal of graduating. The ethnic diversity, the socioeconomic factor, and the population distribution in an entity like Oaxaca represent an unavoidable opportunity to recognize the concurrence of inequalities in the territories, for whose inhabitants migration or daily mobility is necessary.

This requires understanding how indigenous women arrive at the headquarters of the institutions, within the broad and complex phenomenon of migration from the communities of origin to the capital city, with all the implications, the background in their formation, the conditions of their departure and arrival at their destination, their initial expectations until their permanence and graduation from the chosen careers, the conditions of their stay in the city and within the institution, the difficulties in remaining studying, the motivations to continue, as well as the experiences that they acquire by being part of a network of supportive relationships that are built into everyday life.

The particularities of Oaxaca are typical of an entity that has one of the highest indigenous population rates at the national level, which refers to cultural heterogeneity, which cannot be the object of stereotypes, discrimination due to racism, analyzed from an exclusive perspective or folkloric domination of reality. These mentioned municipalities have been governed by Internal Normative Systems, where the system of leadership positions stands out as a necessary element for the social and political organization of the community. This system involves all residents of the community, regardless of educational level, socioeconomic status, or gender. The participation of women, especially those with higher education or professions, as well as the younger population, is important. This is because women, in particular, had not been considered in these leadership positions previously (Díaz Robles, 2023). These organization and service systems focus on various common activities, such as security, political representation, and the organization of civic and religious festivities that ensure community cohesion.

Leadership positions vary, ranging from lower-ranking roles like "topiles," responsible for community security, to the positions of Municipal President and Constitutional Sole Mayor, which are the highest-ranking roles in the community system. There are topiles for the town council and for the church, involved in tasks like safeguarding community property, cleaning the main square, and maintaining the Catholic church. The relationship between the adult population and the youth aims at mutual learning and promoting community values of respect and responsibility on both sides. The persistence of the youth's efforts is a factor driving the constant transformation of normative systems, even in the face of resistance from conservative adults.

Since land ownership in indigenous communities is communal, there are agrarian authorities responsible for mediating conflicts among families, land disputes between neighboring communities, the distribution of roads, electrification, and the construction of communal projects. In terms of the status of women, while women are generally segregated into roles and professions associated with caregiving, such as health and education, indigenous communities have seen teachers pave the way for many more women to obtain jobs and continue their professional education. This has been achieved through equal recognition of their work. "It is important to recognize that indigenous teachers are the ones who have paved the way for the presence of indigenous women in the labor market in rural areas. They have been pioneers in negotiating with husbands and fathers to allow them to work. However, they have also fought for respect and recognition for their work, dealing with sexual harassment and fighting for fair compensation" (Gallardo, 2004: 43).

As part of local strategies, we can also highlight some proposals for intercultural, inclusive education linked to community training. Indigenous youth strive for the recognition of their culture, life, and an awareness of their relationship with nature and their territories. In Tlahuitoltepec, a process of revaluing Mixe culture began through various projects. One example is the Comprehensive Community Polyvalent Baccalaureate (Bachillerato Integral Comunitario Ayuuk Polivalente or BICAP) for upper secondary education. This community proposal imparts knowledge of organization, local knowledge recovery, environmental care, and a sense of education's relevance. Another project is the Center for Musical Training and Mixe Culture Development (Centro de Capacitación Musical y Desarrollo de la Cultura Mixe or CECAM), which offers musical education and training with community principles to strengthen the musical culture of Mixe communities. This creates responsible youth capable of valuing and enhancing their cultural heritage through language preservation and cultural dissemination (Comboni et al., 2005).

These projects have taught youth respect, care, and protection of nature for the benefit of the environment and the community itself. The pedagogical approach aims to ensure the harmonious development of indigenous youth, striving for constant economic improvement through productive activities that optimize the use of natural resources in their territories.

The community of Tlahuitoltepec is known as the birthplace of musicians, where music has become essential—a way of organization and life within the community, considered a backbone for community coexistence. It serves as a mechanism for preserving community history, oral traditions, and ethnic identity. This community proposal aims to continue the tradition of philharmonic bands within the locality and foster musical education and training with community principles for children, youth, and adults who want to learn to play a musical instrument.

These dynamics bring with them a greater openness of families, towns, and communities so that women can be inserted in processes of cultural preservation through music, but this does not mean that they leave aside the fulfillment of roles; on the contrary, the demands are greater on them, who, in addition to the burdens of their musical training, continue with tasks and responsibilities imposed by gender stereotypes. Musical training with community principles acquires relevance from a double perspective: first, the personal trajectories of women and the structural conditions that have led them to occupy important training spaces even with the problems they have had to overcome, and second, the position to transform their circumstances and become, with their example, a reference for other young people and women in unfavorable situations thanks to deploying their capacity for agency as people who build their present to transform their future, decide their own training and their community participation in groups and solidarity networks, from which they can reflect and propose initiatives and actions, based on their life experiences.

In the first decades of this millennium, we can observe the emergence of other intercultural education projects. One such project is the Ayuuk Intercultural Higher Institute (Instituto Superior Intercultural Ayuuk or ISIA), which is noteworthy for providing training to address environmental issues, common goods, sustainable food, and the role of agricultural economy. According to researcher Flor Marina Bermúdez (2020), the training of professionals with an intercultural focus allows for reflecting on the connection between different branches of knowledge, culture, and economics.

These branches form a body of knowledge that observes daily life and the connection to land, water, forests, and territories, as they are vital spaces for collective life. It's also essential to acknowledge that indigenous women, through their academic training as professionals or scientists, generate leadership that guides community participation (Briseño, 2023)

The search for communication alternatives

Indigenous youth understand the need to strengthen the preservation of their language, the defense of natural resources, and the cultural dissemination. This has become a driving force for the emergence of various outreach projects aligned with the needs of indigenous municipalities. One such example is the community radio station "Jën Poj" (Wind of Fire), which began in 2001 in Tlahuitoltepec as an initiative of young professionals. Its purpose was to establish a community communication medium aimed at promoting the culture, customs, and traditions of the Mixe people, while also reviving the Mixe language among the new generations. The radio broadcasts regional music, compositions, and announcements in the native Ayuuk language.

This project began its transmissions without legal permission, leading to government persecution as their platform was considered "illegal." However, they were soon supported by the World Association of Community Radio Broadcasters (AMARC), which stood by them until they obtained legal permission in 2005. Currently, the radio station is a member of AMARC and has hosted continental summits with committee members, municipal authorities, as well as elders' councils and professional and teacher associations from the municipality. The community has developed a strong identity bond with the radio and covers the costs of electricity and production.

It's important to emphasize that Oaxaca has the highest number of community radio stations in Mexico. More of these stations have been established in various towns and indigenous communities, complemented by websites and other forms of self-managed intercultural communication.

In Guelatao de Juárez, the "Estéreo Comunal" project stands as a result of struggles and efforts by indigenous youth, intellectuals, teachers, and individuals within the community. This radio was born as a freer medium in a context where the development of community media has faced greater restrictions. The significance of this project lies in the active involvement of young people who are constantly engaged in addressing their needs and experiences.

The collective efforts of intercultural education and communication projects prove valuable in guiding the daily activities of young people. These initiatives allow them to reclaim ancestral knowledge while viewing it through a modern lens. They engage in dialogue with external actors, including those who return after experiencing migration, and creatively utilize information and communication technologies. However, it's crucial to acknowledge that the outcomes are uncertain in an adverse context of inequality, racism, and the persistence of conservative groups within their own communities (Bautista and Juárez, 2016).

Conclusions

The intercultural education and communication undertaken by indigenous communities do not address isolated matters, but rather strategies aimed at achieving cultural reclamation in the face of systematic denial of their knowledge and processes. According to Martínez Luna (2016), a Zapotec philosopher, teachers, as a sector of intellectuals driving alternative projects, are called to learn from the community about the essence of their work. Teachers must realize their commitment to each community, as the community is the space for constructing present and future horizons. The task of educating is everyone's responsibility, as in this world we are all both learners and teachers, and we can all learn and teach.

The self-managed experiences of young people in reshaping community life warrant the renewal of analytical categories, drawing from other knowledge and theoretical, methodological, and epistemic tools to question the power dynamics that persist in uniform and centralized state strategies, which often view indigenous peoples' responses as mere anachronisms, asserting that they "don't know what's best for them," and attributing persistent inequality in communities to attitudes of "rejection of modernity."

The involvement of young people is expressed in a multitude of tasks and challenges tied to their territories, defending natural resources, becoming increasingly active in the political organization of their communities, and engaging in education and intercultural communication projects. Young people are not passive recipients of public policies that often operate in a top-down manner, disregarding the cultural diversity characterizing Oaxaca.

It's crucial to adopt a complex perspective, recognizing young people as agents of social change with a decentralized vision and a horizon of equality and social justice, where their communal actions, unfolding in the nooks and crannies of the country and the world, are celebrated. It's important to note that experiences in reshaping community life, as illustrated in this chapter, depend on the specificities of local histories and contexts, but it's feasible to thread together different projects as strategies of broader scope. This includes the diverse responses of indigenous peoples in Latin America to the persistence of colonialism, along with the possibilities young people possess for crafting "other worlds," as they say in the southeastern region of Mexico.

Expanding our understanding of indigenous peoples as historical and autonomous entities present, resisting, and transforming is necessary. In particular, as indicated by Pérez (2019), young people are a pivotal sector in both rural and urban domains, vividly expressing the conflicts and dilemmas regarding the change and continuity of their cultural groups, as well as the impact of mass media and the consumption of cultural goods. This impact is intensified by economic globalization and the global spread of culture. Within the theoretical framework of decolonization, the research agenda of indigenous youth can be translated into experiences that affirm identities, construct projects of intercultural education and communication, foster participation, and update their mechanisms of community organization. These mechanisms defend their territories and natural resources, and strive for a dignified life despite adversity, inequality, and racism.

References

Bautista, E. and I. Juárez (2016), Formas emergentes de participación comunitaria. Los jóvenes indígenas en dos municipios de Oaxaca El Cotidiano, núm. 197, mayo-junio, pp. 102–12 Universidad Autónoma Metropolitana Unidad Azcapotzalco, México.

Briseño M. M. L. (2023), Mujeres indígenas en las ciencias. Narrativas situadas y construcción de horizontes emancipatorios. Scriptus- UABJO, México.

Bermúdez, U. M. (2020), Agencia social y educación superior intercultural en Jaltepec de Candayoc, Mixe, Oaxaca. Universidad de Ciencias y Artes de Chiapas, Instituto Superior Intercultural Ayuujk https://cutt.ly/klX3VqF

Butler, J. (2006), Deshacer el género. Paidós

Cervantes, M. A. and R. E. García (2020), Ser mujer e indígena: una propuesta teórica para estudiar la interseccionalidad en la educación superior. Memorias del III Congreso Internacional de Género. Instituto Tecnológico de Sonora. www.academia.edu/61532946/Ser_mujer_e_ind%C3%ADgena_una_propuesta_te%C3%B3rica_para_estudiar_la_interseccionalidad_en_nivel_superior_p

Comboni, S., N. Juárez, and M. Tarrío (2005), Resurgimiento cultural indígena: el pueblo Ayuujk de Santa María Tlahuitoltepec, Mixe, Oaxaca. El Bachillerato Integral Comunitario Ayuujk Polivalente. Revista Latinoamericana de Estudios educativos, Vol. XXXV, núm. 1–2. Universidad Iberoamericana, Pp. 181–208, México. www.redalyc.org/articulo.oa?id=27035207

Consejo Nacional de Evaluación de la Política de Desarrollo Social (CONEVAL) (2018), Medición de la pobreza, 2008–2018. www.coneval.org.mx/Medicion/PublishingImages/Pobreza_2018/Serie_2008-2018.jpg

Comisión Nacional para el Desarrollo de los Pueblos Indígenas (CDI) (2016), Indicadores socioeconómicos de los pueblos indígenas de México, 2015," Coordinación General de Evaluación y Planeación. www.gob.mx/cms/uploads/attachment/file/239921/01-presentacion-indicadores-socioeconomicos-2015.pdf

Cubillos, J. (2015), "La importancia de la interseccionalidad para la investigación feminista. Oxímora," *Revista Internacional de Ética y Política*, (7), 119–37. https://revistes.ub.edu/index.php/oximora/article/view/14502/17834

Díaz Robles, Tajëëw (2023), Santa María Quiegolani: acomodos y resistencias de la organización comunitaria. Ichan Tecolotl, año 34, número 374, CIESAS, México. https://ichan.ciesas.edu.mx/santa-maria-quiegolani-acomodos-y-resistencias-de-la-organizacion-comunitaria

Díaz Ruiz (2011) Retos para la investigación de jóvenes indígenas. En Alteridades No. 42, Universidad Autónoma Metropolitana-Iztapalapa. México. https://alteridades.izt.uam.mx/index.php/Alte/article/view/114

Dirección General de Población de Oaxaca (DIGEPO) (2018), Radiografía demográfica de la población indígena de Oaxaca. Nueva época. Num. 41. www.digepo.oaxaca.gob.mx/recursos/revistas/revista42.pdf

Gallardo, S. (2004), *El ser mujer indígena profesionista y el mercado de trabajo* (Documento de trabajo).

Hettne, Björn (1982), *La Teoría del Desarrollo y el tercer mundo*, Swedish Agency for Research Cooperation with Developing Countries (SAREC), Estocolmo Suecia.

Instituto Nacional de Estadística, Geografía e Informática (INEGI) (2015), Censo de Población y Vivienda, México.

Lugones, M. (2008), "Colonialidad y género," *Tabula Rasa*, (9): 73–101. www.redalyc.org/articulo.oa?id=39600906

Martínez Luna, J. (2016), Conocimiento y comunalidad. Bajo el Volcán, año 16, número 23, septiembre 2015-febrero 2016. México.

Muñoz, R. and G. Cortez (2020), "Impacto social y epidemiológico del COVID-19 en los pueblos indígenas de México," *Debates indígenas*. https://debatesindigenas.org/notas/54-impacto-social-covid-19.html

Pérez-Islas, J. (2006), Trazos para un mapa de la investigación sobre juventud en América Latina. Centro de Investigación y Estudios sobre Juventud. México: Instituto Mexicano de la Juventud. México.

Pérez, M. (2019), Jóvenes indígenas en América Latina: reflexiones para su investigación desde la antropología. Anuario Antropológico, v.44 n.2. En http://journals.openedition.org/aa/4003 DOI: https://doi.org/10.4000/aa.4003

Rostow, Walt (1960), *Las etapas del crecimiento económico*, Fondo de Cultura Económica, México.

18

"Not about us, but with us!" Perspectives of Insurgent Research with Children of the Global South

Manfred Liebel

Introduction

Since the turn of the millennium, social childhood studies have not only become more interested in children and childhoods of the Global South but have also critically reflected on the Eurocentric focus of previous childhood studies. This critical perspective has been taken up recently in a special issue of the international journal *Childhood*, claiming to "develop perspectives that de-construct northern, dominant scholarships, and facilitate generative thinking on the study of childhood and children using southern epistemologies" (Abebe, Tar and Lyså, 2022: 2). Part of this (self-)critical view is that until now, a large part of research has been characterized by an approach that uses children of the Global South as a source of knowledge without asking what the knowledge generated means for the children themselves (Ansell, 2019). When I discuss this here, I want to encourage reflection on forms of knowledge production in childhood and children's rights research that benefit children themselves and contribute to recognizing them as equal human beings and to strengthening their social position in society. The following is inspired by, reflects, and intensifies my experience as a popular educator and action researcher with working and indigenous children in Central and South America since the 1990s (Liebel, 2000; 2004; 2012; 2023a; 2023b).

I will focus my reflections on the question of what kind of knowledge is required to reach this aim and how it can be generated. It must be knowledge that reveals children's perceptions and perspectives, that is, how they think, feel, and deal with everyday experiences. And it must be knowledge in which become visible not only the marginalization, disregard, oppression, exploitation, and violence experienced by children but also their potential to defend themselves against it. In my opinion, such knowledge, sometimes called "situated knowledges" (Haraway, 1988) or "liberatory knowledge" (Giri, 2013), can only emerge together with the children concerned. It must be legitimized by them and accessible to them. The generation of such knowledge requires ethnographic methodologies that do not separate the researcher from the researched but see research as a joint project in which there are horizontal and equal relationships between all participants. And it requires methodologies that do not build on cognitive and linguistic premises usually attributed to adults but give equal weight to mimetic and emotional skills and ways of communicating. Here, I refer primarily to research approaches that have emerged in Latin American and Indigenous contexts and represent what Argentine semiotician Walter Mignolo (2009) has called "epistemic disobedience."

These research approaches have been discussed and practiced since the 1970s under the terms *Participatory Action Research (PAR)* and since the 1990s as *Epistemologies of the South, Indigenous Methodologies* or *Horizontal-Dialogical Methodologies*. The core here is not to define certain methods in advance but to ask ourselves as researchers together with all participants—that is, also with children—which methods are appropriate and applicable in a particular context and in what way the cognitive limits of established academic science can be overcome. The choice of methods is itself part of a communication process in which professional researchers and non-academic co-researchers are equally influential and ultimately arrive at decisions together. The research approaches that have emerged under the terms mentioned here have so far rarely taken children into account, but they can also open up new avenues of research with children and on childhoods and children's rights.

This is particularly important in the case of children who are denied basic conditions of a dignified life. While human rights apply to them as they do to all children, they have little opportunity to claim them. As "children of the Global South" (Diana, 2020; Liebel, 2023) or "children out of place" (Invernizzi et al., 2017), they embody ways of life that are far removed from dominant Western notions of a "true childhood" and are therefore often even denied having a childhood at all. To do justice to these children's life experiences and make them visible, research approaches are needed that are not guided by a predefined pattern of childhood and that contribute to what could be called the decolonization of childhoods (Liebel, 2020; Araneda-Urrutia, 2022). The perspective of decolonization refers primarily to children living in the southern or postcolonial regions of the world, but it is also relevant for children in other regions who are affected by social inequality, poverty, violence, war, and forced human mobility.

I present different dialogical and action-oriented research approaches, highlight the importance of the research attitude and physicality of those involved in the research, and finally discuss pitfalls and possible perspectives in dealing with the relationship between rationality and emotions. Ultimately, I show why these research approaches are also relevant for emancipatory childhood and children's rights research and how they can be applied in this context.

Dialogical and action-based research approaches

Dialogical and action-based research approaches are seen as a critique of the ideology of scientific universalism. This assumes that only scientific knowledge represents true, valid, representative, or objective knowledge and is therefore superior to all other forms of knowledge. The corresponding research defines the researcher as an expert who approaches the people, social conditions, or nature to be investigated from the outside and turns them into an object of knowledge. Thus, according to the critique, the reality experienced by people is only expressed in a distorted form. The knowledge contained in it is devalued and rendered invisible. A logic oriented toward an instrumental and monocultural rationality is imposed, which turns the produced knowledge into a commercial object. It becomes a commodity in the hands of those who have put knowledge into a scientific form and/or have the power to use it in their interests.

In the concept of participatory action research, which was formulated in the 1960s by the Colombian sociologist Orlando Fals-Borda (1970; [1979]2009), research, on the other hand, arises from joint action against living conditions that are perceived as unacceptable, or, on the other hand, it serves to empower excluded and oppressed people.[1] Researchers do not limit themselves to participant observation, as in other forms of participatory research, but see themselves as partners acting in solidarity with population groups whose emancipation and liberation are at stake. They see their research as partisan and consciously take the point of view of these population groups. It is obvious that such an attitude is neither easy to win over nor to apply without contradictions, since the life situation of the researchers in the research period does not become identical to that of their co-researchers. However, since research is an integral part of transformative practice, the knowledge gained is not separated from action but can be used by oppressed and marginalized population groups in their struggles to improve living conditions. In this sense, there is also talk of "insurgent research" (Gaudry, 2011; Corntassel and Gaudry, 2014; Strega and Brown, 2015).

The theoretical outline of *Epistemologies of the South*, which was introduced by the Portuguese sociologist Boaventura de Sousa Santos (2014; 2018), considers it essential to bring to light the knowledges that have been made invisible during colonization. De Sousa Santos calls this a "sociology of absences" and strives for an "ecology of knowledges." By this he means that "the logic of the monoculture of scientific knowledge and rigor must be confronted with the identification of other knowledges and criteria of rigor and validity" (De Sousa Santos, 2014: 176). The ecology of knowledges "aims to create a new kind of relation, a pragmatic relation, between scientific knowledge and other kinds of knowledge" (ibid: 190). According to De Sousa Santos, "throughout the world, there are not only diverse forms of knowledge of matter, society, life and spirit, but also many and very diverse concepts of what counts as knowledge and the criteria that may be used to validate it" (ibid: 192). This does call for "a deeper reflection on the difference between science as monopolistic knowledge and science as a part of an ecology of knowledges" (ibid: 193).

However, the revalorization of non-scientific forms of knowledge does not mean a total rejection of scientific forms of knowledge. Rather, it is about their "counter-

[1] Earlier approaches to action research can be traced back to the social psychologist Kurt Lewin (1946) and have been further developed since the late 1960s in the USA, Australia, and some Western European countries. They had a socio-critical intention, above all to generate an alternative to research standards such as objectivity and neutrality within the social sciences. It was criticized that by claiming neutrality, the social sciences entered into an implicit alliance with the ruling social powers and definitional monopolies, which did not change social structures but affirmed and reproduced them (see Kemmis and McTaggart, 2005).

hegemonic use" (ibid: 207), or "its pretention to being the only valid kind of knowledge" (De Sousa Santos, 2018: 45). In this sense, the ecology of knowledges "focuses on the relations among knowledges, on the hierarchies and power emerging among them" (De Sousa Santos, 2014: 207). In the same circumstances, preference shall be given to knowledge "that guarantees more participation to the social groups involved in the conception, execution, control, and function of the intervention" (ibid: 208). As De Sousa Santos consequently advocates for a "polyphonic epistemology involving poetry and science" (ibid: 208), new possibilities are also opened to establish relations with and give recognition to children's knowledges and modes of articulation. In this sense, the ecology of knowledges could also be understood as "a precondition for achieving cognitive justice" (ibid: 212), from which children are no longer excluded.

Indigenous methodologies are understood as research approaches devised and practiced by the people directly affected or under their direction. They emerged as a critique of research methods, particularly in anthropology and ethnology. These had regarded indigenous peoples primarily as a source of knowledge and were conducted according to standards based exclusively on Western assumptions of scientific truth. On the other hand, Indigenous methodologies are understood as a way for Indigenous individuals and peoples to re-appropriate the knowledge embedded in their cultures. In this way, they oppose research practices that are understood as a continuation of colonial subjugation or the coloniality of power and knowledge (Quijano, 2008) and thus have a decolonial perspective.

The concept of Indigenous methodologies can be traced back to, among others, Shawn Wilson (2008), Margaret Kovach (2010; 2015), Linda Tuhiwai Smith (2005; 2012), and Bagele Chilisa (2012; 2020), feminist-oriented social researchers of indigenous origin from Canada, New Zealand/Aotearoa, and Botswana. This has since given rise to a broad movement of researchers in the continents of the South who are themselves of Indigenous origin, and some of whom work in a kind of diaspora in "Western" universities. These authors do not claim a monopoly on the study of their cultures but insist that their non-Indigenous allies in academia reflect self-critically on their own privileged status and traditional ways of thinking and researching, question them, and adopt a decolonial perspective.

Indigenous methodologies should not only contribute to bringing indigenous knowledge systems and epistemologies to light but also to overcoming the "positional superiority" of Western knowledge (Tuhiwai Smith, 2012: 62). Her starting point is the critique that the "knowledge gained through our colonization has been used [. . .] to colonize us" (ibid.: 62). This corresponds to what the African writer Ngugi wa Thiong'o (1986) has called the "colonisation of the mind." According to Tuhiwai Smith, "the globalization of knowledge and Western culture constantly reaffirms the West's view of itself as the centre of legitimate knowledge, the arbiter of what counts as knowledge and the source of 'civilised' knowledge" (Tuhiwai Smith, 2012: 66). The author therefore stresses:

> For indigenous peoples universities are regarded as rather elite institutions, which reproduce themselves through various systems of privilege. Even those universities, which are state-funded are considered major bastions of Western elitism. It is not

surprising, then, that many indigenous students find little space for indigenous perspectives in most academic disciplines and most research approaches.

Ibid: 132

Margaret Kovach highlights the emancipatory and decolonizing intention of Indigenous methodologies in the following words: "Indigenous researchers make research political simply by being who we are" (Kovach, 2015: 46). She locates the beginning of research based on Indigenous methodologies in the 1990s. At this time, Indigenous knowledge systems began to be considered as part of the discourse of qualitative research that saw itself as transformative. "As a result of decolonization efforts exposing extractive research practices involving Indigenous peoples, there was a focus on the recognition and protection of Indigenous knowledges" (ibid: 48). According to Kovach, it is not possible to practice indigenous methodologies without understanding indigenous knowledge systems from the ground up.

> In applying Indigenous methodologies, researchers are putting force an identity standpoint (whether they desire or not) and there is an expectation for them to engage in anti-colonial work. [...] There must be a deep, abiding respect for Indigenous knowledge systems and Indigenous experience. This often requires unpacking one's understanding of what respect means in this context.
>
> Ibid: 57

Shawn Wilson emphasizes the difference between the prevailing academic research paradigms and the indigenous research paradigm in the following way:

> The major difference between those dominant paradigms and an indigenous paradigm is that those dominant paradigms are built on the fundamental belief that knowledge is an individual entity: the researcher is an individual in search of knowledge, knowledge is something that is gained and therefore knowledge may be owned by an individual. An indigenous paradigm comes from the fundamental belief that knowledge is relational. Knowledge is shared with all of creation. It is not just interpersonal relationships, or just with the research subjects I may be working with, but it is a relationship with all of creation. It is with the cosmos; it is with the animals, with plants, with the earth that we share this knowledge. It goes beyond the individual's knowledge to the concept of relational knowledge. [...] You are answerable to all your relations when you are doing research.
>
> Wilson, 2008: 56

Bagele Chilisa highlights the need for a decolonial perspective and looks at the indigenous research paradigm "as a framework of belief systems that emanate from the lived experiences, values, and histories of those belittled and marginalized by Euro-Western research paradigms" (Chilisa, 2020: 23).[2] The indigenous research paradigm

[2] I quote from the second edition of her book, which has been significantly expanded from the first edition published in 2012.

focuses not only on individuals but also on the "communal forms of living." It creates a space "for inquiries based on relational realities and forms of knowing that are predominant among the non-Western Other/s still being colonized" (ibid: 2). For indigenous methodologies, narrative methods are important, but, according to Chilisa, they go beyond standard qualitative research approaches by "employing indigenous languages, metaphors, spirituality, and relations with the cosmos, symbolism, and art as in dance, song, poetry, and artifacts. Research questions asked are guided by indigenous worldviews" (ibid: 21).

Chilisa also refers to indigenous methodologies as "third-space methodologies" with reference to Homi Bhabha's (1994) concept of "the space in-between." "Thus, in the third space, indigenousness is interrogated to include the voices of those disadvantaged on the basis of gender, race, ethnicity, ableness, health, socioeconomic status, sexual orientation, age, and so on" (Chilisa, 2020: 32). With this enumeration, Chilisa gives an indication that "Indigenous" is not understood, at least by her, in an essentialist way as an ethnic category but tends to include all people who are discriminated against, disregarded, marginalized, and socially disadvantaged, or, in other words, who "have no place" in a social system or world order based on domination and social inequality.[3]

Researchers who advocate and apply *dialogic and horizontal methodologies* in their research share similar principles with representatives of indigenous methodologies. They are rarely of Indigenous origin and can be understood as the "non-indigenous allies" of indigenous researchers. They reject any form of authoritarian monopoly claim to scientific truth in a radically democratic way. They see such claims as rooted primarily in the dominance of White masculinity and the instrumental reason it embodies. They see horizontal methodologies as closely linked to political activism for a socially just society that strives for social equality with respect for cultural diversity and corresponding ways of life. The differences between the claim to academic superiority and the emancipatory aims of research "do not always arise from a gap between Western academic culture and the radical otherness of the researcher/indigenous person, but rather from the gap between the academic and the activist" (Rappaport and Ramos Pacho, 2005: 61–2). In this they are like participatory action research.

Research based on horizontal methodologies cannot avoid questioning the institutionalized hierarchies of the academic enterprise. These make it difficult to conduct research that is not based on hierarchical relationships between researchers and what is researched. Horizontal methodologies are based on the principle of overcoming this separation and of creating communities in which all participants equally shape and control the research process (Corona and Kaltmeier, 2012; Corona Berkin, 2020a; 2020b; Cornejo and Rufer, 2020). In this context, it would be of particular

[3] Nevertheless, the term Indigenous always implies the danger of attributing certain characteristics to people in an essentialist way or of assigning them to an ethnically defined group. It should therefore be recalled that here the term indigenous is understood in the context of decolonial critique as a collective self-designation of persons and peoples who were and are particularly affected by colonial and postcolonial violence and exclusion. Their ways of life and worldviews are by no means homogeneous, and they speak different languages.

importance to maintain equitable relations with population groups that are particularly affected by the consequences of colonization and the current postcolonial constellation. Olaf Kaltmeier, a German historian of Latin American history, points out:

> Horizontal methodologies emerge at this juncture of decolonization and democratization, based on dialogue with groups of subaltern actors who, despite racism, discrimination and lack of recognition, are bearers of multiple knowledges. Several researches working with horizontal methodologies enter into dialogue with indigenous [. . .] or Afro-American peoples promoting "a critical and decolonial collaborative narrative methodology" (Cortés-Gonzales, Leite-Méndez & Rivas-Flores, 2020: 96). At the intersections between the Global South and the Global North, horizontal methodologies have also been applied in the *multi-sited ethnography* of migration processes.
>
> Kaltmeier, 2020: 95; emphasis in original

In the dialogically designed research process, these researchers do not necessarily see a new "we" emerging between researchers and co-researchers, but new social ties of a "with." This "with" requires a self-critical reflection by academic researchers on their own privileges and their own dominance in the research process. Horizontal relations in this process cannot be established by goodwill alone but also require structural changes in the academic system. It also seems important to ask what attitude horizontal relations have in the research process.

Research attitudes and corporeality

Mexican social researcher David Bak Geler (2020) distinguishes between "a conception of tolerant horizontality," which he associates with the ideology of liberalism, and an "agonistic horizontality," which he locates in the history of republicanism. "Unlike tolerant horizontality, the agonistic variety does not start from the supreme rule of respecting the conceptions of others but is open to the criticism and debate of such principles and doctrines" (ibid: 142). In the sense of the latter conception, he always understands horizontality as provisional or improvised, that is, open to self-critical reflection and change.

> Improvising horizontality would insist on the radical impossibility of finding anything but provisional solutions to political conflicts or disagreements. It would also insist that the accumulation and fragmentary visibility of examples, where conflict has been provisionally resolved, establishes precedents and exemplary figures of the political: that is, reminders, inspirations or models that the work of practical improvisation need not always start from nothing.
>
> Ibid: 142–3

In the debate on horizontal and Indigenous methodologies, the connection between knowledge production and the body is also significant. Feminist-oriented researchers point out that the recognition of knowledge as scientific knowledge depends largely on whether it originates from and is represented by White men.

> Scientific authority is almost immediately associated with White, male, heterosexual bodies, which must be recognized as bodies. This does not imply that "feminized," "non-heterosexual" and "disadvantaged racialized" bodies are totally oppressed by the logics of science: there is heterogeneity among people who produce science, even official science, just as there is heterogeneity in colonial bodies.
>
> Nogueira Beltrão, 2020: 246

According to this author, some bodies are recognized as producers of science, while others are not. There is also a corresponding hierarchy within universities and in the academic writing market.

> It is enough to check the curricula of universities considered as recognized to corroborate that most of the knowledge identified as indispensable and classical refers to books and theories written by men, white and heterosexual . . . and preferably European. Assuming a "body-to-body" knowledge could be a way of "provincializing" (Chakrabarty, 2009) knowledge and understanding that any and all science, made by bodies, has limited scope, and could not solve all the world's problems.
>
> Ibid: 246–7

Nigerian sociologist of Yoruba origin, Oyèrónké Oyěwùmí (1997; 2016), also stresses that the recognition of knowledge is largely dependent on the visual perception of bodies. She therefore speaks of a "bio-logic" in which the body is essentialized in binary categories of gender and race, which privileges the knowledge of white men over that of Black women.

Argentinean social scientists Rosana Rodríguez and Sofía da Costa (2019) consider that the hierarchy of forms of knowledge represents an unequal social order that, in the name of neutrality, objectivity, and universality, excludes multiple forms of knowledge, especially the knowledge of women of the Global South. Their interest is to promote the debate on the epistemic decolonial turn in its multiple aspects. They advocate research from a feminist perspective from the South, the production of situated knowledges, the relationship between research and political action, the sensitive selection of methods and techniques, and ethical engagement in research. This requires a methodological strategy that transcends the separation of knower and what is researched, as well as the separation of mind and body or rationality and emotion. Rodriguez and da Costa call the methodology they seek "bodily biographies" or "corpo-biographies":

> The corpo-biographies are senti-corpo-perspective reconstructions of the life trajectories and bodily itineraries of women based on their lived bodily experiences. They are conceptual re-elaborations of the experience of corporeality, which implies the subjectivity and the vital trajectory of the lived experience of the body.
>
> Ibid: 14

The authors define their methodological proposal as "research-action-creation," which they understand as a creative activity "that comes from social, cultural and subjective experience and allows us to know aspects of lived and perceived reality, through sensations, emotions and feelings. Knowing by knowing oneself" (ibid: 15). It is a methodological strategy that is nourished by the desires of the subjects; "it includes sensitive criteria that involve the experience of the body, combining rational intelligence with perception, emotion, intuition and imagination" (ibid: 16). When the authors speak of intuition, they refer to an action that engages the interior, the subjectivity, which is installed in the "mythical capacity" of the subjects, "which is woven with memory, activating all the layers of the body" (ibid: 16). According to them, their proposal coincides with what is understood by "participatory action research" (PAR) and especially by "feminist activist research" (FAR).

With their proposal, Rodríguez and da Costa explicitly oppose the practice of "hegemonic academic feminists who define other women as marginal, poor, illiterate, victims, without recognising their valuable contributions to scientific productions" (ibid: 25). According to them, these "logics of appropriation are sustained by strategies of depoliticisation and decontextualisation, of a profoundly racist matrix, whose purpose is to strengthen the hierarchization of knowledge and the bodies that produce it" (ibid: 25). In this way, they see geopolitics of control and domination of knowledge, through the unequal number of citations of scholars of the Global South, which erases the right to authorship of its true producers.

All the earlier-mentioned authors see a Eurocentric and colonial construction in the claim to scientific authority, as both the term science and the emergence of universities as we know them today have their origins in colonizing countries (see, from an autobiographical perspective, Perrin, 2020).

Sensing-thinking

To produce situated or liberatory knowledge, the divisions between the apparently objective and the subjective, rationality and sense, body and soul, thinking and feeling, and similar opposites, which are deeply rooted in the universalist understanding of science, must be questioned without, however, ignoring the binary distinctions that are made in life and in the dominant languages. In Latin American Spanish, a term has come to express this claim well: the verb *sentipensar* and the noun *sentipensamiento*, an integral combination of being sensitive and thinking.

The term, or what it presumes to express, has a long history in different parts of the world. One might think, for instance, of the cognitive theories of emotion formulated by the psychologist Magda Arnold (1960) in the 1960s and the philosopher Martha Nussbaum (2001) four decades later. These are about how cognitive processes are influenced by emotions. The connection between emotions and thinking has been discussed in the Western tradition since Greek antiquity, especially by Aristotle. Here, however, I refer mainly to the reflections of the Colombian anthropologist Arturo Escobar (2014). The author takes up this concept proposed by the Colombian sociologist Orlando Fals Borda (1970; [1979] 2009), to underline that the "lived life" of people cannot be conceived with the category of rationality on which modern Western science is based, but that the senses, the "heart," also play an important role. By his own admission, Fals Borda had encountered the word *sentipensante* with a peasant on the Caribbean coast of Colombia.

In the book *Libro de los abrazos* (Book of Embraces) by Uruguayan writer Eduardo Galeano, published in English with the title *Celebration of the Marriage of Reason and Heart*, we find the following sentence:

> Why does one write, if not to put one's pieces together? From the moment we enter school or church, education chops us into pieces: it teaches us to divorce soul from body and mind from heart. The fishermen of the Colombian coast must be learned doctors of ethics and morality, for they invented the word *sentipensante*, feeling-thinking, to define language that speaks the truth.
>
> Galeano, 1991: 121

Patricio Guerrero Arias (2010; 2016) introduced the term *corazonar* (synthesis of heart and reasoning) with a similar meaning. According to this author, *corazonar* or *corazonamiento* represents "the affective, epistemic and ethical-political potential of the insurgent wisdoms of all the diverse peoples that inhabit the continent" (Guerrero Arias, 2010: 17). De Sousa Santos (2018: 99–102) takes up this term in his theoretical outline of "Epistemologies of the South" and translates it as "warming up the reason."

The verb *sentipensar*, as Escobar (2014) understands it, refers not only to feelings, but to all bodily emotions such as love, affection, concern, or intuitions.[4] He understands *sentipensar* as a term that comes from below, from the community, and is directed against those who believe they are superior and claim power over others. In this sense, it is a "subaltern" way of making oneself felt. We find it in the narratives of various non-modern worlds, such as indigenous peoples, Afro-descendants, or, in a broader sense, in people and populations subjugated and excluded by power. According to Escobar, the meaning of *sentipensamiento* is always situational. It is also present in social movements that claim their rights within their territory and the disposition of it. Space or territory in this sense includes not only the earth and its ecosystems but also the processes of "territorialisation" that generate identities and appropriations. Escobar conceives of territory as a "life project" in which the socio-political project, autonomy, and future perspectives are combined.

The concept of sensing-thinking is opposed to the Western model of rationality, in which concepts such as economy, body, mind, individual, culture, and nature are separated from each other, with the now widely observed consequence of ecological and socio-political crises and catastrophes. The ways of perceiving and thinking of Indigenous and Black peoples, on the other hand, are relational. From their perspective, the human is not separate from the non-human, nor is culture separate from nature,

[4] In an abridged English publication of Escobar's text, *sentipensar* is translated as "thinking-feeling" (Escobar, 2016). I find it more appropriate to translate the term as *sensing-thinking*, since it is not rational thinking that is in the foreground, but rather sensing or empathizing.

nor is the mind separate from the body. In this perspective, an object, person, or situation cannot be understood isolated from its context, as it does not exist in itself. Similar concepts can be found in the ontologies and ethics of *Ubuntu* in southern Africa, *Sumak Kawsay or Buen Vivir/Vivir Bien* of Indigenous peoples in the Andes, and *mino-bimaadiziwin* of First Nations in Canada (Pérez, 2016; Gruner, 2018: 269–274). Similar ideas have been taken up in feminist thought, according to which no individual can exist on his or her own; all depend on each other in mutual care. However, while the feminist-inspired ethic of care emphasizes compassionate care for others who are perceived as vulnerable (particularly infants), the concept of *sentipensar* emphasizes the aspect of ownership and liberation from imposed and one-sided dependencies. The aim is an intersubjective encounter of equal weight and value between people with different personal characteristics who understand themselves as connected to and responsible for each other.

I return here to the concept of sensing-thinking because it allows us to take seriously children's specific modes of existence and subjectivities and to recognize them as differently equal. In the dominating Western model of thinking or reasoning, which makes the rationality of emotion and the body appear separate, children's specific abilities are denied or devalued as irrational and subordinated to the supposedly more developed, matured, and perfected abilities of adults. The concept of *sentipensar*, on the other hand, is open to recognition of the diversity of capacities children have. It allows them to be recognized and made visible at all. This applies especially to children who are excluded and discriminated against in several and many ways.

Conclusion: Yield for childhood and children's rights studies

The research approaches, methodologies, and ethical concepts presented here have hardly been taken up in childhood and children's rights studies so far. Although they repeatedly emphasize the necessity of not only using children as objects of research or as sources of information but also involving them as partners or actors in the research process, this is rarely done with the intention of empowering children as actors and supporting them in their daily lives (Budde and Markowska-Manista, 2020). Also, participatory research with children usually produces knowledge for those persons or institutions that have access to children—with whatever intentions. Thus, in the sense presented earlier, these are "extractivist" varieties of research that do not question the subordinate or colonized status of children and childhoods.

Basically, childhood and children's rights studies must be re-conceptualized by freeing themselves from their academic corset and opening themselves to a knowledge production that includes all who can contribute to this production. The process of gaining knowledge must be oriented to questions, modes of perception, and modes of articulation that are interesting and useful for all research participants. Thus, scientific research must be transformed into research processes that create equal opportunities for all participants, including children, to engage with their own questions, ways of knowing, and ways of expressing themselves. This also includes the perspective of recognizing children themselves as independent researchers and, if necessary, advising and supporting them in their research (Alderson [2000] 2008; Kellett, 2005; Mason and Watson, 2014).

The research methodologies presented here enable taking the specific modes of existence and subjectivities of children seriously and recognizing children as different but equal. Under the dominance of the Western model of thinking, which makes rationality appear separate from emotion and body, the specific abilities of children are denied or devalued as irrational and subordinated to the supposedly more developed, matured, and perfected abilities of adults. In contrast, the methodological concepts of dialogic and action-oriented research, namely the concept of *sentipensar/sensing-thinking*, are open to recognizing the diversity of abilities found in children. They can help to make visible and better understand the lived realities of children, especially those children who are marginalized and discriminated against in many ways. This is indispensable for a children's rights practice of solidarity with these children that strives for social justice and a life in dignity.

References

Abebe, Tatek, Anandini Tar, and Ida M. Lyså (2022), "Southern theories and decolonial childhood studies," *Childhood*, 29(3): 255–75.

Alderson, Priscilla ([2000]2008), "Children as researchers: the effects of participation rights on research methodology," in P. Christensen and A. James (eds.), *Research with Children: Perspectives and Practices*, London: Routledge, pp. 241–57.

Ansell, Nicola (2019), "Global south research in children's geographies: from useful illustration to conceptual challenge," in T. Skelton and S. Aitken (eds.), *Establishing Geographies of Children and Young People. Geographies of Children and Young People 1*, Singapore: Springer Nature, pp. 51–70.

Araneda-Urrutia, Carlos (2022), "The invention of the 'weird' Southern child: mapping coloniality in the political problematization of disadvantaged children's lives in the Global South," *Childhood*, 29(3): 292–306.

Arnold, Magda B. (1960), *Emotion and Personality. Vol. 1 & 2*, New York: Columbia University Press.

Bak Geler, David (2020), "Alternativas de lo plural: pluralismo liberal, pluralidad republicana e improvisación práctica," in I. Cornejo and M. Rufer (eds.), *Horizontalidad. Hacia una crítica de la metodología*, Buenos Aires: CLACSO and Guadalajara: CALAS, pp. 123–43.

Bhabha, Homi K. (1994), *The Location of Culture*, London and New York: Routledge.

Budde, Rebecca and Urszula Markowska-Manista (eds.) (2020), *Childhood and Children's Rights between Research and Activism*, Wiesbaden: Springer VS.

Chakrabarty, Dipesh (2009), *Provincializing Europe: Postcolonial Thought and Historical Difference*, Princeton, NJ: Princeton University Press.

Chilisa, Bagele (2012), *Indigenous Research Methodologies*, Los Angeles: Sage Publishing.

Chilisa, Bagele (2020), *Indigenous Research Methodologies* (2nd ed.), Los Angeles: Sage Publishing.

Cornejo, Inés and Mario Rufer (eds.) (2020), *Horizontalidad. Hacia una crítica de la metodología*, Buenos Aires: CLACSO and Guadalajara: CALAS.

Corntassel, Jeff and Adam Gaudry (2014), "Insurgent education and Indigenous-centred research. Opening new pathways to community resurgence," in C. Etmanski, B. L. Hall, and T. Dawson (eds.), *Learning and Teaching Community-Based Research. Linking Pedagogy to Practice*, Toronto: University of Toronto Press, pp. 167–85.

Corona Berkin, Sarah (2020a), *Producción horizontal del conocimiento*, Guadalajara: CALAS and Bielefeld: Bielefeld University Press.

Corona Berkin, Sarah (2020b), "Investigar en el lado oscuro de la horizontalidad," in I. Cornejo and M. Rufer (eds.), *Horizontalidad. Hacia una crítica de la metodología*, Buenos Aires: CLACSO and Guadalajara: CALAS, pp. 27–58.

Corona, Sarah and Olaf Kaltmeier (eds.) (2012), *En diálogo. Métodos horizontales en Ciencias Sociales y Culturales*, Barcelona: Gedisa.

Cortés-González, Pablo, Analía E. Leite-Méndez, and José I. Rivas-Flores (2020), "Una experiencia disruptiva en educación secundaria: El caso del IES Esmeralda," *Revista Educación, Política y Sociedad*, 5(1): 85–103.

De Sousa Santos, Boaventura (2014), *Epistemologies of the South: Justice against Epistemicide*, Boulder and London: Paradigm Publishers.

De Sousa Santos, Boaventura (2018), *The End of the Cognitive Empire: The Coming of Age of Epistemologies of the South*. Durham, NC and London: Duke University Press.

Escobar, Arturo (2014), *Sentipensar con la tierra: Nuevas lecturas sobre desarrollo, territorio y diferencia*, Medellin: UNAULA.

Escobar, Arturo (2016), "Thinking-feeling with the earth: territorial struggles and the ontological dimension of the epistemologies of the south," *Revista de Antropologia Iberoamericana*, 11(1): 11–32.

Fals Borda, Orlando (1970), *Ciencia propia y colonialismo intellectual*, Mexico City: Nuestro Tiempo.

Fals Borda, Orlando ([1979] 2009), "Cómo investigar la realidad para transformarla," in M. Moncayo (ed.), *O. Fals Borda. Una sociología sentipensante para América Latina*, Bogotá: Siglo del Hombre Editores and CLACSO, pp. 253–301.

Galeano, Eduardo (1991), *The Book of Embraces*, New York and London: W.W. Norton & Company.

Gaudry, Adam J. P. (2011), "Insurgent research," *Wicazo Sa Review*, 26(1): 113–36.

Giri, Ananta Kumar (2013), *Knowledge and Human Liberation. Towards Planetary Realizations*, London and New York: Anthem Press.

Guerrero Arias, Patricio (2010), *Corazonar. Una Antropología Comprometida con la Vida. Miradas otras desde Abya-Yala para la descolonización del poder, del saber y del ser*, Quito: Ediciones Abya-Yala.

Guerrero Arias, Patricio (2016), *Colonialidad del saber e insurgencia de las sabidurías otras: Corazonar las epistemologías hegemónicas, como respuesta de insurgencia (de)colonial*, Quito: Universidad Andina Simón Bolívar, Sede Ecuador, Área de Letras y Estudios Culturales.

Gruner, Sheila (2018), "Territorio y el ser decolonial: Pervivencia de las mujeres y los pueblos en tiempos de conflicto, paz y desarrollo," in P. López and L. García Guerreiro (eds.), *Movimientos Indígenas y Autonomías en Anmérica Latina: Escenarios de Disputa y Horizontes de Posibilidad*, Buenos Aires: El Colectivo and CLACSO, pp. 259–83.

Haraway, Donna (1988), "Situated knowledges: the science question in feminism and the privilege of partial perspective," *Feminist Studies*, 14(3): 575–99.

Invernizzi, Antonella, Manfred Liebel, Brian Milne, and Rebecca Budde (eds.) (2017), *"Children Out of Place" and Human Rights. In Memory of Judith Ennew*, Cham: Springer International Switzerland.

Kaltmeier, Olaf (2020), "Horizontal en lo vertical. ¿O cómo descolonizar las metodologías en contextos de extrema desigualdad y de la crisis planetaria," in I. Cornejo and M. Rufer (eds.), *Horizontalidad. Hacia una crítica de la metodología*, Buenos Aires: CLACSO and Guadalajara: CALAS, pp. 93–122.

Kellett, Mary (2005), *How to Develop Children as Researchers: A Step by Step Guide to the Research Process*, London: Paul Chapman.

Kemmis, Stephen and Robin McTaggart (2005), "Participatory action research: communicative action and the public sphere," in N. Denzin and Y. Lincoln (eds.), *Handbook of Qualitative Research* (3rd ed.), Thousand Oaks, CA: Sage Publishing, pp. 559–603.

Kovach, Margaret (2010), *Indigenous Methodologies: Characteristics, Conversations and Contexts*, Toronto: Toronto University Press.

Kovach, Margaret (2015), "Emerging from the margins: Indigenous methodologies," in S. Strega and L. Brown (eds.), *Research as Resistance: Revisiting Critical, Indigenous and Anti-Oppressive Approaches*, Toronto: Canadian Scholars' Press and Women's Press, pp. 43–64.

Lewin, Kurt (1946), "Action research and minority problems," *Journal of Social Issues*, 2(4): 34–46.

Liebel, Manfred (2000), *La Otra Infancia: Niñez Trabajadora y Acción Social*, Lima: Ifejant.

Liebel, Manfred (2004), *A Will of Their Own: Cross-Cultural Perspectives on Working Children*, London and Philadelphia: Zed Books.

Liebel, Manfred (2012), in collaboration with Karl Hanson, Iven Saadi, and Wouter Vandenhole, *Children's Rights from Below: Cross-Cultural Perspectives*, Basingstoke: Palgrave Macmillan.

Liebel, Manfred (2020), *Decolonizing Childhoods: From Exclusion to Dignity*, Bristol and Chicago: Policy Press.

Liebel, Manfred (2023a), in collaboration with Rebecca Budde, Urszula Markowska-Manista, and Philip Meade, *Childhoods of the Global South: Children's Rights and Resistance*, Bristol: Policy Press.

Liebel, Manfred (2023b), *Kritische Kinderrechtsforschung: Politische Subjektivität und die Gegenrechte der Kinder*, Opladen, Farmington Hills and Berlin: Barbara Budrich.

Mason, Jan and Elizabeth Watson (2014), "Researching children: research on, with, and by children," In A. Ben-Arieh et al. (eds.), *Handbook of Child Well-Being*, Dordrecht: Springer, pp. 2757-96.

Mignolo, Walter D. (2009), "Epistemic disobedience, independent thought and de-colonial freedom," *Theory, Culture & Society*, 26(7-8): 1–23.

Nogueira Beltrão, Beatriz (2020), "El conocimiento cuerpo a cuerpo como forma de resistencia ante el racismo/sexismo epistémico," in I. Cornejo and M. Rufer (eds.), *Horizontalidad. Hacia una crítica de la metodología*, Buenos Aires: CLACSO and Guadalajara: CALAS, pp. 231–50.

Nussbaum, Martha (2001), *Upheavals of Thought: The Intelligence of Emotions*, Cambridge: Cambridge University Press.

Oyěwùmí, Oyèrónkẹ́ (1997), *The Invention of Women: Making an African Sense of Western Gender Discourses*, Minneapolis: University of Minnesota Press.

Oyěwùmí, Oyèrónkẹ́ (2016), *What Gender Is Motherhood? Changing Yorùbá Ideals of Power, Procreation, and Identity in the Age of Modernity*, New York: Palgrave Macmillan.

Pérez, Diego (2016), "Ubuntu y Sumak Kawsay. Paradigmas urgentes para una paz incluyente y sostenible en Colombia," in S. Gruner, M. Blandon, and C. Mina-Rojas (eds.), *Des/Dibjando el país/aje: Aportes para la paz con pueblos afrodescendientes e indígenas. territorio, autonomía y buen vivir.* Medellín: Poder Negro.

Perrin, Chris "Kikila" (2020), "The hall of mirrors and a landscape of multiple layers: insurgent research, Indigenous resurgence, and challenging the university structure," *Verges—Germanic & Slavic Studies in Review*, 3(1).

Quijano, Anibal (2008), "Coloniality of power, Eurocentrism, and social classification," in M. Moraña, E. Dussel, and C. Jáuregui (eds.), *Coloniality at Large. Latin America and the Postcolonial Debate*, Durham, NC and London: Duke University Press, pp. 181–224.

Rappaport, Joanne and Abelardo Ramos Pacho (2005), "Una historia colaborativa: retos para el diálogo indígena-académico," *Historia Crítica*, 29: 39–62. Universidad de Los Andes Bogotá, Colombia.

Rodríguez, Rosana Paula and Sofía da Costa (2019), "Descolonizar las Herramientas Metodológicas. Una Experiencia de Investigación Feminista," *MILLCAYAC—Revista Digital de Ciencias Sociales*, 6(11): 13–30.

Rufer, Mario (2020), "No vamos a traducir. Instalar un secreto, negar la dádiva, redefinir el juego," in I. Cornejo and M. Rufer (eds.), *Horizontalidad. Hacia una crítica de la metodología*, Buenos Aires: CLACSO and Guadalajara: CALAS, pp. 277–302.

Strega, Susan, and Leslie Brown (eds.) (2015), *Research as Resistance: Revisiting Critical, Indigenous and Anti-Oppressive Approaches*, Toronto: Canadian Scholars' Press and Women's Press.

Thiong'o, Ngugi wa (1986), *Decolonising the Mind: The Politics of Language in African Literature*, London: James Currey.

Tuhiwai Smith, Linda (2005), "On tricky ground: researching the native in the age of uncertainty," in N. Denzin and Y. Lincoln (eds.), *The SAGE Handbook of Qualitative Research* (3rd ed.), Thousand Oaks: Sage Publishing, pp. 85–107.

Tuhiwai Smith, Linda (2012), *Decolonizing Methodologies: Research and Indigenous Peoples* (2nd ed.), London and New York: Zed Books (3rd ed. 2021).

Wilson, Shawn (2008), *Research as Ceremony: Indigenous Research Methods*, Manitoba: Fernwood.

19

From Alternative Epistemologies to Alternative Aesthetics: A Plea for a Decolonial Environmental Aesthetics

Werther Gonzales León

Introduction

In a critical attitude toward modernity and from the notion of *alternative epistemologies*, the Latin American proposal of *buen vivir* (good living) promotes a different relationship between human beings and nature that highlights the need for the cultivation and protection of nature. For this purpose, the theoretical elaboration of this proposal is often based on approaches such as an epistemology of the South, the decolonization of knowledge, or epistemic disobedience.

Insofar as scientific knowledge and theoretical thinking—both of an *epistemic* character—result from a vision of the world essentially different from one based on sensitive perception and artistic representation—both of an *aesthetic* character—this article attempts to show why it is important to also take into account *alternative aesthetics* in favor of the theoretical construction and practical application of *buen vivir*. Following the trail of a decolonial aesthetics, the possibilities of this approach are extended here to the open field of natural beauty and the question of environmental ethics on the value of nature; the aim is to generate a productive interaction between aesthetics and environmental ethics and to advocate for a genuinely decolonial environmental aesthetics.

The model proposed for the promotion and development of such aesthetics is the festival as a mediating reality between community and nature. An exemplary case of this model is the Pachamama Festival, which is celebrated in different places in the Andean region. The moments and elements of this aesthetic event graphically clarify the type of relationship established between human beings and nature in the Andean culture, the relevance of art as a way of humanly inhabiting the world as legitimate as the scientific positioning regarding the phenomena, and the significance that this relationship acquires for the achievement of a good life. The emphasized aspects of this celebration will show why, within the framework of the discussion of *buen vivir*, it is also necessary to think of a decolonial aesthetics with an environmental orientation as an alternative aesthetics.

Demand for alternative epistemologies

Among the contributions to *buen vivir*, we call *alternatives* to those epistemologies that critically position themselves and assume assumptions different from those contained in epistemology understood as a philosophical discipline of European origin, whose subsequent theoretical development took and takes place in different ways and in different parts, especially in the so-called global North. The critical position of these epistemologies is grounded in different ways, as will be seen below. In this context, the term *epistemology* generally indicates reflection on the nature of knowledge, both the most immediate, such as that of common sense, and the most elaborate, such as that of science. Therefore, a strict distinction is not formulated between epistemology and gnoseology, as is usually the case in Latin America, but rather the theoretical domains of the former are extended over the latter.

As a promoter of a civilizational alternative, *buen vivir* questions the epistemic hegemony of modern science's own form of knowledge and its corresponding technical application. Already in the implementation of the idea of *buen vivir* in the National Plan of Ecuador, it is mentioned that, in the discussion around the concept of development, Andean indigenous peoples contribute to this debate from *other epistemologies* and cosmovisions (SENPLADES, 2013: 32). These other epistemologies (*epistemologías otras*) are generated from other knowledge scorned by Western hegemonic knowledge (Acosta, 2014: 53). *Buen vivir* is critically positioned against the epistemic hegemony of the West, not only questioning the basis and limitations of that model but also promoting the development of alternative *indigenous* epistemologies (Ranta, 2016: 20). In geopolitical terms, it is ultimately about the construction of *a new epistemology of the south* (Ribadeneira, 2020: 48).

Among the contributions to *buen vivir*, the *epistemology of the south* is shown, in fact, as the epistemic alternative most in line with the proposal. This epistemology is understood as the search for knowledge and criteria of validity of knowledge that give visibility and credibility to the cognitive practices of classes, peoples, and social groups that have been historically victimized, exploited, and oppressed by global colonialism and capitalism (Santos, 2009: 12). Santos proposes the "paradigm of prudent knowledge for a decent life," an emergent paradigm comprising scientific and social aspects, which is based on five theses justified by the limitations of modern science, namely "all natural-scientific knowledge is social-scientific," "all knowledge is local and total," "all knowledge is self-knowledge," and "all scientific knowledge aims at becoming common sense" (Santos, 1992: 30–43). These are the characteristics of a self-proclaimed oppositional postmodern knowledge outlined by an alternative epistemology. The concrete application of these theses seeks to contribute to the development of a good life in society.

Another epistemic alternative in tune with the postulates of *buen vivir* lies in the idea of *coloniality of power* coined by Quijano. This expression refers to a pattern of world domination that, in the case of our America, orders and structures social relations on the basis of the categories of *race* and *capital* (Quijano 2000). Historically, European hegemony, as a pattern of power, has concentrated all forms of the control of subjectivity, culture, and especially knowledge and the production of knowledge" (Quijano, 2000: 540). Therefore, a decolonization of power begins with the decolonization of knowledge, that is, of Eurocentric perspectives of knowledge. A decolonized knowledge is the recognition of our particular historical experience, the author continues, and knowledge born of a transformative social praxis (Quijano, 2005).

Finally, the *epistemic disobedience* promoted by Mignolo also harmonizes with the theoretical demands of *buen vivir*, since it advocates "other epistemologies, other principles of knowledge and understanding and, consequently, other economy, other politics, other ethics" (Mignolo, 2007: 453). In agreement with Quijano, Mignolo proposes an epistemic decolonization as a process of detachment, that is, as an awakening from the dream and the hegemonic illusion of knowledge and understanding—and the horizon of life—led, in modernity/coloniality, by certain theological and egological policies (Mignolo, 2010b: 34).[1] In other words, this awakening means nothing more than an awareness on the part of racialized and marginalized actors, both epistemically and socially, an awakening that must become a border epistemology.

This outline of the three approaches mentioned is simply intended to show the coherence and, at the same time, the insufficiency of the demand for alternative epistemologies.[2] This demand is coherent, since it is assumed that the hegemony of modern knowledge, especially that of a scientific character, must be questioned and surpassed (not *subdued*, because it is not about recolonizing what was previously colonizing) by knowledge and ideas of the periphery, knowledge invisibilized by capitalism and historically undervalued. What is deficient in it lies, however, in alternating epistemology with epistemology, that is, presupposing a certain privileged function of knowledge—whether it is cognitive practices, ancestral knowledge, situated knowledge, or knowledge of intersubjective origin; *knowledge*, after all—to organize from the foundations and give meaning to our vital world. This assumption could paradoxically also be considered colonial and Eurocentric, since it attempts to radically alter the content of the discourse, but it does not question the very orientation that is decisively followed in modernity: the epistemic path. Therefore, my intention in what follows is to argue in favor of an aesthetics with decolonial character that serves as an *incipient alternative* to alternative epistemologies, and, in accordance with the ethical-environmental aspirations of *buen vivir*, this could thematize the fundamental relationship between human beings and nature.

The path to a decolonial aesthetics

Let us imagine that we are now in front of a mountain, for example, in front of Winikunka, the so-called Mountain of Seven Colors located between the districts of Cusipata and Pitumarca in Cusco, Peru.

[1] Mignolo basically refers to the politics of Christian theology since the Renaissance and the politics of ego-centered philosophies since the Enlightenment. Among the latter, Cartesianism is at the starting point (see Mignolo, 2011).

[2] For a more detailed presentation of classical versus alternative epistemologies from the perspective of a critical pedagogy in the context of Native American cultures, see Knowles and Lovern (2015).

At first glance, one may wonder about the natural reason for its color tonalities, its altitude and the location of its summit, or the age of its current structure. In this case, a *theoretical* attitude is assumed toward the mountain; we seek to know something about it and its presence is interpreted conceptually.

At a second glance, one can try to conjecturally calculate the amount of minerals it hides and the possible profit from its extraction and sale, or simply estimate its surface in order to carry out a certain activity. In this case, a practical attitude is assumed toward the mountain, its usefulness is evaluated, and action is taken accordingly.

At a third glance, one can stop at Winikunka's appearance itself, admire the elevation of its figure and the combination of its colors, and contemplate how its serene aura contrasts with the liveness of its marked lines. In this case, an *aesthetic* attitude is assumed toward the mountain; its attractiveness is carefully perceived, and pleasure or displeasure is experienced in the observation.

These three views constitute three different attitudes toward the same object. Of course, they do not have to be the only possible attitudes, nor does the description of each one have to be the same in all cases. I simply want to point out that these are three *different* attitudes. By this I mean that, although in the same person or in the same community the meaning of one of the attitudes may be more or less in line with the meaning of the other, these attitudes are *irreducible* to each other. The experience contained in the aesthetic attitude cannot be derived from an experience of knowledge; it can certainly coexist and be linked to it in a coherent way, and it could eventually even be enriched or impoverished by it, but it cannot be *derived* from it, since both correspond to two different ways of being in the world, of perceiving and understanding what is given. The same can be said of the relationship between the practical attitude and the theoretical attitude. In fact, we might think about a unity between theory and praxis in relation to an object and thus conceptually and factually realize this thought, but this thought arises precisely from the realization that these are two distinct spheres of human life and of dealing with things.

In philosophy, these three attitudes inaugurate three large areas of reflection referring to three different aspects of the world: theoretical philosophy (with epistemology ahead), practical philosophy (with ethics ahead), and aesthetics.[3] Aesthetics forms a separate reflection since the type of experience at stake does not lead to any knowledge that can be studied, for example, by epistemology, nor to any goal that can be considered, for example, by ethics. As an independent discipline, aesthetics is commonly understood as a philosophical reflection on art, on the fine arts. This understanding dates back to the Enlightenment. Since the middle of the last century, however, this discipline tends to be understood globally in a more radical sense, that is, as a theory on *sensible perception* (*aísthēsis*), in which art is presented as an aesthetic manifestation among others. This return of aesthetics to its etymological

[3] Although the philosophical disciplines belonging to these spheres of reflection were born with modernity—see the introduction to Kant's *Critique of the Power of Judgment* (2000)—Aristotle already distinguished in a way these three types of understanding in *Metaphysics* 6.1.1025b18–28 (see Ross, 1997).

sense and the continuous aestheticization of the various facets of reality has led to talk of a new aesthetics, an aesthetic turn, or simply an *aisthetics*.[4]

In the Latin American decolonial turn, traditional aesthetics have also been questioned and reformulated in a certain way. Mignolo argues in this regard that *aesthesis* (sensation) was colonized by aesthetics as a philosophical discipline, since *aesthesis* is a phenomenon common to all living organisms with nervous systems, *aesthetics* is a particular version or theory of such sensations related to beauty (2010a: 14). Decolonial aesthetics are understood, then, as ways of *dismantling* imperial aesthetics, of displacing them in view of other sensibilities and the future construction of decolonial subjectivities.[5] Continuing Mignolo's thesis, I would like to outline here the preamble of a decolonial aesthetics that, in addition to critically positioning itself against the dominant aesthetic production of modernity, consequently delves deeper into the invisibilized aesthetic experience of our indigenous communities. This deepening leads to the constructive moment that decolonial aesthetics requires, but unlike that developed so far by Mignolo, which refers centrally to the individual contemplation of works of art exhibited in museums or similar installations, the decolonial aesthetics proposed here recovers the force of natural beauty and explores chiefly the joint action that takes place in relation to it. This action is a mode of aesthetic experience of nature, which is exemplarily illustrated in the phenomenon of the *fiesta* (festival). With the inclusion of this phenomenon—not just any festival, but the festivals of indigenous communities in honor of nature—I try to broaden the research field of decolonial aesthetics and mainly of the studies on *buen vivir*. Insofar as the latter has a clear ethical-environmental orientation, since it emphasizes the value of nature and the need for a harmonious relationship between it and human beings, the proposed aesthetics should be not only decolonial but also foremost an *aesthetics of nature*—that is, a decolonial environmental aesthetics.

Aesthetic experience at the Pachamama Festival

In modern society, experiencing nature aesthetically usually means standing in front of a landscape or striking-looking specimens of flora or fauna and contemplating their uniqueness, their shapes, and their colors. It is also considered an aesthetic experience of nature, the sensual and emotional satisfaction when, for example, we walk through a forest or climb a mountain. In fact, these are the two most widespread models of aesthetic experience in German environmental ethics today: *contemplation* of natural realities (defended by Seel) and *satisfaction* in natural spaces (defended by Böhme).[6]

4 This is the case, for example, of Böhme (2001), who understands aesthetics as *aisthetics* from the rediscovery of human corporality, the expansion of artistic possibilities, and the theory of atmospheres. For a comprehensive presentation of the aesthetic turn, see Raffnsøe (2019).

5 In a recent paper, Mignolo (2019) carries out a certain ontologization of the term *aesthesis* and understands the concept of the aesthetic no longer solely as what is linked to sensations but also as that which underlies the various doings in which a colonial matrix of power operates, that is, as a situated *subjectivity*.

6 This article is confined to the aesthetic positions of environmental ethics in German-speaking academia. The two mentioned models play a predominant role, however, also in English-speaking environmental aesthetics: contemplation of natural beauty predominates in cognitive perspectives (e.g., Carlson, 1981, 2000), while emotional and bodily satisfaction by nature predominates in non-cognitive perspectives (e.g., Berleant, 1992, 2013).

Seel (1991) bases the aesthetic value of nature on the contemplative attitude. To contemplate an object aesthetically means to actively perceive its form without instrumentalizing it, without using it as a means for something. Although in principle there are also objects in the artificial world that are susceptible to aesthetic contemplation—objects that can even be usually qualified as anti-aesthetic—nature has elements and spaces that invite mainly contemplative activity. Böhme (1989), for his part, argues that in nature we are able to experience a variety of pleasant sensations both mentally and physically. Let's think, for example, of a quiet stay in the park of our choice. It is no longer about what we see or actively perceive about us but about the way in which that natural atmosphere stimulates our body and our senses, and provides us a state of receptive satisfaction. Regarding the ethical-environmental consequences of these two models of aesthetic experience of nature, for both authors, this experience constitutes *a basic option of good living*.[7]

Now, returning to the idea of a decolonial environmental aesthetics, I would like to highlight another way of aesthetically experiencing nature, a way that currently seems outdated in the Western world. This is the Pachamama Festival in the Andean world.[8] With this I want to emphasize the intersubjective and corporeally active dimension of the aesthetic experience of nature, which is not reduced to the two previously mentioned models.[9]

The festival at issue is celebrated in the context of what is currently known as *Día de la Pachamama* (Pachamama Day), a commemoration that takes place annually on the first day of August in different parts of the Andes mountain range; for example, Zamora Chinchipe (Ecuador), Cusco (Peru), La Paz (Bolivia), Tarapaca (Chile), and Salta (Argentina). Although in recent decades it has been popularized—and sometimes even degraded as a tourist attraction—by regional governments, this *raymi* (*festival* in Quechua) dates back to pre-Hispanic times.[10] Generally, the religious background of this celebration, otherwise undeniable, is the main focus in social science research. This time I would like to draw attention to the aesthetic experience that is at the base of this peculiar way of meeting and relating to nature, the *fiesta*. Although certainly in each place this celebration presents differences in its realization and staging, two constitutive moments can be identified in it, and they are repeated in all cases.

7 For a schematic presentation of both aesthetic perspectives in environmental ethics, see Krebs (1997, 1999).

8 The modern concept of *nature* does not coincide semantically with *Pachamama*, just as neither of them coincide, for example, with *physis* of Ancient Greece, *natura* among the Romans, or the current ways of conceiving nature in *ubuntu* or in Shinto. Without taking this into consideration, it would be imprudent to exchange one term for another. Prudent would be *to highlight* what is distinctive in each case when it comes to understanding that particular region of reality that is at stake, beyond nominal disparities. Regarding Andean culture, a historical documentation of *pacha* and *Pachamama* can be found in di Salvia (2013), as well as an interesting philosophical interpretation of *pacha* in Depaz (2015).

9 The fact that the Pachamama Festival is typical of Andean culture does not mean that a festival related to nature takes place only here. Similar festivals are *still* celebrated in other latitudes, although in a different way and with different conceptions of *nature*.

10 See the contribution of Rozas Álvarez and Calderón García (2000) on the August rituals in the province of Calca, Cusco. In spite of the relatively distant year of its publication, the text is still highly topical.

The first moment of the festival is an opening ceremony: participants share a meal with the Pachamama by burying cooked food, drinks, and coca leaves. This is a way of calling Mother Earth to the table, as if she were a special guest (Alvarado & Mamani, 2007: 87). This activity is known as *the cult of Pachamama*. It is often said that, in exchange for protection and food, the inhabitants of the Andes, especially farmers, feel obliged to offer part of their goods or crops and thus pay a certain tribute.[11] But if this were the only reason, this attitude would reveal a rather instrumental relationship with nature, in which it is used as a means to satisfy personal interests. With this, the mother-child relationship between nature and human beings would be lost, something that the term *Pachamama* denotes: a relationship in which an attitude of love and respectful and selfless treatment prevail. Therefore, more than a payment to the earth, this ceremony essentially represents an act of *recognition*, in the sense of gratitude, but mainly in the sense of seeing nature as a living being, as the bearer of a certain dignity. This is the only way to explain why she—it is also said—*rests* during these days, since she needs rest and food.

The second moment is the festival itself, whose relevance seems to have been overshadowed by the initial rite, and it is often forgotten that it is also part of the *raymi*. The festival is characterized mainly by the presence of music and dance. These two artistic activities configure the festive reality of the aesthetic experience of nature. The aesthetic space is filled with typical dances in which the participants dress in their most colorful attire and perform movements to the sound of the music, while in the perceived time there are songs full of joy and life, especially regional ones. The unity of both becomes a free play of sensations and forms, displayed in the interaction of bodies and a shared feeling of association. The playful manifestations of the festival replace for a few days the annual working days. Work is followed by a well-deserved play break. The series of artistic performances is also accompanied by food and drinks shared by the attendees. In some cases, the festival takes place in the context of a large fair or a sort of open market set up for the occasion. All of this is lived and celebrated in connection with nature, or rather, from a way of understanding, celebrating, and relating to nature as Mother Earth.[12]

Especially this second moment of the festival reveals a peculiar mode of aesthetic experience, highly beneficial for a decolonial aesthetic aligned with *buen vivir*. Unlike traditional arguments based on aesthetic contemplation and satisfaction, the Pachamama Festival allows us to move from reception to group aesthetic production, that is, to dances and music in community. The aesthetic relationship that inhabitants of the Andes establish with nature through the festival is not merely contemplative and receptive, but fundamentally practical and participatory. Our senses move from perceiving natural beauty (Seel), from feeling the pleasant stimuli of nature (Böhme),

[11] This is how the ceremony is often reported in the Peruvian capital's newspapers (e.g., Marcelo, 2022).

[12] For a more detailed approach to the festive dynamics in Andean communities, the saturation of the senses that takes place there, as well as the indissoluble connection between music, dance, and costumes, see Martínez (2014) and Stobart (2002).

to creating and producing beauty communally and artistically with their bodies *for love of* nature.[13]

With a view to the theoretical construction of *buen vivir*, the following three consequences can be drawn from this aesthetic experience of nature represented in the Pachamama Festival.

Firstly, the *communal* character of the festival is ethically and aesthetically valuable. Models of contemplation and satisfaction are primarily individual experiences, but the nature festival of Andean origin develops on a rather interpersonal level. The inclusive dimension of this festival, the exchange and collaboration among the participants in a cohesive environment, are important factors for the achievement of a good life in community. Thus, in complement with the models of contemplation and satisfaction, the festival as a *joint aesthetic performance* represents an extension of the aesthetic experience and also has an important ethical effect with social projection and constitutes a basic option in the idea of a good life.

Secondly, in the Pachamama Festival, there are also two concrete consequences for environmental ethics and the matter of the value of nature. On the one hand, the initial recognition of nature as an animate being and subject of will immediately confirms the attribution of non-instrumental but *intrinsic* value to nature. Nature is not an inexhaustible source of resources but a living being with which a filial relationship is established, a relationship that brings with it feelings of love, respect, protection, and gratitude. On the other hand, nature also possesses a eudemonic value *for us* insofar as it promotes the achievement of a good life on an interpersonal level and, with it, the need to protect and conserve it for our own benefit. This festival shows that it is not contradictory to think that nature can have both a value in itself and a value for us.

Thirdly, a highly profitable consequence for the possibility of an environmental aesthetic of a decolonial character lies in the particular connection established between art and nature in this festival. The aesthetic experience is not expressed here through forms or objects privileged in traditional aesthetics, for example, descriptive judgments or literary statements (Seel), static artistic works, or gardening work where the subject's intervention is limited to producing something external to him (Böhme). Rather, it is expressed through corporeal, dynamic, and intersubjective artistic forms that come from societies historically invisibilized by hegemonic power. Not an individual, but an association of people belonging to a community expresses firsthand aesthetically and with temporal regularity.

With these three consequences, I attempt to outline programmatically the bases of a decolonial environmental aesthetics that can position itself as an alternative aesthetics against the alternative epistemologies demanded by *buen vivir*. Such aesthetics focuses, on the one hand, on a decisive issue for the development of this Latin American proposal, that is, the (harmonious) relationship between human beings and nature,

[13] Meyer (2003: 124–9) questions the categories used by aesthetic positions in environmental ethics and argues in favor of a *love of nature* in order to clarify the type of value that nature gains in the aesthetic experience. Although her proposal presents more complications than solutions, I think that it can be an interesting recourse to interpret the *filial* attitude of Andean inhabitants toward Mother Earth, as well as the different forms of aesthetic expression linked to nature.

and, on the other hand, it does not limit itself to questioning or recreating forms historically privileged by traditional European aesthetics but seeks, in the vital world of indigenous cultures, different aesthetic experiences that have a vindicating and transformative potential.

Conclusion

In order to continue the theoretical construction of the Latin American proposal of *buen vivir*, it would be appropriate to include in the research focus more aesthetic phenomena of the vital world of the Andean culture and give them the importance they deserve, since here a sui generis conception of nature is revealed, able to relativize our interpretation of the world dominated by the modern scientific paradigm. The Pachamama Festival offers us the possibility of questioning this technical, purely scientific representation of nature, and of thinking of another type of nexus with it from another type of perception, that is, aesthetic perception. Of course, this does not mean that these Andean practices should be repeated, imitated, or applied indiscriminately to other social realities and thus utopically save the planet. It would be absurd to want to mechanically implement an identical nature festival in European countries or other latitudes. The idea is simply to show that, just like aesthetic contemplation and satisfaction, the festival, the aesthetic action, and the artistic activities involved in it also represent legitimate ways of experiencing and expressing the way in which nature challenges us, and the way in which we relate aesthetically to it.

Thus, the possibilities of a decolonial environmental aesthetics lie not only in the criticism of the so-called imperial aesthetics and in the promotion of aesthetic productions invisibilized by coloniality but also in the exploration of sensibilities and perceptive and productive experiences typical of communities that have been and are relegated by this pattern of power. As an alternative to alternative epistemologies, a decolonial environmental aesthetics has the potential to highlight and describe the way of perceiving nature as Mother Earth, the type of cultural and artistic manifestations linked to it, and the festive interaction of bodies in performativity. In contrast to and in complementation with a *cognitive* attitude toward the environment (epistemology) and an individual or community *praxis* that configures a moral sphere (ethics), aesthetics is capable of showing the type of particular sensibility that perceives and creates without establishing any theoretical or functional purpose. More specifically, the content of the decolonial environmental aesthetics outlined here offers three aspects of the relationship between human beings and nature that cannot be sufficiently deepened either from epistemology or from ethics as branches of philosophy: 1) the idea of a *community* subject in the festival as an aesthetic experience of nature, unlike the individual who contemplates natural beauty; 2) the idea of a *performative* body in the festival as an aesthetic experience of nature, unlike the thesis of a purely receptive body of aesthetic-natural qualities; and 3) the idea of art as an aesthetic *action* in nature, unlike classical models based on the autonomy of works of art, isolated from any social and natural sphere. The deepening of these three aspects leads to understanding why, in the Andean culture, living well means mainly living in harmony with nature.[14]

References

Acosta, A. (2014), "El Buen Vivir, más allá del desarrollo [Good living, beyond development]," in G. C. Delgado (ed.), *Buena vida, buen vivir. Imaginarios alternativos para el bien común de la humanidad*, Mexico City: UNAM, pp. 21–60.

Alvarado, M. and M. Mamani (2007), *El origen de las fiestas andinas*, Paka-Illa.

Berleant, A. (1992), *The Aesthetics of Environment*. Temple UP.

Berleant, A. (2013), "What is Aesthetic Engagement?" *Contemporary Aesthetics, 11*, 5.

Böhme, G. (1989), *Für eine ökologische Naturästhetik*. Suhrkamp.

Böhme, G. (2001), *Aisthetik. Vorlesungen über Ästhetik als allgemeine Wahrnehmungslehre*. Fink.

Carlson, A. (1981), "Nature, Aesthetic Judgment, and Objectivity." *Journal of Aesthetics and Art Criticism, 40*, 15–27.

Carlson, A. (2000), *Aesthetics and the Environment. The Appreciation of Nature, Art and Architecture*. Routledge.

Depaz, Z. (2015), *La cosmo-visión andina en el Manuscrito de Huarochirí*. Vicio Perpetuo Vicio Perfecto.

di Salvia, D. (2013), "La Pachamama en la época incaica y post-incaica: una visión andina a partir de las crónicas peruanas coloniales (siglos XVI y XVII)." *Revista Española de Antropología Americana, 43*, 89–110.

di Salvia, D. (2015), "Pachamama." In H. Gooren (Ed.), *Encyclopedia of Latin American Religions*.

Gonzales León, W. (2023), *Ética ambiental y estética decolonial: la fiesta de la Pachamama como práctica de un buen vivir*. (Working Paper Series-Puentes Interdisciplinarios 01/ Serie 2). Interdisciplinary Latin America Center (ILZ), University of Bonn.

Kant, I. (2000), *Critique of the Power of Judgment*. Cambridge University Press.

Knowles, F. E. and L. L. Lovern (2015), "Classic and alternative epistemologies," in *A Critical Pedagogy for Native American Education Policy*. Palgrave Macmillan.

Krebs, A. (1997), *Naturethik*, Berlin: Suhrkamp.

Krebs, A. (1999), *Ethics of Nature*, Berlin: Walter de Gruyter.

Marcelo, B. (2022, July 30), "Día de la Pachamama: ¿por qué se celebra el 1 de agosto y cómo surgió esta tradición de pago a la tierra?," *La República*. https://larepublica.pe/datos-lr/respuestas/2022/07/30/dia-de-la-pachamama-por-que-se-celebra-el-1-de-agosto-y-como-surgio-esta-tradicion-de-pago-a-la-tierra-peru-argentina-mundo-andino-evat

Martínez, R. (2014), "Músicas, movimientos, colores en la fiesta andina. Ejemplos bolivianos," *Anthropologica*, 33, 87–110.

Meyer, K. (2003), *Der Wert der Natur. Begründungsvielfalt im Naturschutz* [The value of nature. Diversity of justification in the protection of nature]. mentis Verlag.

Mignolo, W. (2007), "Delinking," *Cultural Studies*, 21(2–3), 449–514.

Mignolo, W. (2010a), "Aiesthesis decolonial [Decolonial aiesthesis]." *Calle 14, revista de investigación en el campo del arte, 4*, 10–25.

[14] An initial Spanish version of this article was published in Gonzales León (2023).

Mignolo, W. (2010b), *Desobediencia epistémica. Retórica de la modernidad, lógica de la colonialidad y gramática de la descolonialidad* [Epistemic disobedience. Rhetoric of modernity, logic of coloniality and grammar of decoloniality]. Ediciones del Signo.
Mignolo, W. (2011), "Geopolitics of sensing and knowing: on (de)coloniality, border thinking and epistemic disobedience," *Postcolonial Studies*, 14, 273–83.
Mignolo, W. (2019), "Reconstitución epistémica/estética: la aesthesis decolonial una década después." *Calle 14, revista de investigación en el campo del arte, 14*, 14–33.
Quijano, A. (2000), "Coloniality of power, Eurocentrism, and Latin America," *Nepantla: Views from South*, 1(3), 533–80.
Quijano, A. (2005), "Of Don Quixote and windmills in Latin America," *Estudos Avançados*, 19(55), 9–31.
Raffnsøe, S. (2019), *The Aesthetic Turn: The Cultivation and Propagation of Aesthetic Experience after its Declaration of Independence* (MPP Working Paper No. 2019/1). Department of Management, Politics and Philosophy, Copenhagen Business School.
Ranta, E. M. (2016), "La compleja construcción histórica del buen vivir en Bolivia: de resistencias indígenas a política del estado [The complex historical construction of good living in Bolivia: from indigenous resistance to state policy]." *Economistas sin Fronteras*, 23, 18–21.
Ribadeneira, K. (2020), "Buen Vivir: críticas y balances de un paradigma social en construcción [Good living: criticisms and balances of a social paradigm under construction]." *Diálogo Andino, 62*, 41–51.
Ross, W. D. (ed.). (1997), *Aristotle's Metaphysics* (2 vols). Oxford University Press.
Rozas, J. and M. C. Calderón (2000), "La renovación: una interpretación de los rituales de agosto," in L. Millones (ed.), *Desde afuera y desde adentro. Ensayos de etnografía e historia del Cuzco y Apurimac*, National Museum of Ethnology, pp. 317–52.
Santos, B. S. (1992), "A discourse on the science," *Review (Fernand Braudel Center)*, 15(1), 9–47.
Santos, B. S. (2009), *Una epistemología del sur* [An epistemology of the South]. Siglo XXI; CLACSO.
Secretaría Nacional de Planificación y Desarrollo (= SENPLADES) (2013), *Plan Nacional para el Buen Vivir 2013-2017* [National Plan for Good Living 2013-2017]. Senplades.
Seel, M. (1991), *Eine Ästhetik der Natur*, Berlin: Suhrkamp.
Stobart, H. (2002), "Sensational sacrifices. Feasting the sense in the Bolivian Andes," in L. Austern (ed.), *Music, Sensation and Sensuality*, London: Routledge, pp. 97–120.

20

Buen Vivir and Sustainability: Relation between Indigenous Knowledge and Western Concepts

Ronald Lutz

Introduction[1]

People always had a multifaceted relationship with what surrounded them and what we call nature: with animals, which were often ascribed a personality of their own; with plants, which likewise had their own characteristics; but also with rivers and mountains, as well as with spirit beings, which people saw everywhere and with which they were in constant contact. People were part of this diversity, they belonged to it; yet it was also alien to them, because dangers came from it, such as storms or other catastrophic events. They dealt with this and formed rituals, which are seen by Indigenous Science as striving for balance, to shape one's own behavior in such a way that a harmonious relationship remains possible (Yunkaporta, 2021).

In addition, a "knowledge" emerged that was constantly changing and expanding as new explanations were sought; it grew over long periods in a relationship of commonality. This was also evident in the fact that people used the fruits of nature, not in excess, but they protected and cared for them at the same time—certainly also out of fear of punishment. Some cultures are reported to have apologized to plants and animals with rituals before killing or picking and eating them.

This relationship between humans and nature, as a result of colonial and global modernization, is now only found in niches, but out of these it is now being re-emphasized as "Indigenous Knowledge" and gaining attention (Lutz, 2021a). However, this should not lead to a general desire to revive it; that would be pure nostalgia. Nevertheless, it can provide ideas for rethinking the completely different relationship the "Global North" has with nature. The place from which many modernization processes originated also led to dramatic climate changes. Given the challenges, a look at "Indigenous Knowledge" can provide arguments for the transformation discourses.

[1] This chapter is a changed version of a german text, published as: Ronald Lutz: Naturverhältnisse: Verbindungen von Konzepten des „Buen Vivir" und starker Nachhaltigkeit", in: Liedholz, Yannick/ Verch, Johannes (Hrsg.): Nachhaltigkeit und Soziale Arbeit, Budrich: Opladen 2023, S. 93–105.

Challenges

The relationship in the "Global North" between nature and humans is an instrumental one due to modernization processes: nature became an "exploitable mine" and is merely a mass of resources to be used for humans and which seem simply available (Beck, 1986; Beck, 2016; Rosa, 2018; Eisenmann, 2019). Nature became the object of mechanistic science and thus also the object of engineers, who use it based on the constructed laws of nature according to people's ideas and at the same time exploit it economically (Scheidler, 2021a, 2021b). Use and care are no longer the unifying factors, but exploitation, transformation, and destruction become the dividing factors.

A dualism of nature and culture has emerged that limits life and is interpreted as the "Great Divide" in the Anthropocene (Eisenmann, 2019; Scheidler, 2021a). This refers to the age developed by humans, with a worldview and diverse actions to use and shape nature, the environment, landscapes, forests, steppes, rivers, seas, plants, and other beings according to their own (human) ideas and needs. This has resulted in an anthropogenically overformed earth, which, in a critical view, is to be understood as a disregard for systemic and processual interrelationships. It has led to massive consequences of human-induced climate change and initiated a tendency toward the collapse of ecosystems (Eisenmann, 2019).

In today's late modernity of the "Global North," hardly anyone knows what an "original state of nature" might look like (Klein, 2017). What people encounter is humanly designed nature. Certainly, a return to an "original state" that we only suspect is not possible, nor should it be intended. Nevertheless, alternatives lie in "Indigenous Knowledge," which is now strongly positioned and receiving attention as to how a different relationship than the instrumental one would be possible and how nature can become a subject of coexistence in a new context (Weber, 2019; Lutz, 2021a).

Against the backdrop of the multiple climate crises, the challenge becomes clear. On the one hand, we must anticipate events that are hardly controllable technologically, economically, and politically and lead to a transformation of the world to which we could all succumb (Beck, 2016). On the other hand, there is also something to counter this: a "new way of thinking" and acting (Rosa, 2018; Göpel, 2019). I tend toward both: we will have to settle into an ecosystemically changed world, but at the same time we can prevent many things.

My optimism stems from the fact that I interpret the Anthropocene and its consequences as a product of us humans. Thus, we are also capable of facing it in a shaping and transforming way and of shaping a (new) different world. In the meantime, many ways out of this are being sought, and "Indigenous Knowledge" should also be reflected. Nevertheless, these concepts must always be questioned according to their significance for the present. Once again, it cannot be about a return to something past but about a departure into the future, which can contribute to other dimensions of natural relations, and also to equality, freedom, equity, and sustainability.

Buen Vivir

Hartmut Rosa called for the relativization of the claim of the Anthropocene to be able to direct and control everything with science, economy, technology, and politics (Rosa,

2018). He contrasts this with the thesis of the unavailable; this means recognizing the limits of the human being and the planet as well as the holistic interconnectedness of life in space and time and daring to think "anew."

"New thinking" is associated with the "Great Transformation" discussed by Polanyi (Polanyi, 1973): economic and social processes that brought about profound change since the mid-nineteenth century. In it, he analyzed two dynamics, market economy and nation-states; he understood capitalism as the driving force of both processes, in connection with the commodification of land, money, and nature. Critical authors call for a similar "Great Transformation," which would counter the prevailing independence of the economy with a new transformation and thereby also reconceptualize our understanding of nature (Beck, 2016; Rosa, 2018; Göpel, 2020; Hanusch et al., 2021; Scheidler, 2021a).

In the meantime, many considerations are condensing into "Real Utopias" (Wright, 2017), which see themselves beyond the capitalist growth model and are reflected, among others, in concepts of a "post-growth society" (Seidl and Zahrnt, 2010) or "strong sustainability" (Konrad and Döring, 2011). The latter is defined by the demand to preserve remaining stocks of nature and to invest in them. Resources should only be used to the extent that they can regenerate. This also means respecting natural environments and a value system that questions growth and profit.

What connects everything and takes it further is the demand to develop a "planetary thinking" (Hanusch et al., 2021) that goes beyond the established political and economic systems of the Anthropocene, not only criticizing them but also intending to transform them. For this, the whole planet must be seen in its systemic, ecological, and social relationships and interconnectedness. However, thinking planetary and sustainable also means reflecting anew on ideas that have been pushed aside in modernization, allowing "other narratives," entering into dialogue, and also opening up to the cosmovisions and knowledge of Indigenous Population Groups (Weber, 2020).

This is not about their cultural appropriation, which would be a new expropriation (Distelhoff, 2021); rather they can be the starting point of a search for similar roots in the "Global North" as well as offer argumentative support for existing alternatives. Especially Indigenous cosmovisions of "Buen Vivir" or "Sumak Kawsay," which come from the Andean and Amazonian regions, have for years experienced special appreciation as "Indigenous Knowledge" and have been described and thought through in many ways. Among many contexts, they express an understanding of nature that is thought-provoking and opens perspectives on post-growth debates, strong sustainability, and ultimately also as a concept of transformation (Lutz, 2022).

The debates on "Buen Vivir" are similar in some aspects to discussions in the Global North, which also demand a fundamental change in thinking and understanding of development and no longer want to follow the anthropocentric ratio. However, the contents of "Buen Vivir" can only be understood if the contexts from which they grew are not ignored. They have a non-colonial origin and want to overcome the "colonial past" seen in the Western anthropocentric development model (Acosta, 2009; Gudynas, 2009; Fatheuer, 2011; Gudynas, 2012; Altmann, 2013; Gann, 2013; Acosta, 2015; Cubillo-Guevara et al., 2018; Lutz, 2021a).

Indigenous people in the Andean and Amazonian regions are reviving that knowledge that was displaced by colonization and turned people into those that colonizers wanted to have. “Buen Vivir” is thus first a rediscovery of the repressed and at the same time represents a new departure toward one’s own Indigenous identity. At the same time, it is a counterweight against the still-existing consequences of colonization and its growth myths. It is a “culture of life” that can suggest ways out of the dilemma of the Anthropocene. It rests on the knowledge of Indigenous People and emphasizes a “way of being” that was suppressed by colonization and is now being revived and further developed (Acosta, 2015).

Starting from Indigenous Movements, these reformulated ideas of “Buen Vivir” were even integrated into the constitutions of Bolivia and Ecuador; the Good Life became a constitutional goal and thus part of the formulation of rights (Acosta, 2015). These include the rights to food, health, education, water, and clean air. Many of these “things” are not only reminiscent of human rights but also of the list of a “Good Life” that Martha Nussbaum drew up in principle (Nussbaum, 1998).

However, “Buen Vivir” distinguishes itself from the ideas of an individual “Good Life” as well as egocentric thinking in the Global North (Guevara et al., 2018). It focuses on the social cohesion of people that communities provide; this also includes recognizing the plurality and diversity of cultures. It is still unclear whether this is not just an attempt to create new worlds with words. The ideas and concepts conveyed cannot be presented as a unified world of thought; rather, there are heterogeneous and contradictory interpretations.

For example, it has led to conflicts in Ecuador and Bolivia over the use of resources and even to neo-extractivism, the proceeds of which were supposed to benefit primarily the indigenous population (Fatheuer, 2011; Acosta, 2015; Cubillo-Guevara, 2018). Buen Vivir” has not produced “islands of the blessed,” and the actual processes are sometimes extremely contradictory (Fatheuer, 2011: 26; Cubillo-Guevara et al., 2018).

According to Acosta, who was involved in the constitutional project in Ecuador, however, “Buen Vivir” clearly sets itself apart from the idea of prosperity and neoliberal capitalism, as it no longer focuses on the exploitation of nature and its accumulation in growth and profit (Acosta, 2015). Rather, it rests on Indigenous cosmovisions that are not about progress as a linear model of neoliberalism but about the production and reproduction of a state of equilibrium.

As a category of the philosophy of life of Indigenous Communities, this concept, without idealization or romanticization, could make an important contribution to giving space to other practices and reflections and to strengthening existing alternatives (Acosta, 2009: 219f). Nevertheless, “Buen Vivir,” like other Indigenous Models, cannot be grasped with simple definitions.

Gudynas pointed out that it is a “concept under construction,” an idea “in the making”; also, Indigenous traditions, in particular, make it difficult to understand for those who do not share these very traditions (Gudynas, 2012: 11). At this point, the breadth and heterogeneity of the debates conducted cannot even be remotely reflected, nor is it possible to reconstruct the political events surrounding “Buen Vivir.” Nevertheless, despite Gudynas’ objection, those positions can be considered that, in the context of a different relationship to nature, force the Global North (and not only there) to reflect and provide new arguments for the ideas of a post-growth society and sustainability thinking.

I see the decisive and unifying element of "Buen Vivir" in the fact that nature is seen as a subject with its rights. It propagates a fundamentally new way of dealing with nature that is unfamiliar from the anthropocentric and Western development model. As a hegemonic paradigm, this has, as already emphasized, above all the exploitation of nature for growth and profit as its goal. This rationality and the mechanistic understanding of nature are challenged by concepts of "Buen Vivir."

There is also a demand to criticize the concept of development in European origins and to show how it can be transformed (Acosta, 2015). Development and nature are thus rethought "to rob the Western understanding of nature, which describes nature as something external to us and breaks it down into individual objects that can be manipulated and appropriated as resources, of its power to act" (Gudynas, 2012: 26).

"Buen Vivir," on the other hand, therefore, emphasizes the complementarity of all beings. The focus is on "life in all its fullness": as harmony with oneself (dimension of identity); with society (justice); and with nature (sustainability) (Cubillo-Guevara et al., 2018: 8ff). A "space" should grow "in which people, animals, and plants live together in a relationship and nature and culture are not in opposition to each other, but complement each other, as one contains the other and is inseparable" (Gudynas, 2012: 13).

This "utopian" relationship with nature comes from the "past" and focuses on the "future." The ideas of balance and harmony are explicitly directed against the dominant implications of the Anthropocene, its accumulation model, and the embedded exploitation of resources (Acosta, 2015). This is not a sustainable solution for life (Cubillo-Guevara et al., 2018). Indigenous thinking and kowledge do not simply oppose anthropocentrism with a nature-centered worldview. Rather, it elaborates a "biocentric worldview" (Cubillo-Guevara et al., 2018: 20), which postulates a "unity of life," "which is precisely not characterized by the opposition of nature and humans" (Fatheuer, 2011: 24).

This includes, above all, recognizing nature's own nature and rights. This does not mean that they lead to nature (animals, water, plants, soils, or mountains) no longer being used; that would be absurd and inconceivable. Nature is rather constituted as a separate good that has a right to exist as a counterpart of humans and thus may no longer be seen mechanically and instrumentally as a "thing" that can be exploited without restraint and dominated and shaped until its destruction.

This does not work without "advocates" and constitutes an "anthropocentric paradox" (Fatheuer, 2011: 24): humans formulate these rights and enforce them at the same time. Their recognition would be an effective change of perspective in the aberrations of the Anthropocene. Nature would be introduced into politics with a new understanding, as shown in the constitutions of Bolivia and Ecuador. This could promote or change existing discourses in the Global North. Especially the debate on the rights of nature could establish a powerful and, at the same time, diverse dialogue between Indigenous theories and alternative thinking in the Global North and give further impetus to the search for ways out of the nature-destructive development model.

The "Laws for the Protection and Right of Nature" passed in Ecuador and Bolivia play a special role in this. The relationship between humans and nature was thereby placed based on primordiality, reciprocity, and dialogue (Fatheuer, 2011: 18; Acosta, 2015). Preserving the balance in nature is defined as a prerequisite for the regeneration of the planet, and respect for nature's own character and preservation of its rights is established. At the same time, these rights are placed in a larger context in the constitutions mentioned, which can be read as a new development model in which respect for diversity and harmonious coexistence with nature are enshrined.

Discourses of "Buen Vivir" also make clear that the Global North, in its hegemonic nature, only recognizes one system, its own, and negates others, as it sees itself as universal (Acosta, 2015; Sousa Santos and Meneses, 2020; Lutz, 2022). "Buen Vivir," on the other hand, calls for the simultaneity of different systems: something is and is not at the same time and includes a third (Fatheuer, 2011; Acosta, 2015). Allowing for contrasts, non-simultaneities, and differences is one of the strengths of this way of thinking; the commitment to pluriversality points far beyond the universal and hegemonic thinking of the Global North and can provide new impulses. It also reinterprets the meaning of the local as a lifeworld in which people locate themselves within the world, find affirmation and identity, and at the same time enter a global network from there.

Once again, it should be emphasized that this also entails a danger of integrating thoughts of "Buen Vivir" too lightly into the Global North's search for "healing" and new ways. The actual demystified image of the "noble savage" is peeking dangerously around the corner. However, as emphasized, this would be cultural appropriation and thus dispossession. The ideas formulated in the concept of "Buen Vivir," however, are not "doctrines of salvation"; it is not about a new romanticism or nostalgia that allegedly wants to restore harmonious living conditions of "original peoples" in the twenty-first century (Cubillo-Guevara, 2018).

However, questions can be derived for a different way of dealing with nature and thus for social and economic concepts. Therefore, looking at the concepts can sharpen debates that have been going on for a long time. Yet this is only possible in a dialogue between actors in the North and South. It is also true here that alternatives to the aberrations of the Anthropocene can only be found if, beyond the national container, injustices of the world order are not neglected.

We are confronted with a global challenge that should lead to diverse, pluriverse, local, and global ways out (Fatheuer, 2011: 31). The reflection of the discourses around "Buen Vivir" can inspire; a platform becomes possible on which many things come together that can promote a dialectical process of exchange. A large-scale experiment seems feasible that enables equal and dialogical communication between North and South. This communication could be the expression of a cultural turn resulting from a paradox: the cultural heritage of Indigenous traditions, historically incompatible with development, becomes the key to its renewal and comes into focus in a changed way (Cubillo-Guevara, 2018: 23).

This could release discursive energies that contribute to an intensification of global transformation discourses and thus to socio-cultural and political change (Cubillo-Guecva, 1918: 23ff; Lutz, 2022). In this, however, "Buen Vivir" always remains an independent and Andean concept that must not be culturally appropriated or appropriated for other interests (Gudynas, 2012: 18). This confrontation means questioning one's own rational and Western thinking in the Global North as well, decolonizing it and opening up spaces for new thoughts that also exist in Western thinking: more and more people in different regions of the world are seeking their roots and a different relationship to the surrounding nature, to life itself, in a process of re-encounter.

"Buen Vivir" is like these critical debates on the capitalist economic system, which have been seeking new ways to turn away from the neoliberal model for a long time. In Western thinking, too, anthropocentrism is questioned, and the rights of nature are pleaded for. However, the ways out for the North will be different from those in the South. The reflection of "Buen Vivir" opens discourses in which different movements, each with its perspective on the future, meet.

In summary, the preliminary result can be stated: Buen Vivir not only bundles the fundamental critique of the anthropocentric understanding of nature and development model, it also opens views beyond its horizons that are quite similar to discourses of a "new way of thinking" in the Global North. Nevertheless, the differences must always be emphasized and acknowledged. Indigenous and Western perspectives nevertheless come into a communication context that can become fruitful.

Consequently, these considerations can be used to discuss ways out of the growth compulsion in a more pointed way. In rudiments, even a different concept of development emerges, "which wants to say goodbye to western paradigms of prosperity" (Fatheuer, 2011: 7).

Sustainability

Recknagel has presented a comprehensive study that traces the context of "Buen Vivir" and strong sustainability in a broad approach that includes nature, economy, politics, society, development, sustainability, and culture (Recknagel, 2019). Only the results on nature and sustainability will be discussed here. These confirm and expand on the considerations that have already been presented.

Overall, "Buen Vivir" provides a new understanding of nature and fundamentally sets itself apart from the anthropocentric understanding of nature. The propagated rights of nature are more than the "protection of species and the environment," as emphasized in the sustainability debate in the Global North; they are directed more strongly against destruction and valorize nature as a subject in its own right. Nature comes to the center, as the worldview is not anthropocentric but biocentric.

This offers a new paradigm and enables a change of perspective that can influence environmental and economic discussions. Despite its contradictions, "Buen Vivir" emphasizes elements of the sustainability debate that are significant for debates in the Global North. This is sometimes reflected in a particular emphasis on aspects: the past is present in the future, everything must always be biodegradable, equal rights for all, and sources of life must be preserved for future generations.

Buen Vivir" confirms discourses in the Global North and provides further argumentation aids. There are similarities between "Buen Vivir" and strong sustainability. This does not mean, however, that both are completely compatible with each other; rather, it becomes clear that both concepts come from different traditions and ways of thinking and stand for these.

Thus, in texts on "Buen Vivir," positions can also be found that criticize the dominant aspect of Western sustainability discourses and interpret them as an attempt merely to modify Western hegemonic thinking without fundamentally changing anything. "Buen Vivir" is altogether more oriented toward transformation processes.

However, both can exchange views in dialogue and sharpen their standpoint. This can lead to critically examining and expanding positions of sustainability as well as the Eurocentric way of thinking and the embedded horizon of values. In this way, the inflationary use of the term "sustainability" can also be questioned to make the actual core, which also lies in a revision of the relationship between humans and nature, clearer and thus support the transformational opportunities of strong sustainability.

Outlook

"Buen Vivir" calls for a "critical negation of the present time with the intention of a better future (Gann, 2011: 90). Future is a core term of the concepts to be seen as normative utopia. Non-transferable, they inspire. Utopianism can be realism since politics always means a narrow focus on constraints and given things that merely need to be improved. Utopian maxims of action, on the other hand, assume that things cannot go on as before: "Fundamental change is needed, not within the framework of existing practices, but a change of the framework itself" (Gann, 2011: 90).

Perhaps a utopian normative view of "Buen Vivir" is what new thinking in the context of transformative approaches needs. Perhaps Western science also needs to engage with it beyond its rational understanding. In any case, a reflection on "Buen Vivir" can show that the rational explanation of the world is not the only one. But this also means opening to the plurality of knowledge and world relations. Boaventura de Sousa Santos argues for a dialogue between different epistemologies today rather than tomorrow, in which southern as well as Indigenous forms are recognized as equally valid (Sousa Santos, 2018). Courage is needed to transcend boundaries, including those of science, to liberate thinking from the brazen shell of the Anthropocene.

References

Acosta, Alberto (2009), "Das 'Buen Vivir.' Die Schaffung einer Utopie," in Juridivum 2009/4.
Acosta, Alberto (2015), *Buen Vivir. Vom Recht auf ein Gutes Leben*, München: oekom.
Altmann, P. (2013), "Das Gute Leben als Alternative zum Wachstum? Der Fall Ecuador," *Sozialwissenschaften und Berufspraxis*, 36(1): 101–11. https://nbn-resolving.org/urn:nbn:de:0168-ssoar-406442
Beck Ulrich (1986), Risikogesellschaft. Auf dem Weg in eine andre Moderne, München.
Beck, Ulrich (2016), *Die Metamorphose der Welt*, Berlin: Suhrkamp.
Böhnisch, Lothar (2020), *Sozialpädagogik der Nachhaltigkeit*, Juventa Weinheim.
Cubillo-Guevera, Ana, Julien Vanhulst, Luis Antonio Hidalgo-Capitan, and Adrian Beling

(2018), "Die lateinamerikanischen Diskurse zu buen vivir. Entstehung, Institutionalisierung und Veränderung," *Peripherie*, 149(2018): 8–28.
Distelhorst, Lars (2021), *Kulturelle Aneignung*, Hamburg: Edition Nautilus.
Eisenmann, Charles (2019), *Klima. Eine neue Perspektive*, Berlin-München-Zürich-Wien.
Fatheuer, Thomas (2011), Buen Vivir. Eine kurze Einführung in Lateinamerikas neue Konzepte zu gutem Leben und zu den Rechten der Natur, Heinrich Böll Stiftung, Schriften zur Ökologie Band 17, Berlin.
Freire, Paulo (1973), *Pädagogik der Unterdrückten*, Hamburg: rororo.
Gann, Anna-Lisa (2013), "Das Konzept des Buen Vivir in der ecuadorianischen Verfassung aus feministischer Perspektive," *Femina Politica*, 11(2013): 81–93.
Gudynas, Eduardo (2009), "Politische Ökologie: Natur in den Verfassungen von Bolivien und Ecuador," Juridicum, 2009, 4.
Gudynas, Eduardo (2012), Buen Vivir. Das gute Leben jenseits von Entwicklung und Wachstum, Rosa Luxemburg Stiftung—Analyse, S. 1–35, Berlin.
Göpel, Maja (2020), Unsere Welt neu denken: Eine Einladung, Ullstein, Berlin.
Hanusch, Frederic, Claus Leggewie, and Eric Meyer (2021), Planetar denken. Eine Einführung, transcript, Bielefeld.
Klein, Naomi (2017), Die Rache der Realität und die Kraft der Utopie. In: Blätter für deutsche und internationale Politik, Heft 9, S. 51–62.
Liedholz, Yannick (2021), Berührungspunkte von Sozialer Arbeit und Klimawandel, Budrich Opladen.
Lutz, Ronald (2021a), Vom Süden lernen? Eine kritische Rezeption indigenen Denkens, in: Fischer, Jörg und Tuider, Elisabeth: Sozialer Zusammenhalt, Sozialmagazion, Sonderband 4, Weinheim, S. 168–86.
Lutz, Ronald (2021b), Krise—und jetzt? WAS PASSIERT MIT DER SOZIALEN ARBEIT DURCH PANDEMIE UND KLIMAWANDEL, in: Forum sozial 4/2021, S. 27–31.
Lutz, Ronald (2022), "Anthropozän und Klimaverwandlung. Skizzen einer „transformativen Sozialen Arbeit," in Tino Pfaff, Barbara Schramkowski, and Ronald Lutz, (eds.), *Klimakrise, Sozialökologischer Kollaps und Klimagerechtigkeit. Spannungsfelder für Soziale Arbeit*; Juventa: Weinheim, pp. 370–94.
Lutz, Ronald and Alexander Stauss (2016), "Internationale Soziale Arbeit," in *neue praxis*, 6(2016): 544–61.
Nussbaum, Marta C. (1998), *Gerechtigkeit oder das Gute Leben*, Frankfurt am Main: Suhrkamp.
Ott, Konrad and Ralf Döring (2011), *Theorie und Praxis starker Nachhaltigkeit*, Metropolis: Marburg.
Pfaff, Tino: Schramkowski, Barbara and Lutz, Ronald (Hg) (2022), Klimakrise, Sozialökologischer Kollaps und Klimagerechtigkeit. Spannungsfelder für Soziale Arbeit; Juventa: Weinheim.
Polanyi, Karl (1973), The Great Transformation. Politische und ökonomische Ursprünge von Gesellschaften und Wirtschaftssystemen, Suhrkamp Frankfurt 1973 (aktuelle Auflage 2021)
Recknagel, Lukas (2019), Buen Vivir als Konzept einer Bildung für nachhaltige Entwicklung?—Eine nachhaltigkeitsorientierte Diskursanalyse, Dissertation, Martin-Luther-Universität Halle-Wittenberg siehe auch: https://opendata.uni-halle.de/bitstream/1981185920/14038/1/2019_Recknagel_Dissertationsschrift.pdf
Ronald Lutz: Naturverhältnisse: Verbindungen von Konzepten des "Buen Vivir " und starker Nachhaltigkeit", in: Liedholz, Yannick/Verch, Johannes (Hrsg.): Nachhaltigkeit und Soziale Arbeit, Budrich: Opladen 2023, S. 93–105.

Rosa, Hartmut (2018), Unverfügbarkeit, Residenz: Wien.
Scheidler, Fabian (2021a), Der Stoff aus dem wir sind. Warum wir Natur und Gesellschaft neu denken müssen, Piper München.
Scheidler, Fabian (2021b), "Die große Trennung. Die Geburt der technokratischen Weltsicht und der planetarischen Krise," in *Blätter für deutsche und internationale Politik*, 4/2021, S. 51–62.
Seidl, I. and Zahrnt, A. (Hrsg.) (2010), Postwachstumsgesellschaft. Konzepte für die Zukunft. Marburg: Metropolis-Verlag.
Sousa Santos, Boaventura de (2018), Epistemologien des Südens: Gegen die Hegemonie des westlichen Denkens; Unrast Verlag: Münster.
Sousa Santos, Boaventura de; Meneses, Paula (ed.) (2020), *Knowledge Born in the Struggle. Constructing the Epistemologies of the Global South*, Routledge, London and New York.
Stamm, Ingo (2021), Ökologisch-Kritische Soziale Arbeit, Budrich Opladen.
Wright, E. O. (2017), Reale Utopien. Wege aus dem Kapitalismus. Berlin: Suhrkamp.
Weber, Andreas (2019), *Indigenialität*, Nicolai Publishing. Berlin.
Yunkaporta, Tyson (2021), Sand Talk. Das Wissen der Aborigines und die Krisen der modernen Welt, Matthes und Seitz: Berlin.

Conclusion

21

Conclusion: Decolonializing Knowledge

Ronald Lutz

The texts in this book provide examples of a pluriversal form of critical thought, which distances itself from a universal and hegemonial science as developed in the Global North and exported in the colonial context of Western Modernism, its development paradigms and programs. Resistance against this grew not only in liberation movements, but also when formulating postcolonial narratives as well as in the discourses of an Indigenous science.

This development requires mutual understanding and the acceptance of each other's way of thinking as equally valid, which can only be achieved through dialogues, in line with the principles outlined by Paulo Freire. He understood dialogue as an "encounter between men . . . in order to name the world," therefore "it must not be a situation where some men name on behalf of others. [. . .] it must not serve as a crafty instrument for the domination of one man by another. The domination implicit in dialogue is that of the world by those who enter into dialogue" (Freire, 1972: 61f).

With the dialectic of encounters and shared thinking that forms in contradictions and diversity, the growing challenge in the North is to decolonize the products of colonial modernism and to reinvent itself, so as to be able to conduct an honest dialogue with the South and the preservers and protectors of Indigenous knowledge. That means the deconstruction of hegemony, the acknowledgment of differences and the acceptance of other life concepts as equally valid.

With all relativity, however, shared basic values are necessary, with a universal core that is subject to continuous negotiation—what I understand as "dialogic humanism." In outline, this aims at a "New Enlightenment," which extends over and beyond the strictures of the Global North.

Opening comments

Faced with erosion and criticism, the ideas of the Enlightenment threatened to fade over time. Writing at the end of World War II, Horkheimer and Adorno showed how the instrumental reason of the European Enlightenment had flipped into the barbarism of the death camps of the twentieth century (Horkheimer and Adorno, 2002). With the ongoing research into European colonialism, it became clear that the philosophies of the Enlightenment included racist views that came together in a radical devaluation of others (Barth and Osterhammel, 2005). In the narratives of the Anthropocene, it was recognized that humankind is in the process of destroying itself and the world in the name of humanism (Noller, 2023). The reality of the wars being waged as I write shows once more the negative side of instrumental reason and technology, which lead unconscionably to the degradation and destruction of human beings.

Nevertheless, European Enlightenment continues to have its attractions. In view of the multiple crises of the present, the immanent ideas of liberty, justice, and autonomy, when viewed in a new light, can contribute to the transformation of what the socializations of the Global North have brought about. In the "self-criticism of the modern" more and more philosophers and economists (Coccia, 2020; Göpel, 2020; Pelluchon, 2021; Neumann, 2022), are calling for a new accentuation of the theses of the European Enlightenment which places thought and action on a new basis, with long-term consequences for the material exchange of humans with the non-human world (Sloterdijk, 2023: 9).

This "New Enlightenment" does not imply the abandonment of existing ideas and the destruction of important universal ideals such as emancipation, as some post-humanist thinkers demand. The motto of Immanuel Kant "*Sapere aude*—have courage to use your own understanding" is still relevant for dialogic humanism, freedom and autonomy. The human rights developed in the wake of the Enlightenment, particularly that all people are equal and endowed with universal rights, not only remain central for democracies and international relations and organizations but are also starting points for emancipatory movements. However, while they continue to provide an essential anchor, they do need to be extended:

- Firstly, collective human rights should be formulated that are not directed to an inviolable individual, but which emphasize the importance of communities.
- Secondly, "human rights from below" should be included which have developed in social and indigenous movements and which emphasize cultural and local particularities. This includes the affinity with the non-human world and the "knowledge" about its "sacredness."

New Enlightenment

Reconceptualization of the Enlightenment means expanding and defending central elements of its legacy such as reason, autonomy, liberty, and justice (Pelluchon, 2021). This is of great importance because there are many powerful opponents of the Enlightenment and its values and ideals who would like to establish a completely different world, one of unfreedom, intolerance, of ideologies, religious fundamentalism, populism, autocracy, and oppression. Under cover of a criticism of Western values and hegemony, often packaged in postcolonial theories, many elites conceal an authoritarianism that would deprive people of the possibilities gained in the course of long emancipatory struggles.

Forces of the counter-enlightenment, according Pelluchon, use their hatred of reason as a "weapon against any emancipation project." The topicality of the imperial war that Russia is waging in Ukraine against all reason, the newly erupted Middle East conflict, in which fundamentalists confront one another irreconcilably, the oppression of women in Iran and Afghanistan, the attacks against abortion in the USA, and the authoritarianism in China confirm this again and again. Critical and humanistically minded people and movements must continue to take a stand against these developments, but to do this they need a basis of common values—worldwide and not only in the North. The New Enlightenment that Pelluchon outlines could provide this basis.

Firstly, the fundamental criticism should be reformulated. An understanding of humans based exclusively on an understanding of their place in this world has changed everything into a sort of war; even the human being has been "colonialized" in the name of civilization (Lutz, 2022). The "great divide" of nature and culture, of humans and animals, which is essential for a humanism fixated only on humans (Descola, 2011), leads to an "amputation of reason," depriving it of its communicative aspect (Pelluchon, 2021). Denial of the fact of not being alone can be seen as "the vice of our civilization."

The domination of humans over nature and other humans made possible by a narrow humanism must be opposed by a different narrative. Pelluchon therefore formulates another image of humanity. This goes far beyond what has previously been understood under Enlightenment; she includes a responsibility for collectives, for other living organisms, for the more than human world and for coming generations. The narrative is expanded beyond its focus on the humans as the dominant beings. Humans not only exist individually but are always born into a communal world, with a rich natural and cultural inheritance. They are always parts of interactions and dependencies. The material exchanges with the non-human world can never be a one-way process.

Pelluchon sees the human as a "relational being," whose actions always have consequences for others, for coming generations, and the entire planet. This gives rise to a far-reaching responsibility that makes humans the "advocates" for the non-human world and opponents of all forms of oppression (Latour, 2021). This is also a clear call to create a more just and more peaceful world. The utopian potential that lies in such a "New thought of the Enlightenment" can in my view contribute to overcoming limited narrow mindedness, accelerating the decolonialization of thought.

Thematic universe of decolonization

A world in a normal state (in the sense of a natural state) does not exist. This is made impossible already by the material exchanges with the non-human world initiated by our deeds. The result is a human construction—everything is always in motion and changing, based on wonderful and possible dreams which can become true if we work continuously to realize them, but which can also collapse into catastrophes.

Anthropocentric thought and action has long been criticized in the discourses about the Anthropocene, neo-colonialism, the imperial ways of living, the approaches of posthumanism, the theses of a "New Enlightenment," the visions of a post-growth society, and demands for "planetary thought." Such approaches provide a foundation for the creative formulation of new ideas, enriched by visions and utopias, which above all demand the decolonization of the established knowledge of Western science.

This legacy knowledge, derived in the colonial contexts of "Western science" and its methods, has claimed the status of a defined objectivity at a distance from things and contexts. And these concepts have spread as colonial exports to the Global South, displacing and devaluing their own stocks of knowledge. This privileged status of epistemic thought of a universally accepted Western science, is not only regarded critically in postcolonial theories, but also in the debates on post-humanism, Indigenization, and the subject-character of nature, which call into question humanism as the product of the West and of anthropocentric thought.

These confrontations are significant because they critically deconstruct a self-perception, anchored in the North, of a generalizable theory and methodology. The analysis of colonial power and dependence also requires critical reflection of European points of view in their context and their integration in colonialism, anthropocentric perspectives, globalization, and imperial ways of living.

Criticism of anthropocentric thought has for some time been radicalized in "post-humanist thought," which not only elevates itself to an explanatory narrative, but also to a philosophy of transformation, indicating ways to escape from the devastations of anthropocentric thinking and acting (Braidotti, 2014; Coccia, 2021). The prominent position of human beings in the Enlightenment and in Anthropocentrism is called into question. Braidotti speaks of the self-glorification of humans in anthropocentrism (Braidotti, 2014: 71). In many cases, the charm of the special is highlighted, which lies in questioning this self-glorification, so as not only to dissolve the previously privileged position of humans, but also to cause them to disappear as the dominant actor of life.

These ideas make "post-humanist thought" particularly attractive for some. It resembles to a certain extent postcolonial and Indigenous thought and in a similar way argues that anthropocentric thought above all consolidates the positions of white European males (Said, 2004). Hierarchical thought and the exploitation and oppression of both human life (in colonialism, in imperial way of living) and non-human life (fauna, flora, and landscapes) are fundamentally called into question, along with the narrative of Western thought in which the human is at the center. The hope is that this criticism will open up new pathways.

However, in particular in the light of the theses of a New Enlightenment and the further extension of Critical Theory that have been formulated, it is necessary to reflect whether this narrative really can provide the basis for transformation processes. With its ideas of a "disappearance" or an "improvement" of humans, as discussed in "transhumanism" (Loh, 2018), it would not, in my opinion, be possible to overcome the crises generated by humans. I doubt whether a "dissolution of humans," which in any event can only be a thought experiment, actually offers a way into the future, which can only be shaped differently by the humans themselves.

A positive aspect of the theses of posthumanism is that human beings or human actions should only be regarded as one possibility of existence among many. Human beings are then no longer at the center of life, as the sole drivers and shapers, but instead are part of a more extensive world in which they play an equal role in a network of exchange. However, this also means that humans have a particular responsibility toward the world of which they are part—and that applies also for their sciences.

Decentering is also demanded by postcolonial thought in order to relativize and displace the dominance of Western thought and to promote "pluriversal thought" (Sousa Santos, 2018). This goes along with an "Indigenization," involving the reformulation of Western theories and methods, and their adaptation to histories and different local conditions of the South (Sousa Santos and Meneses, 2020) and also the upgrading of "Indigenous knowledge" as well as "Native science" (Cajete, 1999).

In these discourses, considerations arise concerning the knowledge and thinking that can be found beyond colonial influences and devastation (Mbembe, 2016), in order to develop questions and answers against historical, political, and social backgrounds. This leads to a re-evaluation of colonial imported science and its images, theories, and methods. Epistemologies of the South must stand on an equal footing alongside the epistemologies of the Global North (Mignolo, 2000; Sousa Santos, 2018; Sousa Santos and Meneses, 2020).

The critical indigenization processes view life as "biocentric"; humans too are seen as biological systems which are integrated in the environment, acting in and with this, rather than existing separately from it. In the Andean concepts of *buen vivir* (Acosta, 2015), but also in other approaches to "Indigenous science," a worldview is developed which postulates a unity of life which is not characterized by the opposition of nature and human beings (Cajete, 1999). This involves above all accepting the uniqueness of nature and its own rights.

If non-human nature has its own rights, this gives rise to an "anthropocentric paradox," because it will call for human "advocates" (Latour, 2021): humans formulate these rights and at the same time enforce them. The debates in a "Parliament of things" can converge in an "organisms democracy."[1] This represents a possible political system in which all living organisms of a country are represented as citizens with equal rights, entitled to political participation and with human "advocates." In the crises and confusions of our times, such recognition could result in an effective and instructive change of perspectives.

Transformation as a project

A decolonization of anthropocentric and hegemonial thought could gradually overcome the trap of Western science defined by Peter Sloterdijk. "As long as we look through the glasses made by the modern sciences, which show us the earth as if from the outside, as if we did not really belong to its wilful outgrowths" (Sloterdijk, 2023: 74), we will not comprehend what we have done and what we urgently have to change.

The debate about the rights of nature could lead to a fruitful dialogue between Indigenous epistemologies and alternative thought in the Global North and provide further impulses in the search for ways to escape from the development model that is destroying nature—thus forming the basis of a "dialogic humanism."

[1] https://organismendemokratie.org

Concepts and contents of "Indigenous knowledge" have an enriching significance and can offer a perspective of transformation, by relativizing the previously elevated status of humans in the world. This necessitates dialogues, the acknowledgment of other ways of thinking and knowledge, along with their local characteristics. At the same time, an exchange at the international level is necessary. A confrontation with the epistemologies of the Global South is unavoidable. However, enlightened criticism requires humility and honesty.

Critical Theory must learn to show humility when facing the knowledge of the others that has hitherto been pushed aside. But it must also be honest toward itself and scrutinize its own contextuality, and self-critically address its own history, which is always linked with colonialism and its hegemonial claim of universally valid knowledge. This history is therefore necessarily full of contradictions, which as a continual challenge have to be emphasized in the dialogue with other systems of thought.

Closing remarks

Science, as a whole, is the result of argumentation and "respects the disputes of hypotheses and theories, the continual weighing up of arguments, the never-ending search for the right convictions" (Nida-Rümelin, 2023: 19). But this does not result in a universal show of the good; instead, pluriversal ways emerge, with intersections and voids as well as open water, plains, hills, and valleys, where utopias grow.

It is time to find new and inspiring pathways to another future. Against this background, Naomi Klein calls on us to go where we have never been before (Klein, 2017: 62). In order to achieve this, we must again draw on utopian traditions that have breathed life into so many forward-looking movements (Klein, 2017: 62). Courage is required to overcome boundaries in order to free thought from the confines of the Anthropocene. Klein reminds us of the words of Oscar Wilde (Klein, 2017: 62): "A map of the world that does not include Utopia is not worth even glancing at, for it leaves out the one country at which Humanity is always landing. And when Humanity lands there, it looks out, and, seeing a better country, sets sail."

References

Acosta, Alberto (2015), *Buen Vivir. Vom Recht auf ein Gutes Leben*, Munich: oekom.

Barth, B./Osterhammel, J. (Eds.) (2005), *Zivilisierungsmissionen. Imperiale Weltverbesserung seit dem 18. Jahrhundert*, Konstanz: UVK.

Braidotti, Rosi (2013), *The Posthuman*, Cambridge: Polity Press.

Cajete, Gregory (1999), *Native Science: Natural Laws of Interdependence*, Santa Fe: Clear Light Publishers.

Coccia, Emanuelle (2020), *Metamorphosen. Das Leben hat viele Formen. Eine Philosophie der Verwandlung*, Munich: Hanser.

Descola, Philipp (2011), *Jenseits von Natur und Kultur*. Berlin: Suhrkamp.

Freire, Paulo (1972), *Pedagogy of the Oppressed*, London: Penguin Books.

Göpel, Maja (2020), *Unsere Welt neu denken: Eine Einladung*, Berlin: Ullstein.

Horkheimer, M. and T. W. Adorno (2002), *Dialectic of Enlightenment*, Stanford. CA: Stanford University Press.
Klein, Naomi (2017), "Die Rache der Realität und die Kraft der Utopie," in *Blätter für deutsche und internationale Politik*, Heft 9, pp. 51–62.
Loh, Janina (2018), *Trans- und Posthumanismus*, Hamburg: Junis.
Latour, Bruno (2021), *Das Parlament der Dinge. Für eine politische Ökologie*, Berlin: Suhrkamp.
Lutz, Ronald (2022), "Anthropozän und Klimaverwandlung. Skizzen einer transformativen Sozialen Arbeit," in Tino Pfaff, Barbara Schramkowski, and Ronald Lutz (eds.), *Klimakrise, Sozialökologischer Kollaps und Klimagerechtigkeit. Spannungsfelder für Soziale Arbeit*, Weinheim: Juventa, pp. 370–94.
Mbembe, Achille (2016), *Ausgang aus der langen Nacht: Versuch über ein entkolonisiertes Afrika*, Berlin: Suhrkamp.
Mignolo, Walter (2000), *Local Histories—Global Design. Coloniality, Subaltern Knowledges, and Border Thinking*, Princeton, NJ: Princeton University Press.
Neumann, Peter (2022), *Die neue Weltunordnung. Wie der Westen sich selbst zerstört*, Berlin: rowohlt.
Nida-Rümelin, Julian (2023), *Cancel Culture—Ende der Aufklärung? Ein Plädoyer für eigenständiges Denken*, Munich: Piper.
Noller, Jörg (2023), *Ethik des Anthropozäns. Überlegungen zur dritten Natur*, Basel: Schwabe.
Pelluchon, Corine (2021), *Zeitalter des Lebendigen. Eine neue Philosophie der Aufklärung*, Darmstadt: Wissenschaftliche Buchgesellschaft.
Said, Edward (2004), *Humanism and Democratic Criticism*, New York: Columbia University Press.
Sloterdijk, Peter (2023), *Die Reue des Prometheus. Von der Gabe des Feuers*, Berlin: Suhrkamp.
Sousa Santos, Boaventura de (2018), *Epistemologien des Südens: Gegen die Hegemonie des westlichen Denkens*, Münster: Unrast Verlag.
Sousa Santos, Boaventura de and Paula Meneses (eds.) (2020), *Knowledge Born in the Struggle. Constructing the Epistemologies of the Global South*, London and New York: Routledge.

Index

The letter *f* following an entry indicates a page with a figure

ABA (Asociación Bartolomé Aripaylla) 27–8
abayas 173
academia 28–9, 30–5, 218–19, 222 *see also* education
  decolonization 39
  research 220
Acosta, Alberto 246
action-based research 216–20
aesthesis 235
aesthetic attitudes 234
aesthetic philosophy 234
aesthetics 231–40
affective geographies 88, 91, 92–3, 95 *see also* place
affective theory 88, 90 *see also* affective geographies
Africa 70, 71, 141, 142 *see also* South Africa; Zambia
Africa Agenda 2063 70
  African Research Survey 83
  AIKS 70, 71, 72, 73–5, 76, 77
  Angola 143, 144
  Botswana 143, 144
  Comoros 144
  DRC 143, 144
  ECA 70
  economy 146–7
  education 72
  indigeneity 84
  ISSS 142–3
  Kenya 86–96
  languages 186
  Lesotho 143, 144
  Malawi 74–5, 143, 144, 145
  Mozambique 143, 144, 145
  Namibia 143, 144, 145, 146
  railways 86, 87*f*, 88, 91, 92, 93, 94, 95, 96
  São Tomé 144
  SAPs 146–7
  Senegal 138
  Seychelles 144
  southern Africa 141–50
  Tanzania 74, 143, 144, 145
  Zimbabwe 143, 144, 145, 146
African Indigenous Knowledge System (AIKS) 70, 71, 72, 73–5, 76, 77
African Union (AU). *See* AU
Afrocentrism 148
Against Purity (Shotwell, Alexis) 48–9
agonistic horizontality 221
agriculture 27–8, 29
Ahmed, L. 170
AIKS (African Indigenous Knowledge System) 70, 71, 72, 73–5, 76, 77
aimu 90–2, 94
aisthetics 235
Akamba people 81, 82–3, 86, 90–3, 94–5
  ethnic region 84
  narratives 96
  resource sustainability 83, 85
Akiwenzie-Damm, Kateri 45
alternative aesthetics 231, 238–9
alternative epistemologies 232–3
AMARC (World Association of Community Radio Broadcasters) 210
Amazonian people 245–6
ancestry 6, 53–4
  ancestral memories 43
  ancestral ways of knowing 53
  Ayoreo cosmovision 107*f*–8
Andean people 26, 27–36, 240
  Buen Vivir 26, 225, 232–3, 237–9, 245–50
  Pachamama Festival 231, 236–9
Andean region 26, 27–36
Angola 143, 144
Antezana, Fernando 28–9, 30–1

Anthropocene, the 244
anthropocentrism 245, 247, 249, 257–8
apartheid 144–5, 183, 184
Asociación Bartolomé Aripaylla (ABA) 27–8
attitudes 234
AU (African Union)
Agenda 2063 196
Australia 84
Ayacucho, Peru 27
Ayoreo people 17–22, 107*f*–9
Ayuuk Intercultural Higher Institute (Instituto Superior Intercultural Ayuuk [ISIA]) 209

Bachillerato Integral Comunitario Ayuuk Polivalente (BICAP [Comprehensive Community Polyvalent Baccalaureate]) 209
backwardness 84, 204, 205
Bak Geler, David 221
Bakhtin, Mikhail 93
Bamba, Cheikh Amadou 137–8
Basso, K. H. 96
Bastien, Betty
Blackfoot Ways of Knowing. The Worldview of the Siksikaitsitapi 3
Bear, Tracy Lee
Power in My Blood 45
Beltrão, Nogueira 222
Bermúdez, Flor Marina 209
BICAP (Bachillerato Integral Comunitario Ayuuk Polivalente [Comprehensive Community Polyvalent Baccalaureate]) 209
biocentrism 247, 249, 259
Blackfoot Ways of Knowing. The Worldview of the Siksikaitsitapi (Bastien, Betty) 3
body/thought disconnect 39
Böhme, G. 235, 236, 237, 238
Böhme, Hartmut and Böhme, Gernot 40
Bolivia 246, 247–8
Bonino, J. M. 69
Bophela, M. J. K. and Khumalo, N. 147
botho (humanness) 186, 187, 188
Botswana 143, 144
Braiding Sweetgrass (Kimmerer, Robin) 48
Braidotti, Rosi 258
Brazil 136, 137, 169–77, 196
BRICS (Brazil, Russia, India, China and South Africa) 196
Buen Vivir/*Vivir Bien* 26, 225, 231, 232–3, 238–9, 245–50
burqas 173–4
Butler, Judith 204

Cajete, Gregory 58
Cameron, Michelle 45
Canada 84
Candre, Anastasia 31–2
capability approach (to human development) 69–70, 72, 73
capitalism 197, 198, 200–1, 245
CECAM (Centro de Capacitación Musical y Desarrollo de la Cultura Mixe [Center for Musical Training and Mixe Culture Development]) 209
Center for Musical Training and Mixe Culture Development (Centro de Capacitación Musical y Desarrollo de la Cultura Mixe [CECAM]) 209
Centro de Capacitación Musical y Desarrollo de la Cultura Mixe (CECAM [Center for Musical Training and Mixe Culture Development]) 209
ceremony 43–4, 125
Cree umbilical cord 42, 47
Chagunda, Chance 74, 75
childhood studies 215–26
Eurocentrism 215
research approaches 215–26
chilimba (informal banking systems) 147
Chilisa, Bagele 218, 219–20
China 86, 88, 92, 93, 95, 196
Chiquitano people 25, 35
Christianity 35–6, 91
Livonian Crusade 154
modesty 178
PCEA 94, 95
Venezuela 134–5
climate change 188–90, 191, 244
cognition 223–4
collectivism 21, 74–5, 77, 83 *see also* community
colonialism/colonization 106, 118–19, 126, 201 *see also* decolonization

Africa 83, 84, 86, 132
amnesia 5–6
colonial amnesia 5–6
colonized minds 5–6
crimes 5–6
eroticism/sex 45
Europe 153
Global North 4
horizontal methodologies 221
indoctrination 5–6
ISSS 143
Kenya 88, 94
land 47
language 5
Malawi 75
migration 130
philosophy 46–7
power axes 130–1
settler 143
social work 187
South Africa 143, 145, 183, 187
southern Africa 141–2, 143
Western science 5
coloniality 195–6, 200–1
coloniality of power 232–3
color 92
coming-to-know process 58
community 75–6, 120, 124, 259
Brazil 136
development 190–1
festivals 231, 235–9
indigenous youth 203–11
leadership 208
Muridia 138
Oaxaca, Mexico 203–11
social development 142
southern Africa 145
Ubuntu 73, 75, 183–91, 225
Umunthu 73, 74, 75–6
Venezuela 134–6
Comoros 144
Comprehensive Community Polyvalent Baccalaureate (Bachillerato Integral Comunitario Ayuuk Polivalente [BICAP]) 209
conflict 110, 208
contemplation 235, 236
corazonar 224
corpo-biographies 222–3
corporeality 221–3
corruption 71
cosmovision 107*f*–9, 117–18
Cree people 41–4, 47–8
Cubillos, J. 205
cultural imperialism 183
cultural roots 3
culture 17, 21, 119–20, 124
Andean 27–8, 231, 236–40
culture/nature relationship 39–40, 44–5, 243, 244, 247, 257
dance 237
diversity 203, 204
exchange 35–6
festivals 231, 235–9
foreign, absorbing 25–6
Livonian 156, 158, 159, 163
mestizaje 25–6
mestizo 26
music 209, 237
Oaxaca, Mexico 203, 209
youth 203

dance 237
de Sousa Santos, Boaventura 217–18, 224, 250
decoloniality 195–201
decolonization 4–5, 39, 232–3, 257–9
aesthetics 233–9
migration 138
of the mind 6–7
of narratives 133–4, 138–9
of philosophy 43, 44, 48, 49
South Africa 183, 185
southern Africa 144, 149–50
Zambia 148
deep listening 59
defaqane 145
Deleuze, Gilles and Guattari, Félix 95
Democratic Republic of Congo (DRC) 143, 144
dependency 199
depersonalization 138
Descartes, René 40, 41
desktop analysis 142
development 205, 247 *see also* human development
SADC 143–4, 145

social development 142
southern Africa 141–50
development cooperation 27–9 *see also* intercultural cooperation
dialogic humanism 255, 259
dialogical based research 216–17, 220–1
dialogue 5, 255
Dikoko, D. v Mokhatla, P. (2006) 185
discrimination 34, 205, 206 *see also* prejudice
disembedding 197
dispersals 145–6
Ditlhake, K. J. 186, 187
Dominelli, L. 190
Doolittle, Edward 54, 55, 56
DRC (Democratic Republic of Congo) 143, 144
Dwa Estate 86, 88, 89*f*, 91, 92, 93, 94, 96

eami (world; realm) 19, 20–1, 22, 108
Earth 117–18
ECA (Economic Commission for Africa) 70
eco-eroticism 46
ecology 83, 91
ecology of knowledge 216–17
Economic Commission for Africa (ECA) 70
economics 144, 146–7
capitalism 197, 198, 200–1, 245
neoliberalism 196, 197, 198–200
Ecuador 86, 246, 247–8
education 28–9, 30–5, 205
Africa 72
Faith and Joy School 135
knowledge transmission 57
mathematics 64
Oaxaca, Mexico 207–9, 211
research 64–5
South Africa 183
transformative/non-transformative 64
Elberfeld, R. 47
emotion 223–4
endogenous concept 142
Enlightenment, the 5, 40–1, 83, 255–7
environment, the 82, 85 *see also* sustainability
aesthetics 231
Aroreo cosmovision 108–9
climate change 188–90, 191, 244
Ermine, Willie 48
ethics 235–6, 238
Native Science 12, 13
railways 86, 87*f*, 88, 91, 92, 93, 94, 95
epic, the 93
epistemic disobedience 216, 233
epistemologies 232–3
epistemologies of the south 4, 216, 217, 224, 232, 259
eroticism 44–6
Escobar, Arturo 223–4
"Estéreo Comunal" project 210
Estonia 156, 157, 159
eSwatini 143, 144
ethic space of engagement 48
ethnoautobiography 7
Ethnoautobiography. Stories and Practices for Unlearning Whiteness, Decolonization and Uncovering Ethnicities (Kremer, Jürgen W. and Jackson-Patton, Robert) 3
Euler, Leonhard 54–5
Eurocentrism 215, 232–3
Europe 6, 153 *see also* Global North; Latvia; West, the
Enlightenment, the 5, 40–1, 83, 255–7
Estonia 156, 157, 159
Eurocentrism 215, 232–3
Finland 155, 156
imperialism 41
philosophy 44
Russia 196
Soviet Union 157–8
experiential knowing 52–6

Faith and Joy School 135
Fals Borda, Orlando 224
Farmers Clubs 74–5
fashion 177–8
fear 132–3, 135, 138
feminism 221–2, 223, 225
festivals 231, 236–9
Finland 155, 156
First Nations Climate Action Fellowship 49
football 169
FPIC (Free, Prior, Informed Consent) 109, 112–13

Free, Prior, Informed Consent (FPIC) 109, 112–13
freedom 122, 197
Freire, Paulo 5, 133, 255
Frontline States 144
fun 125
functioning 72
Fundación Encuentro 31

Galeano, Eduardo
  Libro de los abrazos (*Celebration of the Marriage of Reason and Heart*) 224
German people 159
Giddens, A. 197
Glasser, William 125
Global North, the 4, 243–4, 248 *see also* Europe; West, the
  colonialism 4
  migration 131
Global South 196, 198, 199, 200–1 *see also* Africa; Latin America
  children 215, 216
  China 86, 88, 92, 93, 95, 196
  India 196
globalization 201
Gough, N. 72
Great Transformation 245
Greenland 84
Grosfoguel, R. 196
Guarani people 111
Gudynas, Eduardo 246
Guelatao de Juárez 204, 207, 210
Guerrero Arias, Patricio 224
Gumilla Leadership Training Center 135
Gutierrez, G. 69
Gutstein, E. 58

Habermas, Jürgen 25
HAD (Human Development Approach) 69–70, 73
Hall, Edward T. 120
Harding, Sandra 29–30
Harms-Smith, L. and Rasool, S. 196
heteroglossia 93, 94
Hettne, Björn 205
hijab, the 169, 170–6, 177, 178
Hirtz, F. 95
Hobbes, Thomas 40

horizontal-dialogical methodologies 216, 220–1
hospitality 117
Hughes, Amber 53
human development 69 *see also* intercultural cooperation
  capability approach 69–70, 72, 73
  freedoms 72, 73
  HDA 69–70, 73
  SDGs 69–77
  UDA 70–1, 75–7
Human Development Approach (HDA) 69–70, 73
human rights 104–5, 109–14, 176, 216, 246, 256
humaneness 186–7
humanism 256, 257, 258 *see also* posthumanism
  dialogic 255, 259
humanness 186
humans
  agency 197
  decentering 258–9
  nature relationship with 239–40, 243–50, 257, 259
  world, place in 257
hybrid societies 106

Iceland 84
identity 173
  documents 176
  Latvia 162–3, 164, 165
  Livonian people 159, 161–6
  two spirited 44
imperialism 41, 44–5, 48
  cultural 183
India 196
indigenous, the 84–5
  definition 119
indigenous methodologies 216, 218–20
Indigenous people 220 n. 3 *see also* Andean people; intercultural cooperation; Livonian people
  Ayoreo people 17–22, 107*f*–9
  Chiquitano people 25, 35
  claiming to be 45 n. 2
  coming-to-know process 58
  Cree people 41–4, 47–8
  diversity 104

indigeneity 84, 93–4, 95
indigenous youth 203–11
inequality 204–6
Inuit people 84
knowing 56, 57, 58
knowledge/knowledge systems 4, 5, 29, 32–3, 243, 244, 245–6, 259–60
livelihoods 106, 108
mathematics 51–2
Mixe people 207, 209, 210
monitoring 18
Native science 9–14
Native spirituality 9
Ndwandwe people 145–6
place 58
protecting the integrity of 17–18
Prusai people 3
rights 104–5, 109–14
rural-urban mobility 29–30
science 4, 9–14, 58, 64, 81, 82, 95–6
social class 147
Sotho people 146
systems 146–7
territoriality 105, 109–10, 111
Wayuu people 34
White Earth Band of Ojibwe 85–6
indigenous social security systems (ISSS) 142–3, 148–9
indigenous youth 203–11
Indigeogy. Invitation to Learning in a Relational Way (Wilson Stand and Schellhammer, Barbara) 47
individualism 7, 48, 69, 120, 124
AIKS 73–4
Ayoreo people 21
capability approach 72, 73
Enlightenemnt, the 83
inequality 204–6
Instituto Superior Intercultural Ayuuk (ISIA [Ayuuk Intercultural Higher Institute]) 209
intellectual property rights 30
intelligence 11
intercultural cooperation 103–5, 111–12
actors 103–4
cosmovision 107*f*–9
FPIC 109, 112–13
land use 106
legal entities 112
topics 104
intersectionality 205, 206, 220
Inuit people 84
ISIA (Instituto Superior Intercultural Ayuuk [Ayuuk Intercultural Higher Institute]) 209
Iskay Yachay (double knowledge) 28
Islam 169–78
hijab, the 169, 170–6, 177, 178
ISSS (indigenous social security systems) 142–3, 148–9
ithembo 91, 92, 94, 95–6

"Jën Poj" (Wind of Fire) radio station 210
Joop, Wolfgang 178

Kakangani 96
Kaltmeier, Olaf 221
Kant, Immanuel 40
Kaupo (Livonian chieftan) 159, 160
Keewatin, Alice 42
Kenya 86–96
Kibwezi 86–8, 89*f*, 90–2, 93, 94, 95
Kiewitt, Karsten 3
Kilundo Meli 94–5, 96
Kimmerer, Robin
Braiding Sweetgrass 48
Kinuwa, Lionel 43
Klein, Naomi 47, 260
knowing 20–3, 56–7
ancestral ways of 53
dialogical 57–8
experiential 52–6
indigenous 56, 57, 58
place 58
knowledge/knowledge systems *see also* knowing; mathematics; science
AIKS 70, 71, 72, 73–5, 76, 77
alternative epistemologies 232–3
Andean 28, 29
Ayoreo people 21–2
childhood studies 215–16
corporeality 221–3
de Sousa Santos, Boaventura 217–18
decoloniality 198
decolonization of 232–3
ecology of 216–17
epistemologies 232–3

epistemologies of the south 4, 216, 217, 224, 232, 259
Global North, the 4
indigenous 4, 5, 29, 32–3, 243, 244, 245–6, 259–60
intellectual property rights 30
Iskay Yachay 28
liberatory knowledge 215
nature 243
non-scientific 216–17
openness 27
research 216–17
Santos, B. S. 232
scientific 222, 223
scientific universalism 216–17
situated knowledges 215
South Africa 183
transmission 57
unity 12
Western 218, 258
Wilson, Shawn 219
WKS 69, 70, 71–3, 74, 77
Kohlberg, Lawrence 41
Königsberg bridges problem 51, 52*f*, 54*f*–5, 64
Kovach, Margaret 218, 219
Kremer, Jürgen W. 3
Kremer, Jürgen W. and Jackson-Patton, Robert
Ethnoautobiography. Stories and Practices for Unlearning Whiteness, Decolonization and Uncovering Ethnicities 3
Kristeva, J. 93
kyondo 90, 94, 95, 96

labor migrants 132
Labrador Métis people 84
Lāčplēsis (Pumpurs, Andrejs) 159–60
land
conflict 110, 208
FPIC 109, 112–13
indigenous territoriality 105, 109–10, 111
memory 86
Oaxaca, Mexico 208
pollution 47
relationship to 42, 43, 47, 90
rights 104–5, 109–14
tekohá 111
use strategies 105–6
language 5, 11
African 186
Ayoreo people 21, 22
Cree people 42–3
Latvian 155, 160, 163, 164
Livonian 155, 156, 158, 159, 160, 161, 162–6
Oaxaca, Mexico 207, 210
of religion 35
semiotics 32–3, 93
ubuntu 186
Wittgenstein, Ludwig 53
Latin America 104–5
Andean region 26, 27–36
Bolivia 246, 247–8
Brazil 136, 137, 169–77, 196
Ecuador 86, 246, 247–8
Mexico 203–11
Peru 233–4
Venezuela 134–6
Latvia 155, 156, 158, 159–61
identity 162–3, 164, 165
language 155, 160, 163, 164
Livonian Coast 155–6, 157–8
"Laws for the Protection and Right of Nature" 247–8
leadership 208
Lesotho 143, 144
liberatory knowledge 215
Liboiron, Max 47
Pollution is Colonialism 39
Libro de los abrazos (*Celebration of the Marriage of Reason and Heart*) (Galeano, Eduardo) 224
lineage 3, 42
Linhart, Jan 33
Livonian Coast 155–6, 157–8
Livonian Chronicle of Henry 154, 159–60, 161
Livonian people 153–66
contemporary 161–5
data about 162
heritage 163, 164–5
history 154–9
identity 159, 161–6
language 155, 156, 158, 159, 160, 161, 162–6

narratives for 161
perception of 159–61
Lloyd, B. et al. 58
Loorits, Oskar and Kettunen, Lauri 161
Lorenz, W. 196, 198–9, 201
love 124
Lugones, María 205
Luna, Martínez 211
Luongo, K. 94

Madagascar 144
Malawi 74–5, 143, 144, 145
Maldonado-Torres, N. 198
mana (spiritual power) 123–4
"Manifiesta no saber firmar: Nacido 31 de diciembre" (Pushaina, Estercilia Simanca) 34
Mansilla, Hugo Celso Felipe 26
mapping methodology 59–61, 62*f*–3*f*
clustering and mapping 60
coding and graphing 59–60
conclusions from the map 60–1
conversations/dialogue 59
how to read the map 60
Marks, S. E. 146
Maslow's hierarchy of needs 125
mathematics 51–2, 56–7, 58, 61–5
Königsberg bridges problem 51, 52*f*, 54*f*–5, 64
spirituality 64
mathembo 91–2, 93, 94, 95
Mathu, Mr. 90–1
Mbororo people 84
memory 5–6
Mercado, Rafael 32–3, 34
mestizaje (racial/cultural mixing) 25–6
methodological nationalism 129
Mexico 203–11
mfecane 145
Mignolo, Walter 216, 233, 235
migration 129–39, 169, 207–8 *see also* mobility
minds
colonization 5–6
decolonization 6–7
modern 11
participatory 11
mino-bimaadiziwin 225
Mixe people 207, 209, 210
mobility. *See* dispersals; migration
modernity 95, 196–8, 199, 201
modesty 178
Moises (Venezuelan asylum seeker) 134–7
Mokgoro, Y. 186
Montes Rodriguez, Maria Emilia 30, 31–3, 34
Moreton-Robinson, A. 58
Mozambique 143, 144, 145
Mugumbate, J. R. and Chereni, A. 185
Munandu Mwololo 94
Muridia 137–8
music 209, 237
mu'unde mu'uoi 90–1, 94
Mzilikazi (king of the Ndebele people) 146

Namibia 143, 144, 145, 146
Natarajan, B. 76
nation-states. *See* state, the
National University of Colombia 32
nationalism 133, 156, 159, 160–1, 203–4
nationality 132
Native science 9–14
Native spirituality 9
natural disasters 188–91
natural world 83, 85–93
aesthetics 231, 234, 235–9
agency 85, 93
art 238
Ayoreo people 20–1
Buen Vivir 246–9
culture/nature relationship 39–40, 44–5, 243, 244, 247, 257
development 247
eroticism 46
estrangement from 11
exploitation 243–4
humans, relationship with 239–40, 243–50, 257, 259
knowledge 243
"Laws for the Protection and Right of Nature" 247–8
Pachamama Festival 231, 236–9
philosophy 40
railways 86, 87*f*, 88, 91, 92, 93, 94
RoN 85, 86
value of 238
Winikunka, Peru 233–4

Ndlovu-Gatsheni, S. J. 196–7, 198
Ndwandwe people 145–6
Nelson, Melissa 46
neoliberalism 196, 197, 198–200
Nestor
 Tale of Bygone Years, The 154
New Enlightenment, the 255, 256–7
New Zealand 118, 123–4
Ngolano 91–2, 94
Nicol, C. et al. 58
9/11 attacks 174
Nissiotis, N. A. 69
non-scientific knowledge 216–17
novel, the 93
Noyoo, N. 144
Noyoo, N. and Sakala, B. M. 148
Noyoo, N. and Sobantu, M. 145–6
Nussbaum, B. 186–7
Nussbaum, M. 72, 246
Nyerere, J. K. 187

"O Clone" TV soap opera 175
Oaxaca, Mexico 203–11
oral traditions 11, 33, 57
 Ayoreo 21
 deep listening 59
 myths 30
 Umunthu 75–6
Oranga Tamariki—Ministry for Children 123–4
organism democracy 259
othering 160
Oyěwùmí, Oyèrónkẹ́ 222

Pachamama Festival 231, 236–9
"Panos Quentes" (Hot Cloths) (Macedo, S. and Damasceno, V.) 169
PAR (participatory action research) 216, 217
participatory action research (PAR) 216, 217
participatory minds 11
Paul, Annie Murphy 39
PCEA (Presbyterian Church of East Africa) 94, 95
Pelluchon, Corine 41, 47, 256–7
perception
 limitations of 18–19
 of Livonian people 159–61
Pere, Rose 126
Pérez, M. 204–5, 206, 211
Peru 233–4
philosophy 234
 colonization 46–7
 decolonization 43, 44, 48, 49
 European 44
 male dominance 41
 Western 39, 40–1
place 58, 88, 90, 91, 96 *see also* affective geographies
 Guarani people 111
 territory 224
planetary thinking 245
Plato 40
Polanyi, Karl 245
political affections (of mobility) 131–4, 135–7, 138–9
political economy 82–3
Pollution is Colonialism (Liboiron, Max) 39
Poole, Roger
 Towards Deep Subjectivity 48
posthumanism 258
Power in My Blood (Bear, Tracy Lee) 45
practical attitudes 234
practical philosophy 234
PRATEC (Andean Project for Peasant Technologies) 26, 28
Pregel River 55*f*
prejudice 173–6 *see also* discrimination
Presbyterian Church of East Africa (PCEA) 94, 95
Príncipe 144
Prusai people 3
purity 48–9
Pushaina, Estercilia Simanca 34
 "Manifiesta no saber firmar: Nacido 31 de diciembre" 34
Pusisuyu 29

Qatar 169
queerness 44, 45
Quijano, A. 130–1, 232–3

race/racism 84, 130, 174, 175–6
 apartheid 144–5, 183, 184
 mestizaje 25–6
radio 210

railways 86, 87*f*, 88, 91, 92, 93, 94, 95, 96
Ramose, M. B. 187
rationalism/reason 40, 41, 126, 224
  responsive reasoning 47–8, 49
Recknagel, Lukas 249
relational accountability 43, 48
relationality 39, 41, 42, 47–8, 224–5
religion *see also* Christianity; Islam
  Muridia 137–8
  Native spirituality 9
  science, integration with 12
Rengifo, Grimaldo 26
research 64–5, 142, 148
  action-based 216–20
  agonistic horizontality 221
  childhood studies 215–26
  corpo-biographies 222–3
  corporeality 221–3
  dialogical based 216–17, 220–1
  epistemologies of the south 4, 216, 217, 224, 232, 259
  horizontal-dialogical methodologies 216, 220–1
  indigenous methodologies 216, 218–20
  PAR 216, 217
  research-action-creation 223
  scientific universalism 216–17
  sensing-thinking 222–5, 226
  sentipensamiento 223–4, 225, 226
  third-space methodologies 220
  tolerant horizontality 221
research-action-creation 223
Research is Ceremony (Wilson, Shawn) 44
responsive reasoning 47–8, 49
revitalization 153–4
rhizome metaphor 95
rice 85–6, 90
Rifkin, Mark
  When Did Indians Become Straight? 45
risk 82
Rodríguez, Rosana and da Costa, Sofía 222–3
RoN (rights of nature) 85, 86
Rosa, Hartmut 244–5
Rose, N. 200
Royal, Te Ahukaramū Charles 119
rural-urban mobility 29–30
Russia 196
SADC (South African Development Community) 143–4, 145
SADCC (Southern African Development Co-ordination Conference) 144
Safatle, V. 132–3
Samaran, Nora 45
Samba (Senegalese migrant) 137–8
San José de Chiquitos 25
Santa María Tlahuitoltepec 204, 207, 209, 210
Santos, B. S. 232
São Tomé 144
SAP (Structural Adjustment Programme) 146–7
satisfaction 235, 236
Scarritt, R. J. 74
Schmid, J. 200
science 9, 61, 260
  Global North, the 4
  indigenous 4, 9–14, 58, 64, 81, 82, 95–6
  integration 29–30
  Native 9–14
  political economy 82–3
  religion, integration with 12
  scientific universalism 216–17
  Western 5, 12, 82, 83, 86, 93, 95–6, 258, 259
scientific knowledge 222, 223
SDGs (Sustainable Development Goals) 69–77
Seel, M. 235, 236, 237, 238
"self-as-relationship" 43
semiotics 32–3, 93
Sen, Amartya 69, 72, 73
Senegal 138
sensing-thinking 222–5, 226
sentipensamiento 223–4, 225, 226
settler colonialism 143
sex 44
Sexkultur ("Sex Culture") (Stangneth, Bettina) 44
sexual violence 45
Seychelles 144
SGR (Standard Gauge Railway) 86, 88, 91, 92, 93, 94, 95
Shaka (king of the Zulus) 145, 146
Shotwell, Alexis
  Against Purity 48–9
Simpson, Leanne 47

situated knowledges 215
Skolimoski, Henryk 11
Sloterdijk, Peter 259
Smith, L. 200
Smith, Linda Tuhiwai 218–19
Sneed, A. 76
social class 147
social development 142
social networks 177
social security. *See* ISSS
social work 119, 187, 197, 199–201
climate change 190, 191
mana 123–4
modernity, 199
natural disasters 188–9, 190, 191
political roles 200
South Africa 183–5, 186–91, 199
Spirited Five-Finger model 120–6, 127
state, the 199
Te Wheke model 126–7
ubuntu 187–9, 190, 191
Socrates 40
Soshangane (king of Gaza) 145–6
Sotho people 146
South Africa 76, 143, 144, 145, 146 *see also* Africa
BRICS 196
decolonialization 183, 185
economics 147
natural disasters 188, 189–90
SADC 143–4, 145
social work 183–5, 186–91, 199
ubuntu 183–91
southern Africa 141–50 *see also* Africa
Southern African Development Co-ordination Conference (SADCC) 144
sovereign erotics 39, 44–6
Soviet Union 157–8
Soydan, H. 200
speaking 133–4, 138
Spirited Five-Finger model 120–6, 127
spirituality 64, 65*f*
Akamba 90, 94
mana 123–4
Split Tooth (Tagaq, Tanya) 46
Standard Gauge Railway (SGR) 86, 88, 91, 92, 93, 94, 95
Stangneth, Bettina
Sexkultur ("Sex Culture") 44
state, the 109–10, 112
capitalism 197, 198, 200–1, 245
migration 132, 137, 138
nation-states, creation of 155–6
social responsibility of 196, 197, 198–9, 200
stokvels (informal banking systems) 147
Structural Adjustment Programme (SAP) 146–7
Sumak Kawsay 225, 245
Sustainable Development Goals (SDGs) 69–77
sustainability 82, 85, 105, 153–4, 245, 249–50 *see also* SDGs
Syetali 96

Tagaq, Tanya
Split Tooth 46
Tale of Bygone Years, The (Nestor) 154
Tanzania 74, 143, 144, 145
Te Wheke model 126
technology 61
tekohá (place of being) 111
territoriality 105, 109–10, 111
territory 224
texts 42
erotic 45, 46
third-space methodologies 220
theoretical attitudes 234
theoretical philosophy 234
thought/body disconnect 39
tolerant horizontality 221
Tomichá Charupá, Roberto Claudio 34–6
Towards Deep Subjectivity (Poole, Roger) 48
transformation 259–60
trauma, 53, 121–2
Treblanche-Greef, A. C. 74
Trudell, John 6
two spirited identity 44

UAINN (Universidad Autónoma Indígena Nasa) 34
Ubuntu (humanness) 73, 75, 183–91, 225
UDA (umunthu developmental approach) 70–1, 75–7
Ujamaa 187

Umunthu (humanness) 73, 74, 75–6
umunthu developmental approach (UDA) 70–1, 75–7
Universidad Autónoma Indígena Nasa (UAIIN) 34
Universidad Indígena del Tawantinsuyo del Ajlla 31
universities 28–9, 30–5, 218–19, 222 *see also* education
University of Cauca 34
uoi 90–1, 92, 93, 94
utopianism 245, 247, 250, 260
uwe 90–1, 93, 94

Vaalgamaa, Edgar 162
veiling 169, 170–6, 177, 178
Venezuela 134–6
violence 88, 92, 118
  migration 132–3
vocero de la comunidad (community spokesperson) 136
Vygotsky, Lev 57

wahkotowin (relationality) 41, 42, 47–8
Waldenfels, Bernhard 41
war 118
Wayuu people 34
West, the 81, 126, 218 *see also* Europe; Global North
  Canada 84
  influence 26, 185, 259
  New Zealand 118, 123–4
  philosophy 39, 40–1
  science 5, 12, 82, 83, 86, 93, 95–6, 258, 259
  sustainability 250
  WKS 69, 70, 71–3, 74, 77
Western Knowledge System (WKS) 69, 70, 71–3, 74, 77
Whanganui River 118
Whatmore, S. 92
When Did Indians Become Straight? (Rifkin, Mark) 45
White Earth Band of Ojibwe 85–6
Wilber, Ken 12, 55
wild rice 85–6
Wilde, Oscar 260
Wilson, E. O. 12
Wilson, Shawn 218, 219
  Research is Ceremony 44
Wilson, Stan 39, 43
Wilson Stan and Schellhammer, Barbara
  Indigeogy. Invitation to Learning in a Relational Way 47
Winikunka, Peru 233–4
Wittgenstein, Ludwig 53
WKS (Western Knowledge System) 69, 70, 71–3, 74, 77
women 42, 47, 91–2
  discrimination 205, 206
  education 207–8, 210
  feminism 221–2, 223, 225
  hijab, the 169, 170–6, 177, 178
  Islam 169, 170–6, 177, 178
  migration 207–8
  modesty 178
  music 209
  Oaxaca, Mexico 205, 206, 207–8, 209, 210
World Association of Community Radio Broadcasters (AMARC) 210
writing 21

Yatiris 31

Zambia 74, 143, 144, 145, 146
  economics 147
  ISSS 148–9
Zimbabwe 143, 144, 145, 146

www.ingramcontent.com/pod-product-compliance
Ingram Content Group UK Ltd.
Pitfield, Milton Keynes, MK11 3LW, UK
UKHW020246290926
15470UKWH00002B/767

9 781350 42520